Basic Telecommunications: The Physical Layer

Basic Telecommunications: The Physical Layer

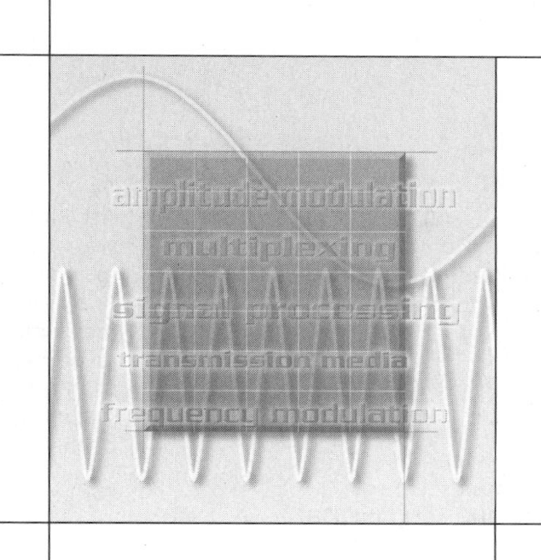

Gary J. Mullett

National Center for Telecommunications Techologies

Springfield Technical Community College
Springfield, Masssachusetts

THOMSON
DELMAR LEARNING

Australia Canada Mexico Singapore Spain United Kingdom United States

Basic Telecommunications: The Physical Layer
Gary Mullett

Executive Director:
Alar Elken

Executive Editor:
Sandy Clark

Senior Acquisitions Editor:
Gregory L. Clayton

Senior Development Editor:
Michelle Ruelos Cannistraci

Executive Marketing Manager:
Maura Theriault

Channel Manager:
Fair Huntoon

Marketing Coordinator:
Brian McGarth

Executive Production Manager:
Mary Ellen Black

Production Manager:
Larry Main

Senior Project Editor:
Christopher Chien

Art/Design Coordinator:
David Arsenault

Senior Editorial Assistant:
Dawn Daugherty

Printed in Canada
1 2 3 4 5 XX 06 05 04 03 02

For more information contact
Delmar Learning
Executive Woods
5 Maxwell Drive, PO Box 8007,
Clifton Park, NY 12065-8007

Or find us on the World Wide Web at
http://www.delmar.com

ISBN: 1-4018-4339-5

NOTICE TO THE READER

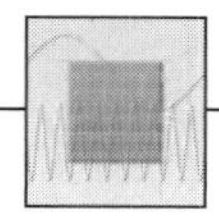

CONTENTS

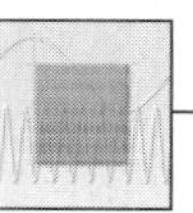

PREFACE

Audience

This text, *Basic Telecommunications: The Physical Layer,* was written primarily for either the two- or four-year college level. It will most likely find its largest audience at the community college level for use by students involved in the study of the technical aspects of telecommunications technology. It certainly could also be used by others who would like to gain an insight into this field or as a reference text for still others who need to know more about the fundamentals of telecommunications.

Today, the telecommunications industry continues to implement new technologies in an effort to provide high-bandwidth, high-speed data transmission capabilities to the consumer. In the end, a combination of sophisticated wireline, fiber-optic, and wireless technologies will most likely be deployed by this industry to satisfy consumer demand. The technology of choice for a particular situation will be dictated by both economic considerations and marketplace preferences. This text provides comprehensive coverage of the fundamental aspects of information transmission by means of industry-standard transport mechanisms over each of these transmission mediums.

This book is the first in a series of texts from the National Center for Telecommunications Technologies that deals with today's telecommunications systems. It will provide foundational theories and concepts for students interested in the physical-layer technology of the telecommunications industry regardless of the type of transmission media used—wireline, wireless, or fiber-optic.

Approach

During the last three decades, there has been an accelerating rate of convergence of computer data transmission and classic electronic communications. At the present time, driven by the steady pace of advances in the field of microelectronics and other technological innovations, the ubiquitous PC has become our gateway to the Internet and all the possibilities it holds for changing how our society functions.

Discussions of advances in fiber-optic communications, with its promise of almost limitless bandwidth and predictions of universal wireless Internet access in the not too distant future, abound in both the popular press and technical journals. This author has observed this ongoing evolution of technology with both fascination and awe. As with most technology, it is hard to comprehend the incredible changes that have occurred in one's lifetime and then observe a small child quickly adopt the latest technology with little need for instruction in its use.

This ongoing technology convergence, which has served to redefine the modern field of telecommunications, has brought together two relatively different groups under a common goal—the transparent transmission of information. One group consists of the computer information systems community and the other group is the electronic communications community. It is my belief that these two communities of telecommunications users/implementers tend to look at the telecommunications field from different viewpoints. These two different viewpoints can be explained quite well by the 7-layer OSI model. The computer systems people tend to view telecommunications from the applications layer (layer 7) perspective downward with the necessary physical-layer hardware being a commodity available from a host of different vendors. On the other hand, the communications hardware systems people prefer to view telecommunications from the physical layer (layer 1) perspective upward with the ultimate computer application being somewhat less important than the hardware with its exacting specifications. This is only natural due to the influence of one's training, education, and occupational status within these fields.

This text was written from the perspective of the "physical layer" of the 7-layer OSI model with an additional goal of endeavoring to bridge the gap between the hardware and the eventual applications of the system. The result of many years of teaching and working in the hardware end of this field, it explains the most important basic concepts in this field. As already mentioned, an additional goal of the text is to tie in the modern applications of telecommunications and how they relate to the system hardware and vice versa. It is my firm belief that the concepts presented will serve the student well in any segment of the telecommunications industry, regardless of the type of transport media used by that sector. The reader of this text should have completed fundamental courses in dc and ac circuit theory, electronics, and an introduction to digital electronics.

This text abounds with numerous screen captures of real waveforms and frequency spectra from standard test and measurement equipment connected to representative hardware systems. This has been done in an effort to help the student visualize and connect points made in the text about theoretical topics with real-world systems and typical operational measurements of these systems. It is the author's contention that numerous equations that model a system's operation can be presented to the student, but few will be as effective as a visual presentation of the actual signals involved in system operation. This type of presentation is also in keeping with the increasing use of sophisticated simulation software available for use in this field.

One might also note that this text is almost totally devoid of any detailed circuit diagrams. Almost all the topics that are presented here are basic fundamental concepts and therefore, transcend the changing technology that may be used to implement them over the course of time. Additionally, one might point out that application-specific ICs presently allow the integration of entire complex telecommunications hardware systems on an integrated circuit chip with limited if any physical access to any of the subsystems that comprise the complete system. Today, one must deal with these systems from a terminal characteristic point of view—the same approach taken by this work. It is also the author's belief that the technician of the future will be primarily tasked with the evaluation of system operation and the possible reconfiguration of programmable hardware/system functions more than the repair of this hardware.

In summation, the style of presentation used in this text closely resembles how the author presents and demonstrates this material in a classroom setting and, most importantly, it furnishes the required systems-based overview of basic telecommunications hardware operation.

Organization

This text is organized in a fairly traditional manner with some additional topics and features built in. The first chapter goes into a relatively detailed history of both the technology and governmental regulation of the telecommunications industry. After an introductory chapter on fundamental concepts and several brief tutorials, the next four chapters deal with modulation schemes starting with traditional AM and FM and ending with pulse and present-day digital modulation techniques. The next chapter takes a look at the important concepts of signal multiplexing and wireless access techniques. The final two chapters are detailed looks at what are presently the most popular transmission media: wireline (copper pairs and coaxial cable), fiber-optic cable, and wireless. Where appropriate, each chapter starts with a historical perspective of the evolution of the particular technology to be discussed. Also included throughout the text are numerous examples, pictures, and screen shots of representive signals and signal spectra.

Chapter 1

Chapter 1 contains presentations of two similar yet different topics. The first topic is the history and evolution of telecommunications from a hardware perspective; the second topic is the history and evolution of the regulation of this industry. These two topics, consisting more of historical facts than technological facts, are valuable in the insight they provide the student about how the telecommunications industry has evolved. The increasing pace of technological innovation, coupled with the acceptance of the computer into everyday life, has driven this industry at a pace seldom experienced by any other field. Governmental regulation has also played a major role in how the implementation of new technology has played out. The importance of the material in this chapter is certainly debatable if one takes the approach that technology is only as good as the next best implementation. This author feels differently! A short section on standards is included at the end of the chapter.

Chapter 2

Chapter 2 sets the stage for the rest of the text by introducing the student to definitions and the basic hardware components of a telecommunications system. The major emphasis is on system hardware—the transmitter, channel or transmission media and switching fabric, and the receiver. The function and purpose of each subsystem is explained and several typical examples are described. The second part of this chapter consists of five short tutorials that again are used to set the stage for the rest of the text. These topics are critical to a student's understanding of the following chapters. Indeed, they are so critical that the author feels compelled to introduce them here with numerous practical examples. In order, the topics are filter theory, dBs, test equipment, noise measurements, and Fourier analysis and its application to signals.

Chapter 3

Chapter 3 covers the first type of classic modulation scheme, amplitude modulation (AM). It is given a fairly detailed coverage with a rather in-depth level of mathematical treatment. The purpose of this detailed coverage is to emphasize the basic concepts of signal spectra and bandwidth. A further grounding in these topics is provided by numerous examples of frequency spectra and oscilloscope time displays. The index of modulation for AM and the resulting power relations are introduced to again emphasize signal bandwidth, signal frequency components, and power relations. The production of AM is briefly covered at a block-diagram level. Mixer concepts and the superheterodyne technique are introduced with coverage of receiver concepts and AM demodulation. Alternative versions of AM and the theory of digital radio are covered at the end of the chapter.

Chapter 4

Chapter 4 covers angle or frequency modulation from the perspective of signal bandwidth and spectral frequency components. Bessel functions are introduced to facilitate this coverage and to provide a basis for how practical transmitter calibration is performed. FM is contrasted to phase modulation and then AM with system advantages and disadvantages chronicled.

FM production is briefly covered at a block-diagram level. The coverage of superheterodyne radio receivers is extended to FM with an explanation of FM receiver differences and FM detection. This chapter also introduces the student to the important concept of multiplexing via FM stereo—a practical system that is familiar to the student.

Chapter 5

Chapter 5 introduces the reader to classic pulse-modulation techniques and its most popular implementation, pulse-code modulation (PCM). After introducing signal sampling theory, analog pulse-modulation techniques, PCM, and some PCM variations, common practical line codes, encoding techniques, and line-code bandwidth for baseband modulation systems are discussed in detail. Eye diagrams and intersymbol interference are introduced and typical mask measurements are detailed. The typical frequency spectra of a pulse-modulated signal is introduced in some detail. The problem of under-sampling is examined from a spectral viewpoint and the demodulation of a pulse-modulated signal is demonstrated in the frequency domain. The chapter concludes with a short overview of digital signal compression and the MPEG standards.

Chapter 6

Chapter 6 introduces the student to the theory of modern digital-modulation techniques. After a short discussion of amplitude and frequency-shift keying, phase-shift keying (PSK), and quadrature amplitude modulation (QAM), are both explained in some degree of detail to the reader. Typical transmitter and receiver implementations are shown in block-diagram form with special consideration given to the balanced modulator and its role in the realization of these forms of modulation. Typical output signals with representative digital coding and their frequency spectra are demonstrated for several systems. Emphasis is again placed on the achievement of bandwidth efficiency for these passband systems. A modern modem is highlighted and contrasted with one of the first modems introduced earlier in the chapter. System bit error rate (BER) is introduced and typical plots of BER versus symbol-to-noise ratio are given for various levels of modulation.

Chapter 7

Starting with an explanation of the need for multiplexing, one of the oldest and most common forms, frequency division, is introduced. Typical examples of this multiplexing technique are examined. Time-division multiplexing is introduced next with the North American T1-carrier system the primary example used to illustrate this technique. Additional detail is provided about the function of the time-slot interchanger in the switch used by the PSTN and higher-order T-carrier systems. Wireless access technologies like code-division multiple access and time-division multiple access are introduced next with spatial and wavelength division multiplexing covered at the end of the chapter. The chapter concludes with an overview of the multiplexing techniques used to implement asymmetrical digital subscriber line (ADSL), over a copper pair in the local loop.

Chapter 8

Chapter 8 covers two related topics: electromagnetic propagation and various transmission media. After an overview of the radio frequency spectrum, a fairly in-depth coverage of electromagnetic propagation is undertaken. EM propagation properties and the effect of the ionosphere on propagation is provided. The next topic, transmission lines, is covered from a systems viewpoint. After considering examples of typical transmission lines, the concepts of reflection coefficient and standing-wave ratio for transmission lines is introduced. Types of transmission-line matching techniques are introduced through example. The chapter concludes with an overview of the typical infrastructure of the local telephone loop, hybrid fiber/coaxial-cable TV systems, and long-haul fiber-optic cable systems.

Chapter 9

Chapter 9 completes our coverage of the telecommunications physical layer with an introduction to the antennas that are used primarily with the free-space or wireless channel. After defining and introducing basic antenna models and properties, a survey of typical practical antennas is presented. Next, the radiation patterns for the half-wave dipole are illustrated and the effect of ground on the antenna's directivity is described. Array antennas are introduced next with several common wire array examples highlighted. The final topics covered in the chapter are typical parabolic reflector antennas, practical microwave and millimeter-wave horn antennas, microstrip patch-antenna arrays, and an overview of cell-site antenna technology.

Features

1. This text makes liberal use of screen shots from digital-storage oscilloscopes and spectrum analyzers to reinforce the theoretical discussions in the text about the operation of practical telecommunications systems hardware.
2. The text is written from a systems perspective with only a few relatively detailed circuit diagrams.

3. Where appropriate, a short summary of the evolution of a particular technology is included.
4. The mathematics in the text is limited to algebra, some trigonometric identities, and the use of Bessel functions, making this text suitable for use at the two-year college level.
5. The topic coverage has been chosen to provide the reader with the necessary fundamental concepts and theories needed to become proficient in the understanding of the practical physical-layer implementation of the telecommunications system hardware. Additionally, it is believed that in reading this text the student will gain some sense of the ultimate relevance of the system to a real-world application.
6. Helpful learning aids such as Objectives, Outlines, Key Terms, Examples, Summaries, Questions and Problems, and Equation Lists are found in every chapter, focusing attention on key concepts to facilitate learning.

Supplements

Instructor's Guide. ISBN: 1-4018-3617-8

Laboratory Manual. ISBN: 1-4018-3207-5

Web Tutor. This student study guide and interactive supplement offers notes, flashcards, web links, quizzes, and discussion-room topics. Web Tutor is available using:

Web CT ISBN: 1-4018-3866-9

Blackboard ISBN: 1-4018-9539-5

Online Companion. Visit the textbook's online companion at http://www.electronictech.com for up-to-date information.

NCTT

The National Center for Telecommunications Technologies (NCTT: www.nctt.org) is a National Science Foundation (NSF: www.nsf.gov) sponsored Advanced Technological Education (ATE) Center established in 1997 by Springfield Technical Community College (STCC: www.stcc.edu) and the National Science Foundation. All material produced as part of the NCTT textbook series is based on work supported by the Springfield Technical Community College and the National Science Foundation under Grant Number DUE 9751990.

NCTT was established in response to the telecommunications industry and the worldwide demand for instantly accessible information. Voice, data, and video communications across a worldwide network are creating opportunities that did not exist a decade ago, and preparing a workforce to compete in this global marketplace is a major challenge for the telecommunications industry. As we enter the twenty-first century, with even more rapid breakthroughs in technology anticipated, education is the key and NCTT is working to provide the educational tools employers, faculty, and students need to keep the United States competitive in this evolving industry.

We encourage you to visit the NCTT Web site at **www.nctt.org** along with the NSF Web site at **www.nsf.gov** to learn more about this and other exciting projects. Together we can explore ways to better prepare quality technological instruction and ensure the globally competitive advantage of America's telecommunications industries.

Acknowledgments

The Author and Delmar Learning would like to thank the following reviewers:

Mike Awwad, DeVry University, N. Brunswick, NJ

Michael O. Beaver, University of Rio Grande, School of Technology, Rio Grande, OH

Joe Gryniuk, Lake Washington Technical College

David Holding, DeVry University, Kansas City, MO

Bill Liu, DeVry University, Fremont, CA

Gilbert Seah, DeVry University, Mississauga, Ontario

About the Author

Gary Mullett is a long-time faculty member at Springfield Technical Community College in Springfield, Massachusetts. For the past decade he has been actively involved in the field of telecommunications as a consultant and co-director of the National Science Foundation, Advanced Technology Education, National Center for Telecommunications Technologies located at Springfield Technical Community College. A principal figure in the implementation of the pace-setting Verizon NextStep Program, he has spent many years as a consultant to local industry.

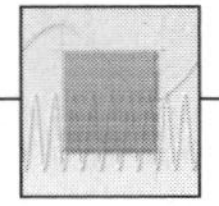

DEDICATION

This book is dedicated to my wife, Robin, for her infinite patience, understanding, and encouragement.

Introduction to Telecommunications: A Technical and Regulatory History

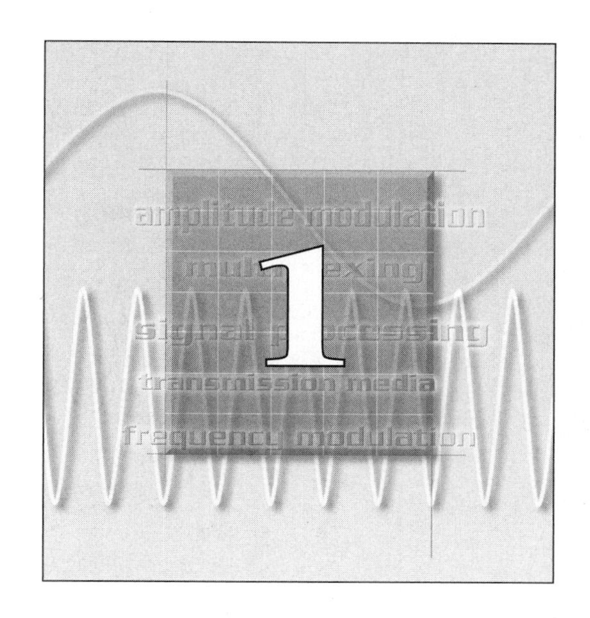

Objectives Upon completion of this chapter, the student should be able to:

- Discuss the general history and evolution of telecommunications technology from a North American perspective.
- Explain the impact of the microprocessor and advances in microelectronics and fiber-optic technology upon the telecommunications industry.
- Discuss the history of the cable-television industry.
- Discuss the history of wireless communications.
- Discuss the history of the public-switched telephone system.
- Discuss the convergence of technological innovation, the use of the personal computer (PC), and the advent of the Internet in the transformation of the world into a networked environment.
- Describe major regulatory events and their effects on the evolution of the public-switched telecommunications network in the United States.
- Describe the purpose and functions of standards organizations.

Outline

Key Terms

communications
deregulation
digital
gigascale integration
Internet
interoperability
local access territory
Moore's Law
multimedia
network
regulated monopoly
regulation
standards
telecommunications
wireless
wireline

Introduction

Practical electrical communications began in the United States over one-hundred-and-fifty years ago with Samuel B. Morse's telegraph system. In turn, the invention of the telephone in 1876 began the era of telecommunications. Shortly thereafter the first manually switched wireline **network** was introduced. Radio (wireless) was invented at the turn of the nineteenth century, adding different forms of **communications**—namely, broadcasting and mobile operation. At the start of the new millennium, advances in microelectronics, computers, fiber optics, and telecommunications technology have put us at the threshold of a networked world unlike anything previously known to man. The rapid evolution of the network from analog to **digital**, coupled with continued increases in bandwidth and accessibility, are turning into reality the promise of "anywhere, anytime" delivery of rich **multimedia** content over the **Internet**.

Since the beginning of the twentieth century, both state and federal governments have been regulating the telecommunications industry. Over the course of time regulation has played an important role in how the industry has evolved, and in how it continues to evolve. As the global economy has become more extensive, the need for global telecommunications standards that offer the promise of worldwide **interoperability** has also become more demanding. This chapter will introduce students to both the technological and regulatory history of the telecommunications industry and to the various standards bodies that are presently involved in the shaping of new and emerging telecommunications technologies.

PART 1

A Technical History of Telecommunications

1.1 Mankind's Desire to Communicate

Since the dawn of civilization, mankind has expressed a desire to communicate; thus, verbal communications developed through the spoken word and the evolution of language. As civilizations spread and interactions between the people of the world increased, counting and number systems were developed to aid in trading and commerce. The written word evolved from pictographs to alphabets, and the modern graphic representation of the printed word soon appeared. From the earliest days to the present times, mankind has shown a desire to extend interhuman communications beyond the limitations of sight and hearing. For example, sound-producing instruments like drums, horns, and bells and visual signals using smoke, flags, and lights have been used since ancient times to overcome the limitations of our senses in our quest to communicate with our fellow human beings.

The Early Days

This chapter is going to introduce the field of telecommunications via a history of electronic communications, digital computers, and electronic devices. The story starts in the 1830s with Samuel B. Morse, pictured in Figure 1-1 on page 4.

An aspiring artist and inventor, Morse had spent time studying portrait painting in Paris. It is said that Morse, while in transit back to America in 1832, had a conversation with a fellow passenger aboard the *Sully* about electricity and its properties. Morse apparently had an "inventor's brainstorm" about this newly discovered topic, for he quickly conceived the idea of using the flow of electrical current through a wire to communicate over a distance. He invented the first practical telegraph in 1837 and later developed Morse code, a means by which he could transmit what are known today as "alphanumerics." These alphanumeric symbols were encoded as combinations of short and long signals (dots and dashes). Interestingly, his code for the most common letter in the English language, "e", consisted of a single dot, while the code for a relatively uncommon letter like "z" consisted of two dashes followed by two dots. It seems that Morse could also be credited with developing the first efficient coding scheme!

Morse received Patent 1,647 for his invention. Then, in the early 1840s, the United States Congress gave Morse $30,000 to fund a prototype of his

FIGURE 1-1 Samuel B. Morse
© CORBIS

telegraph system. This first North American telegraph system was constructed between Baltimore, Maryland and Washington, D.C. It consisted of approximately 700 telegraph poles spaced at 300-foot intervals (a quick calculation will yield a distance of about 40 miles, the approximate distance between the two cities). On May 24, 1844, Morse demonstrated his telegraph by sending a quote from the Bible. The phrase "What hath God wrought" was transmitted from the Supreme Court room in the District of Columbia to Morse's assistant, Alfred Vail, in Baltimore. By 1846 the first commercial United States telegraph systems were operational. (Do an Internet search on *Samuel B. Morse* for a more in-depth history of the early telegraph.)

The Mid-1800s

It should be noted that concurrent to the development of the first telegraph system in the United States, William Cooke and Charles Wheatstone, coinventors of another form of the telegraph, started to construct telegraph systems between towns and cities in Great Britain. During the middle of the 1800s the basic concepts of electricity and electrical circuits were known by a relatively small group of people. Some of the other early contributors to the development of the theories of electricity and magnetism were Georg Ohm, Charles Wheatstone, Gustav Kirchhoff, Michael Faraday, Andre Ampere, Christian Oersted, and Carl Gauss. An Internet search on any of their names will lead to Web sites that discuss their individual contributions to the fundamental understanding of electricity and magnetism.

Interestingly, direct current (dc) circuit theory is the same now as it was in the mid-1800s; however, the number of people with knowledge of this topic is certainly much greater now than in those early days. Anyone who has ever taken a course in basic electricity most likely knows almost as much about the topic as the early scientists in the field.

The 1850s brought about the era of transatlantic telegraph cables, the expansion of telegraph systems between cities, and the desire to transmit more than one signal per telegraph wire (a concept known today as multiplexing). Figure 1-2 shows a picture of the transatlantic cable station at Heart's Content, Newfoundland, in 1872. This is the site of the North American side of the first transatlantic telegraph cable (go to **http:\\www.atlantic-cable.com/NF2001/hc.htm** for numerous photos of the present-day site, which is now a museum).

Presently, another transatlantic telegraph museum also exists on Cape Cod in Orleans, Massachusetts. The French Cable Station Museum has an interesting display of early transatlantic telegraph technology on exhibit from its early days as a transatlantic telegraph relay station from 1891 to 1898 and then as a cable station from 1898 until 1959 (see Figure 1-3). This station served as the North American side of the longest telegraph cable (christened *Le Direct*) laid between Europe and America up until that time (1898).

The 1860s brought the completion of a transcontinental telegraph line in 1861 that linked together the east and west coasts of the United States. Then, in 1865, Maxwell published his theory of electromagnetic waves, linking together electricity, magnetism, and light through mathematical equations.

FIGURE 1-2 Transatlantic Cable Station at Heart's Content, Newfoundland (From *Proceedings of the IEEE*, Volume 64, Number 9 © 1976 IEEE)

FIGURE 1-3 Telegraph Museum in Orleans, Massachusetts

The Late 1880s

The pace of technological innovation sped up during the final decades of the 1800s. The telegraph had linked most of the great eastern cites of the United States when, in 1876, Alexander Graham Bell, who began his technical career working on ways to improve the telegraph, invented the telephone. Bell's invention was not initially embraced due to the popularity of the telegraph. However, improvements to the telephone, such as Edison's carbon microphone, and demonstrations to the public eventually improved its popularity. Figure 1-4 shows Bell demonstrating the operation of the telephone in a crowded lecture hall in Salem, Massachusetts in 1877. The audience was able to hear the voice of Watson, who was located in Boston eighteen miles away.

In 1879, Edison invented the incandescent lightbulb, which was a precursor to the vacuum tube (see Figure 1-5).

In 1887 Heinrich Hertz performed laboratory experiments that proved the existence of electromagnetic waves, just as Maxwell predicted back in 1865. During the 1880s and 1890s geographically limited telephone systems were constructed in countries around the world. Early telephone systems usually consisted of a single iron wire mounted on poles, connected to a central manual switchboard with additional manual exchanges to connect calls from one part of town to another. Bell patented the idea of using two wires in 1881. This improvement, which reduced system noise, was slow to be implemented; however, the innovation allowed for the possibility of long-distance telephone calls. Further, in the late 1890s loading coils that were inserted into the wire circuits to improve frequency response also enhanced the maximum distance of operation.

(Go to **http:\\www.privateline.com** for more historical information about the telephone and the development of that industry.)

FIGURE 1-4 In a crowded lecture hall in Salem, Massachusetts in 1877, A.G. Bell demonstrates the operation of the telephone. © Bettmann/CORBIS

From 1895 until 1901 Marconi experimented with a **wireless** telegraph. He initially started his experiments at his family's villa in Bologna, Italy. He then moved to England in 1896 to continue his work. He built several

FIGURE 1-5 A replica of Edison's lightbulb (*Courtesy of Edison National Historic Site, National Park Service, U.S. Department of the Interior*)

radio telegraph stations there and started commercial service between England and France during 1899. However, the defining moment for wireless is usually considered to have taken place on December 12, 1901, when Marconi sent a message (the signal was a repetitive letter "s" in Morse code) from Cornwall, England to Signal Hill, St. John's, Newfoundland, the first transmission across the Atlantic Ocean. One should note that at this time there were no "electronic devices" in existence—no vacuum tubes to amplify weak signals or to generate high-frequency signals for transmission.

Historical Footnote Located within the National Sea Shore in Wellfleet, Massachusetts is a historical site known as Marconi Station where further transatlantic radio experiments were conducted by Marconi at the turn of the nineteenth century. It is claimed that the first transatlantic telegraph message from the United States to England was sent from this site on January 19, 1903. During the same time period, Reginald Fessenden was conducting experiments with wireless radio on the Outer Banks of North Carolina.

The 1900s and the Age of Electronics

The 1900s brought the world into another technological era, for it was not long into the twentieth century before the advent of electronic devices. In 1904, John Fleming invented the diode vacuum tube (see Figure 1-6).

In 1906, Lee de Forest expanded on this, inventing the triode vacuum tube by inserting an additional control element (eventually known as the "grid") into Fleming's valve. Also in 1906, Reginald Fessenden conducted the first radio broadcast on Christmas Eve. Figures 1-7a and 1-7b show the birthplace of radio broadcasting—the National Electrical Signaling

FIGURE 1-6 Fleming's diode vacuum tube (From *Engineers & Electronics* New York: IEEE Press © 1984 IEEE)

FIGURE 1-7 (a) Brant Rock Station (*Courtesy of the North Carolina Office of Archives and History*); (b) The antenna of the Brant Rock Station

Company (NESCO) radio station located at Brant Rock, Massachusetts, the site of Fessenden's first radio broadcast. An Internet search of *Reginald Fessenden* will yield many interesting Web sites about his work. One such site can be found at **http:\\www.kwarc.org/hammond/fessenden/html**.

Early wireless transmitters and receivers used the limited technology available at the time and operated at low frequencies (LF). Wireless telegraph transmitters were usually "spark-gap" devices augmented with resonant circuits (see Figure 1-8 on page 10) or high-power, high-frequency Alexanderson alternators. If one looks carefully at the bottom-right corner of Figure 1-7a, one can see the high-frequency alternator used by Fessenden.

Wireless radio receivers initially used crude, unstable coherers made from powdered metal filings or liquid mercury. These early mechanical detectors were bistable devices that exhibited different levels of resistance when subjected to a dc or an ac voltage. Later, crystal detectors that used galena with a "cat's whisker" contact were used to detect the received signal. At the time it was thought that only low frequencies could be used to transmit over long-distances and that propagation losses increased with frequency. Therefore, most wireless systems operated at frequencies below 200 kHz or at wavelengths longer than 1500 meters.

The Vacuum Tube Comes of Age

The 1910s brought many improvements in vacuum tube technology (see Figure 1-9 on page 10). In 1913, with the new ability to produce a very

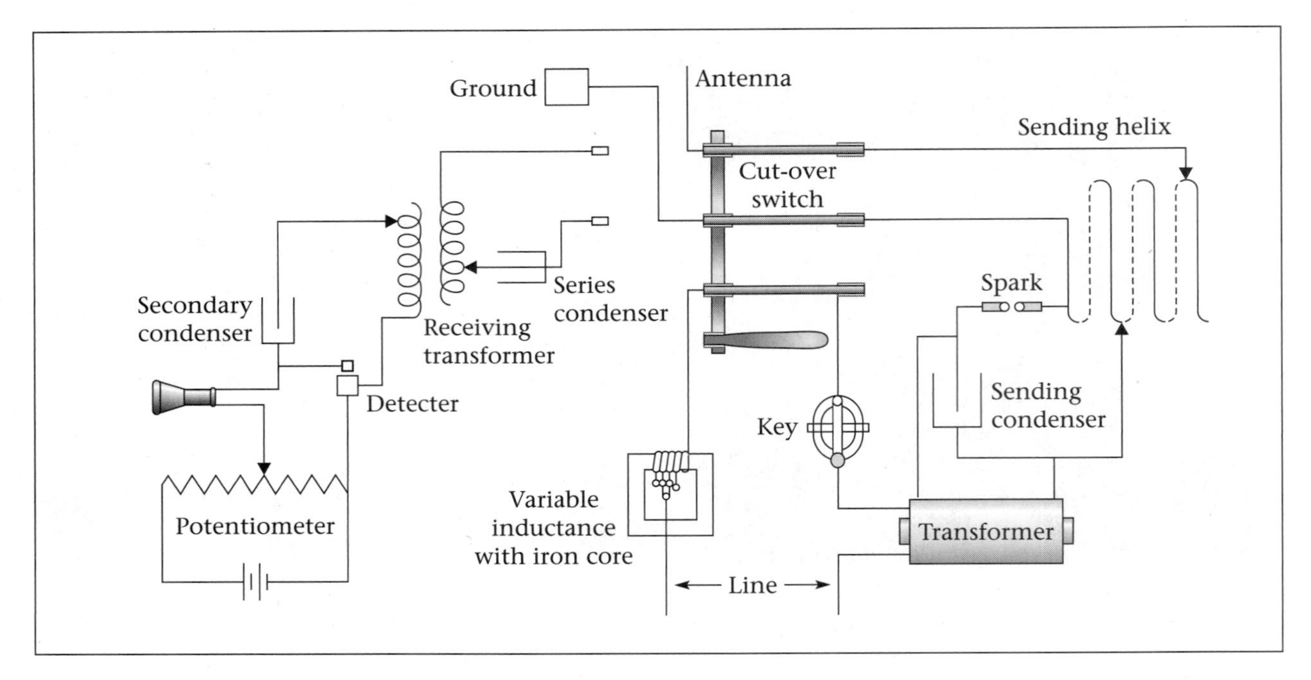

FIGURE 1-8 Wireless telegraph station schematic

high vacuum and the invention of the oxide-coated filament, early triodes made practical the cascade amplifier and triode oscillator.

The use of vacuum tubes as amplifiers and oscillators in conjunction with resonant circuits was exploited by wireless radio engineers of the day in both transmitters and receivers. Edwin Armstrong, an early radio

FIGURE 1-9 Early vacuum tubes designed by van der Bijl (*Courtesy of Vermeulen*)

pioneer, was instrumental in the development of the regenerative radio receiver in 1913. Armstrong is also credited with the invention of the superheterodyne receiver in 1918, even though the heterodyning technique was first introduced by Fessenden earlier in the century.

During this same decade, the United States Navy led a major effort to develop wireless radio for ship-to-ship and ship-to-shore communications. Historical accounts of the sinking of the Titanic on April 12, 1912, tell of the transmission of futile "SOS" distress messages by the ship's wireless operator.

In 1913, Western Electric tested prototype vacuum tube repeaters for long-distance telephone calls (see Figure 1-10). By 1915, the first transcontinental telephone line was completed (see Figure 1-11). This system used 8-gauge (1/6 inch) copper wire, repeater amplifiers using van der Bijl's type-101A vacuum tubes, and reportedly had a bandwidth of approximately 900 Hz.

The start of World War I during the last part of the decade was also a major driver of the development of radio technology by the United States military. See Figure 1-12 on page 12 for a typical transmitter-receiver pair from that era or go to **http:\\www.users.erols.com/oldradio/index.htm** for examples of old radio equipment.

FIGURE 1-10 Vacuum tube repeaters used on the transcontinental telephone link (*Courtesy of Lucent*)

FIGURE 1-11 Connecting together the east and west coasts of America (*Courtesy of Lucent*)

FIGURE 1-12 Western Electric airplane transmitting and receiving set (From *Transactions of the AIEE*, vol. XXXVIII, 1919)

The 1920s: The Shift to Short Waves

The 1920s might well be characterized as the decade of high-frequency radio development. During this era, Marconi's research on radio-wave propagation revealed that transatlantic radio transmission was feasible at frequencies much higher than had been thought possible. At the same time, vacuum-tube technology had improved to such an extent as to increase the upper-frequency limit of their operation. Propagation studies by Sir Edward Appleton demonstrated that ionospheric layers could be used to reflect high-frequency waves back and forth between the Earth's surface and the ionosphere, hence allowing the propagation of radio waves around the world! Pioneering work by Ralph Bown in the field of wave propagation and the use of multiple antennas to prevent signal fading helped to make transatlantic communications a practical reality. By 1926, transoceanic telephone calls were available via high-frequency radio transmission.

The 1930s: Radio and Television Come of Age

During the early 1930s there was renewed interest in the transmission of pictures via television, prompted by the earlier invention of the iconoscope and the kinescope by Zworykin (see Figures 1-13a and 1-13b).

During this decade the facsimile machine was developed and Armstrong started to experiment with the use of wideband frequency modulation (FM) for broadcasting. (An excellent Web site about Armstrong and his work is located at **http:\\www.users.erols.com/oldradio/index.htm**.) Another innovation of note was the development of automated electromechanical telephone switches and the deployment of these switches in the central offices of the public-switched telephone network (PSTN).

The 1930s saw the introduction of several different versions of television or "TV" systems, the development of the low-distortion negative-feedback amplifier, and a push to higher frequencies of operation. In the

FIGURE 1-13 (a) Prototype iconoscope (From *Proceedings of the IRE*, vol. 21, © 1933 IRE); (b) Prototype kinescope (From *Proceedings of the IRE*, vol. 21, © 1933 IRE)

United States, 1934 brought the Federal Communications Commission (FCC) and its regulation of radio broadcasting and long-distance telephony.

World War II

World War II started in the last half of the 1930s and proved to have a significant influence on technology. The invention of radar (*ra*dio *d*etecting *a*nd *r*anging) for the detection of enemy forces was essentially a joint effort by both the United States and England, and brought with it many technological advances. Some of these can be traced back to World War II and radar, including the development of practical pulse modulation systems, new high-frequency antenna arrays, practical waveguide transmission lines and components, and the invention of both the Klystron and Magnetron vacuum tubes for operation at what were called "microwave" frequencies. The first FM broadcasting band from 42 to 50 MHz was opened in 1938, and further research and experimentation was done on both the very-high frequency (VHF) band (30 to 300 MHz) and the ultra-high frequency (UHF) band (300 to 3000 MHz). Additionally, the PSTN saw the implementation of the crossbar switch at both central offices and exchanges.

The 1940s: The Digital Computer and the Transistor

The 1940s saw further development and refinement of television technology in the United States with the adoption of the National Television Standards Committee's (NTSC) technical recommendations in 1941. Coaxial cable networks that had their beginnings in the late 1930s were used to transport "network" TV programs to other distant TV stations. The PSTN started to utilize the same technology to deliver multiple long-distance phone calls over coaxial cables using frequency division multiplexing (FDM). 1943 saw the invention of the traveling wave tube (TWT), the same device that makes present-day direct satellite broadcasting possible. World

War II continued to drive the development of technology. Funding from the United States government hastened the development of the first practical electronic digital computers during the mid-1940s and ushered in the age of digital technology. During 1947, the FM band was shifted to its present 88- to 108-MHz range and the first wideband microwave relay system was installed. Finally, during late 1947 the transistor was invented and the age of solid-state technology arrived. See Figure 1-14 for a picture of the first point-contact transistor (the junction transistor appeared on the scene in 1950). Due to manufacturing problems and reliability concerns, it took approximately ten years to turn this invention by Shockley, Bardeen, and Brittain into a technical innovation.

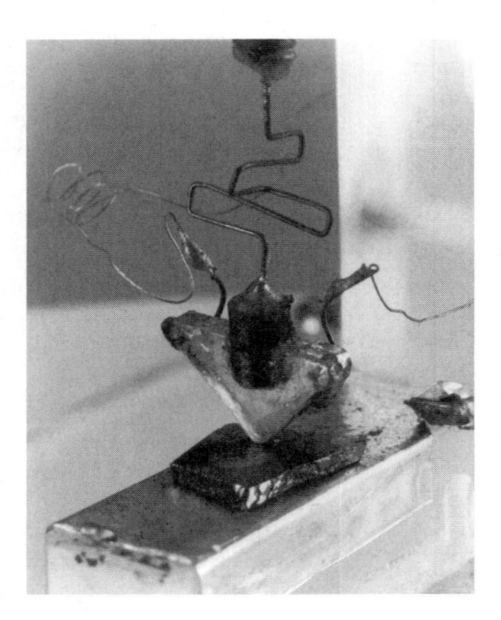

FIGURE 1-14 The first transistor (*Courtesy of Lucent*)

The 1950s: The Pace Quickens

During the 1950s, television became extremely popular, the UHF television band was opened to provide for more TV stations, and public-safety agencies (police and fire departments, etc.) began to adopt the use of high-frequency mobile radio for their fleet communications. In an effort to bring TV to areas of poor reception, community antenna TV (CATV) systems were implemented for the first time. By 1951, the first transcontinental microwave-relay network was operational and was used to deliver long-distance telephone calls and network television programming. In 1956, TAT-1 (Transatlantic Telephone Cable #1), a 36-circuit submarine cable was put into operation. One might be surprised to know that it did not use transistors, but instead utilized highly reliable vacuum tubes for the cable-repeater amplifiers.

The 1960s: New Frontiers

The 1960s saw a proliferation of cable TV systems (over 800 systems with close to one million subscribers), U.S. television broadcasting transitioned to color programming, and television receivers became 100 percent solid state, meaning that every active device was a semiconductor except the cathode-ray tube (CRT). In 1961 the telephone company deployed the first "T1" trunk line, which uses time division multiplexing (TDM) to transmit twenty-four simultaneous pulse-code modulation (PCM) phone calls on the same cable or pair of wires. In 1962 FM stations began broadcasting in stereo and Telstar, a commercial orbiting communications satellite, was launched. Like the live televised broadcasts of the early United States space-program launches, Telstar, a low-earth-orbit satellite (LEOS), was tested during prime time while the entire nation watched the amazing technological advance. The world had suddenly grown a little bit smaller! 1964 brought about the first deployment by AT&T of dual-tone multifrequency (DTMF) signaling, also known as touch-tone dialing, to the PSTN. In 1965, Intelsat I or "Early Bird," the first geosynchronous communications satellite, was put into operation. In 1968 the last transatlantic telephone cable utilizing vacuum-tube repeaters was placed into service. Also during this decade AT&T started to deploy the #1 Electronic Switching System (1ESS), a stored-program digital electronic switch, into the PSTN.

1.2 The Modern Era of Telecommunications

The period of time from 1970 to the present can be referred to as the modern telecommunications era. During this period there have been many advances in telecommunications; however, the driving force behind many of these changes has been other technological innovations. Presently, the build-out of a worldwide broadband telecommunications infrastructure has become the largest industrial undertaking in the history of mankind. To explain why this has come about it is necessary to first consider several key technological innovations and their evolutions.

The Driving Forces of Technological Innovation

First and foremost one must look at the development of computer and digital technology. Recall that the first digital computers appeared during the mid-1940s. Built based on vacuum tube (first generation) technology, they were large, expensive, and difficult to maintain and run. Later generations of the computer became more affordable, and mini-computers based on off-the-shelf digital integrated circuits (third-generation computer technology) changed the landscape of computing during the 1960s due to their widespread use and improved software (advances occurred in both operating systems and high-level application languages). During this era, timesharing and remote computer access over telephone lines also became popular. The desire to transmit data from computer to computer gave birth to the concept of packet-data switching and the initial development of data networks that linked organizations together.

The Microprocessor and the PC

In the early 1970s, Intel and Busicom, a Japanese electronic calculator company, had a business agreement that Intel would develop an integrated circuit (IC) "chip-set" that would reduce the parts count and hence the size of Busicom's products. In November of 1971, Intel developed four ICs—the 4001, a read-only memory (ROM) chip; the 4002, a random-access memory (RAM) chip; the 4003, an input/output (I/O) chip; and the 4004, a central-processing (CPU) chip. This chip-set could manipulate four binary bits at a time. Interestingly, due to financial difficulties Busicom agreed, for financial consideration, to allow Intel to secure the rights to their designs and to sell their chip-set to other customers. This was the start of the microprocessor age.

Intel quickly introduced the 8008, an 8-bit machine on a single IC, in 1972. The 8008 evolved into the 8080 in 1974 and the 8085 was introduced in 1976. The 8080 needed three different power-supply voltages to operate, while the 8085 only needed a single +5-V supply. The birth of the modern microprocessor began in 1978 with the introduction of the 8086, a 16-bit machine capable of accessing 64K of memory. In quick succession, Intel introduced the 8086 family of processors. The 8086 and its successors, the 80286, 80386, and 80486 were used in the highly successful line of IBM personal computers (PCs) first introduced in 1981 (based on another Intel microprocessor, the 8088). The 80186 was incompatible with other members of this family, and was therefore rarely used in PCs. The IBM offering legitimized the PC, which until this time had not been taken seriously by the business community. During the same timeframe Apple Computer introduced the Apple 2e home computer; however, it was based on the 6800, a 16-bit microprocessor from Motorola. Various faster and more powerful versions of the 80386 and 80486 were produced by Intel before the introduction of the Pentium I processor in 1993. As of 2002, the Pentium family has evolved to include the Pentium IV, which operates at speeds as high as 2.53 GHz. On the horizon is a 64-bit microprocessor, the Itanium processor, introduced by Intel during 2002. See Intel's Web site **http:\\\\www.intel.com**, for more information.

With the lower cost of computer hardware, businesses and schools could afford to equip offices and classrooms with PCs. It was not long before the required technology evolved to interconnect these PCs into local-area networks (LANs) that allowed the PCs to share resources such as printers and databases stored on servers. This innovative application of telecommunications technology, sporting unusual names like "Token Ring" and "Ethernet," spawned a new field of computer networking.

Present-day PCs, usually costing less than $1000, have many times more processing power than the mainframe computers from the 1970s and 1980s, which cost hundreds of thousands to millions of dollars! Furthermore, technological and software innovations have made the PC capable of manipulating, storing, and displaying rich multimedia content in a networked environment.

Moore's Law

What has led to this rapid development of the PC? The answer to this question brings us to another technological innovation, the integrated circuit, or IC. The advent of integrated-circuit technology, or microelectronics, that began in the late 1950s has contributed greatly to the rapid advances in the PC field. Once initial fabrication problems were overcome, Moore's Law applied and the integration of more and more devices on a single piece of silicon has continued unabated for over four decades (see Figure 1-15).

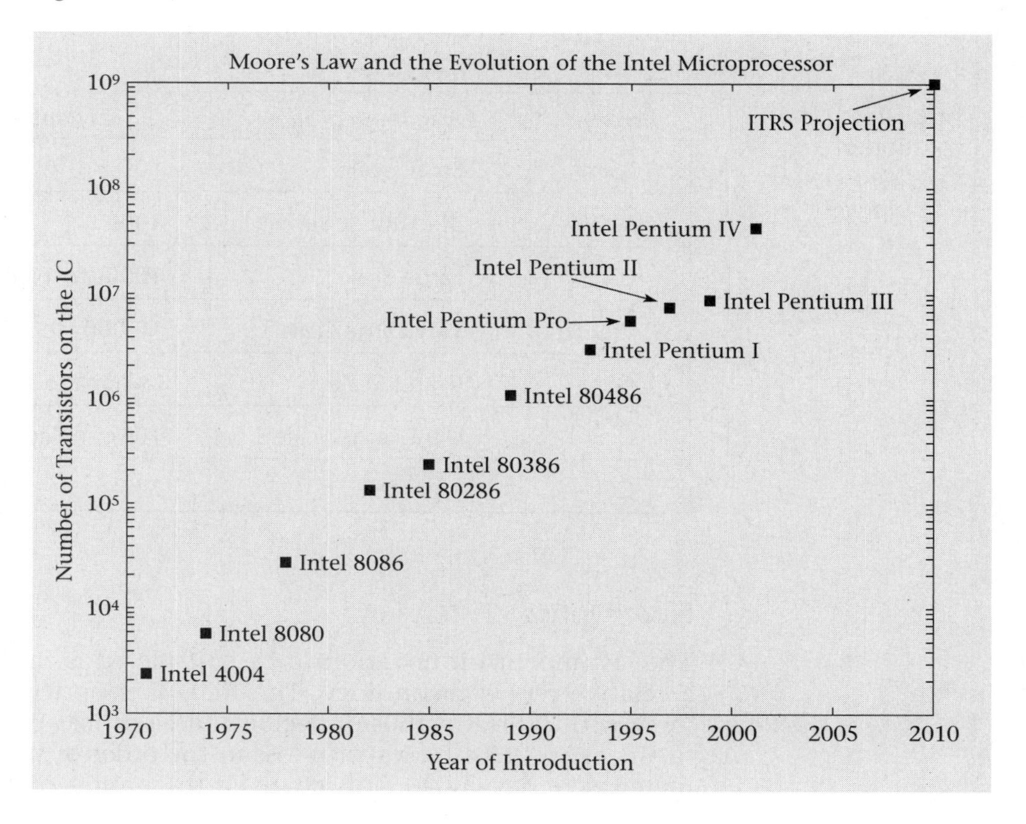

FIGURE 1-15 Depiction of Moore's Law using the evolution of the Intel Microprocessor as a reference

Moore's Law states that the number of devices on an IC chip doubles every eighteen months to two years. Figure 1-15 illustrates this phenomenon by plotting the number of transistors used in the various microprocessors developed by Intel Corporation over the last three decades. The future of Moore's Law may be in danger as a result of the approach of fundamental limits in silicon technology. However, the International Technology Roadmap for Semiconductors (ITRS) project still predicts that by 2010 approximately a billion devices will be integrated on a single chip! This is represented by the top-right point on Figure 1-15.

Table 1-1 shows the development of the IC in terms of the names or acronyms used to describe the levels of integration. Beginning with small-scale integration (SSI), the technology has advanced to today's grand scale or ultra-large scale integration (GSI and ULSI). Today's ICs can contain tens of millions of devices (the Pentium IV contains 42 million transistors). The future holds the promise of **gigascale integration** (also GSI) levels! One might argue that most electronic technological innovations have occurred simply because our microelectronics manufacturing processes have been improving. It should also be noted that IC speed has also increased dramatically during this timeperiod and that discreet analog devices for RF applications are capable of operation at frequencies of tens of GHz, a vast improvement over earlier devices.

TABLE 1-1 The evolution of integrated-circuit technology

Acronym	*Level of Integration*	*Number of Transistors*
SSI	Small Scale	10s
MSI	Medium Scale	100s
LSI	Large Scale	1000s to 10,000s
VLSI	Very Large Scale	10,000s to Millions
GSI	Grand Scale	Millions to 10s of millions
ULSI	Ultra-Large Scale	10s of millions to 100s of millions
GSI	Gigascale	Billions

Fiber Optics

The last modern innovation to be considered is that of fiber-optic cable, a relatively new technology. The first fiber-optic cables were extremely lossy with typically 100s of decibels (dBs) of loss per kilometer. However, in the early 1970s cables with loss in the order of several decibels per kilometer were developed; this changed the nature of "guided" information transmission. Fiber-optic cables have an almost unlimited bandwidth available for data or information transmission. Presently, cable losses have been reduced to a few tenths' of a decibel per kilometer. Additionally, we have moved from the use of multimode to single-mode cables with their inherently higher data-transmission rates. Development of reliable solid-state light sources at 1380, 1510, and most recently 1400 nanometers has increased the data capacity of fiber-optic cables. Additional new technology to more fully utilize the cable's bandwidth is now available through wavelength division multiplexing (WDM) and dense-wavelength division multiplexing (DWDM) techniques that have recently been commercialized. The use of WDM and DWDM allows older fiber-optic cable infrastructure to be upgraded with the opportunity to increase data-transmission rates by a factor of one hundred or more. Numerous submarine cables have been installed linking the world's continents together and in-cable rare-earth (erbium) amplifiers have been developed, which in essence provide loss-free fiber-optic cables. To date, the only disadvantage of fiber-optic cable is that it is still relatively expensive to install. Recently, inexpensive plastic fiber-optic cable has been developed which might be used in the local-area network (LAN) environment.

1.3 Modern Telecommunications Infrastructure

With these three innovations—the microprocessor, the PC, and fiber optics—as a backdrop, let's return to our chronicle of modern telecommunications by examining the current major transmission infrastructure. Some of these systems have a long history, but in each case technology has been evolving these systems into something totally different from what they were only ten years ago.

The Cable-TV System

In starting a discussion about cable TV, it should be noted that the original cable systems consisted of twelve channels (channels 2 through 13 in the VHF band) compatible with the average subscriber's TV receiver. Today cable TV penetration has evolved to a subscriber base of over sixty percent of the population of the United States, and the newest cable technologies can provide over one-hundred channels or close to 1 GHz of bandwidth. In recent times most cable-systems have undergone a fiber "build-out" to upgrade their cable system hardware and are transitioning to two-way systems, allowing for an upstream path as well as the present downstream path.

Cable operators now provide "set-top" converter boxes that can deal with new digital-modulation formats (primarily used to deliver pay-per-view content) and, depending upon location, have offered cable modems for fast Internet access. Major corporations such as AT&T and MediaOne (now known as AT&T Broadband) have merged together believing that broadband cable will be the medium that will deliver the telecommunication services (telephone, Internet access, and high-definition video entertainment) desired by the consumer in this new century. A recently announced merger will occur shortly between AT&T Broadband and Comcast to form an even larger corporation with over 20 million customers.

Wireless Mobile Radio

The next innovations to consider are fixed and mobile wireless communications. Although wireless mobile radio has a history that goes back many decades, the modern era started in the United States in 1983 when the first modern commercial system went into operation. The first commercial cellular telephone system was known as the advanced mobile-phone system (AMPS). It was a simple analog FM system that used frequencies originally assigned to UHF television (formerly channels 70 through 83 on the UHF dial) in the 800- to 900-MHz range. The cellular concept made use of frequency reuse plans to reduce co-channel interference between cells and different uplink and downlink frequencies, thus making it possible to simplify transceiver design. In the early 1990s D-AMPS (digital AMPS) technology was introduced in an effort to increase the capacity of the systems.

Cellular radio has evolved in others parts of the world using different access technologies and frequencies. At the present time, with over one billion subscribers, there are plans to bring high-speed data (Internet access) to cellular telephones. This initiative is known as 3G (Third Generation). Standards bodies from around the world are attempting to harmonize the various standards in existence so that one might use their cell phone anywhere in the world regardless of the technology in use. Meanwhile, new frequency allocations and new technologies such as wideband code-division multiple access (WCDMA) and variations of time-division multiple access (TDMA) have recently emerged, which adds to the confusion. During the period between 2000 and 2002, most providers of cellular service began to offer low-speed data transmission on their systems and began to introduce intersystem text messaging interoperability. These are two of the first steps toward high-speed data access for mobile networks.

Other Wireless Services

Other wireless services exist and have recently become more popular. Wireless LANs operating at 2.4 GHz and running at 11 Mbps are available, and like modern modems these LANs have fallback rates if excessive interference occurs. Newer technology will extend the data rates to approximately

50 Mbps at frequencies in the 5-GHz range. Bluetooth™ technology will create personal-area networks (PANs), with which one's laptop or desktop computer will be able to do such things as talk to a printer or scanner without the need for interconnecting cables. Other emerging wireless technologies such as local multipoint distribution systems (LMDS) and microwave multipoint distribution systems (MMDS), offer broadband fixed wireless access (BFWA) at millimeter wave frequencies from 10 to 66 GHz or in the industrial, scientific, and medical (ISM) and unlicensed national information infrastructure (U-NII) bands at 2 and 5 GHz. Direct satellite broadcasting has become popular and has augmented cable TV in rural and not so rural areas of the United States. DirecPC (high-speed Internet access via satellite) is now being offered by DirecTV. Finally, low-earth-orbit satellite (LEOS) systems providing high-speed Internet access are closer to becoming operational, albeit after several system redesigns that have resulted in fewer total satellites. The Teledisic system, Bill Gates' "Internet in the Sky," is poised for start-up sometime in 2004. However, it should be pointed out that other start-up satellite systems for personal communications have met with failure—Iridium failed in 2000 and as of 2002 Globalstar is having financial difficulties.

The PSTN

What about the traditional telecommunications providers of the PSTN, the local and long-distance telephone companies? Although not obvious to the individual consumer, the long-distance providers and the local telephone companies have been constantly upgrading their facilities with fiber-optic cables and new high-speed broadband digital-transmission technologies as they have been developed over the past three decades. Transmission facilities using digital transport schemes like frame and cell relay, asynchronous transfer mode (ATM), and synchronous optical network (SONET) are all in use throughout the Public Data Network. Modern switching systems including #5ESS and DMS-100 digital switches, routers for Internet Protocol (IP), and digital cross-connect systems (DCS) can be found in the central offices of the local telephone provider and the competitive local exchange carrier (CLEC).

On the other hand, the local loop of the telephone company (that infamous last mile) still consists of copper pairs with limited bandwidth. In an effort to compete with the broadband cable systems and satellite providers, local telephone systems have started to offer asymmetrical digital subscriber line (ADSL) service for high-speed data transmission (Internet access) over existing copper pairs. However, there are strict distance limitations, and sometimes copper pairs that seemingly meet the criteria needed for ADSL do not work. It remains to be seen how successful ADSL or some of its variations will be in the local loop compared to the cable modem.

1.4 The Convergence of Technology

At this point the convergence of all of these technologies should be discussed. In the early 1990s, delivery of entertainment (video on demand) was perceived to be the driving force behind the build-out of broadband telecommunications systems. Since the last half of the 1990s, the Internet has been the major driver of this effort. During this time period the world has experienced a technological revolution unmatched by any other in the history of mankind. As previously stated, the building of today's telecommunications infrastructure is the largest industrial endeavor ever to be undertaken. There are, however, historic parallels in the building of the telegraph, railroad, telephone, and interstate highway systems in the United States. As history has witnessed with the introduction of previous technologies, our society will forever change as we enter the "information age" facilitated by broadband access to the Internet.

Technology and the Internet

The question might be asked, But why now? Because of the convergence of the enabling technologies previously discussed. With the advent of PCs that can deal with rich multimedia content (i.e., video, high-definition pictures, CD-quality audio, and so forth) people have the desire to be connected to other computers that are sources of this information or to share the content of their own computers with others. This type of communication is facilitated by the Internet. The desire for connectivity is presently extremely strong and has been the driving force behind the PC industry, the business community, and the telecommunications industry. All are striving to satisfy the public's thirst for broadband connectivity to the Internet "at anytime and from anywhere."

Summary of Part I

As historians look back at the early days of the Internet, they will most likely note the relatively slow development of technology leading up to the creation of our present-day telecommunications infrastructure over the course of the last century. To be sure, there will be rapid changes and vast improvements to that infrastructure during the next decade, the effects of which we can only guess. What will be the ultimate effect of a broadband Internet upon society? Like other technological innovations, the full story is yet to be written. We can only hope that the best is yet to come!

PART II

A Regulatory History of Telecommunications and Telecommunications Standards

1.5 The Beginnings of Regulation

To start a discussion of **regulation** requires a definition of telecommunications. Presently, **telecommunications** is the communication of voice or data (including multimedia content) over a distance using the Public Switched Telephone Network (PSTN) or privately owned networks. It is important to note the use of the word *public* when describing the existing telephone network, as it intimates government involvement.

Why is regulation of the telecommunications industry necessary? In the beginning, during the era of the telegraph, there was no regulation of the industry. However, in a related matter, the rapid development of the railroad industry in the nineteenth century led to the formation of monopolies, and it soon became apparent that the federal government had to step in to protect the consumer from unfair business practices.

In 1887, with the passing of the Interstate Commerce Act, federal regulation of businesses officially began. Another law regulating corporations, the Sherman Antitrust Act, was passed in 1890. Initial legislation concerning the telecommunications industry permitted a **regulated monopoly** so as to provide what is known as "economies of scale" (i.e., the ability to provide services at lower costs due to a larger business volume). In general, a regulated monopoly is allowed to charge a rate for its services that will provide it with a certain fixed return (percentage) on its investment. The rate is usually set by the appropriate government regulatory agencies. A valid question is whether this is beneficial or harmful. The answer to this is often dependent upon whether the industry is newly evolving or mature at the time when the question is asked. The truth usually lies somewhere between the two extremes.

Regulation and the Early Telecommunications Industry

Let us recap the early history of the telecommunications industry. Morse invented the telegraph in 1837 and formed a telegraph company based on his new technology in the mid-1840s. The Western Union Telegraph Company was established in 1856 and within a decade bought out most of its competitors (an early monopoly?). Early long-distance telegraph systems required many relay points because signals had a limited maximum range. In 1867, an improved telegraph relay was invented by Elisha Gray. Gray's company was bought out by another company that in 1872 became the Western Electric Manufacturing Company.

Now Comes the Telephone!

Bell received a patent in 1876 (he filed his patent application for the telephone only hours before Elisha Gray) and formed the Bell Telephone Company in 1877. In 1882, Bell bought the Western Electric Company and in 1885 formed what became the American Telephone & Telegraph Company (AT&T). Bell's patent expired in 1893, leading to the establishmeent of numerous "independent" telephone companies. Most of these independents provided service to rural areas where Bell had not done so because he felt that it would not be profitable. However, some independents competed directly with Bell's system in larger cities. By 1900, the Bell system served approximately sixty percent of telephone subscribers in the United States; the rest were served by the approximately 4,000 independents in existence at the time. During the 1900s and early 1910s, AT&T, under J.P. Morgan's direction, purchased large amounts of Western Union stock and also bought out many of AT&T's competitors, in essence forming a monopoly provider of telecommunications.

Increasing Regulation

Due to perceived misdeeds by the Bell telephone system, the Mann-Elkins Act was passed in 1910, the first United States government step in regulating the early telecommunications industry. However, the type of regulation imposed was basically the same as that used to regulate the railroad industry, an inadequate approach for regulation of telecommunications. The Interstate Commerce Commission (ICC) was placed in charge of enforcing the regulations of the Mann-Elkins Act. Shortly thereafter AT&T made what is called the "Kingsbury Commitment" of 1913 (Kingsbury was the AT&T vice president who wrote the commitment letter to the government). The Commitment stated that AT&T would not buy out any more independents. Furthermore, the Commitment would allow independents to connect to AT&T's network and AT&T would sell its interest in Western Union. Why did all this happen? At the time, the Department of Justice (DOJ) and the Sherman Antitrust Act were the driving forces threatening to further regulate AT&T. An interesting twist to all this was that AT&T required the independents to pay long-distance fees. So, in essence, the Bell system became a monopoly anyway.

Telecommunications Regulation in the Twentieth Century

More regulation—or deregulation, depending upon one's viewpoint—occurred due to the 1921 Graham Act. This act, lobbied for by AT&T, called for the exemption of the telecommunications industry from the Sherman Antitrust Act (recall the economies-of-scale idea) and monopoly service providers like AT&T, the Bells, and local independents resulted. The Graham Act prompted AT&T to expand its network nationally, as desired by the government at the time. However, there were no controls built into the Graham Act to ensure that these telecommunications providers would keep down the fees to be paid by the consumers. Thirteen years later the United States Congress passed the 1934 Communications Act. Designed to regulate radio broadcasting and the interstate telecommunications provided by AT&T, the purpose of this act was to correct the shortfalls of the Graham Act. Taking this increased regulation one step further, the Federal Communications Commission (FCC) was established, taking over from the ICC the duties of policing the ever-expanding telecommunications industry.

The Rural Electrification Act

Another government initiative, The Rural Electrification Act of 1936, created the Rural Electrification Administration (REA). This act helped to provide electrical power to rural America where power companies, due to economics, refused to expand their presence. This act was expanded in the late 1940s to cover telephone service in rural areas. At the time, it was felt that the telephone was a household's lifeline to public safety agencies and that every household should be connected to the PSTN. These programs exist to this day as Federal Domestic Assistance programs.

Regulation at Different Government Levels

What about regulation at the local or international level? The regulation of telecommunications within a state (known as intrastate telecommunications) became vested in the state. This led to the creation of Public Utilities Commissions (PUCs) in each state. State regulations allowed only one local telephone company to serve a **local access territory** (LAT), though recent (1996) legislation has changed this. Regulation at both the local and federal level required telephone companies to file documents called tariffs detailing any new service and its associated charges.

At the other end of the spectrum are the International Telecommunications Regulations designed to facilitate the most efficient operation and interoperability of worldwide telecommunications across national and international boundaries. Generally, in the interest of their own well-being, nations voluntarily abide by these regulations.

1.6 The Beginnings of Deregulation

In 1949 the DOJ sued AT&T for alleged violations of the Sherman Antitrust Act. At the time, the Bell Operating Companies (BOCs) only purchased equipment from Western Electric, which was then a part of AT&T. From the government's point of view there appeared to be little incentive for Western Electric to charge competitive prices for their equipment, for the cost of the equipment could be passed on to the consumer by the BOCs. The DOJ believed that Western Electric was selling the same equipment to independent telephone companies at lower prices, therefore employing anticompetitive practices against other, smaller equipment manufacturers. In 1956, to prevent further action by the government, AT&T agreed to sell equipment to the BOCs at competitive prices, but was allowed to keep Western Electric. This agreement, known as the "Consent Decree of 1956," allowed companies like Northern Telecom (parent company of Nortel) and others to enter the marketplace and become major sources of telecommunications equipment for both independents and the BOCs. This action was the beginning of the **deregulation** of the telecommunications industry, putting the wheels in motion for free competition and a self-regulating market.

The Carterfone Decision

Another important point in the history of telecommunications deregulation is the 1968 Carterfone Decision. This decision effectively allowed private mobile radio systems to be connected to the PSTN. This FCC decision led to the ability of the consumer to buy phone equipment from someone other than Western Electric. The competing companies allowed in this market are called "interconnecting companies." AT&T fought this deregulation, claiming that equipment attached to their network from interconnecting companies would be inferior to their own equipment, thus causing harm to the PSTN. At first, in an interesting ruling, the FCC required an "interconnect" device between the network and the other companies' equipment. However, this requirement was struck down in 1978; instead, the FCC required registration of all equipment intended for connection to the PSTN.

The MCI Ruling

The pace of deregulation picked up with the "MCI Ruling of 1969." This ruling allowed businesses to use companies other than AT&T for private telecommunications networks. Additionally, the "Specialized Common Carrier Decision of 1971" allowed any qualified "common carrier" to compete with AT&T for private point-to-point business networks. In 1975,

Microwave Communications Inc. (MCI) began providing long-distance service to the public. The FCC issued a cease and desist order to MCI at AT&T's request. A Federal Appeals Court reversed the FCC ruling, resulting in the "MCI Decision of 1976," which opened up the long-distance market to competition.

The Effect of Deregulation

At this point, one might wonder what the effect of all the deregulation of the telecommunications industry has been. As with most controversial issues, there appear to be two schools of thought. The proponents of deregulation believe that it is a good thing and will eventually lead to lower rates and more services as a result of marketplace competition. The opponents believe that this deregulation will lead to inferior service and higher prices. What has really happened? Most knowledgeable observers believe that over time telecommunications service has improved at the same time as long-distance rates have decreased. Local rates, however, have tended to increase. Why has this happened? Prior to 1984, AT&T shared its long-distance profits with local phone companies. After 1984, to make up for the lost revenues, local companies were permitted to raise rates and charge network access fees to long-distance companies. However, since the Telecommunications Reform Act of 1996 local telephone companies have been forced to lower access fees to long-distance carriers, and are therefore filing tariffs to increase local rates to make up the difference. Presently, Regional BOCs like Verizon have filed for permission to offer long-distance service within their territory, thus competing with the long-distance carriers. This is but another chapter in the evolution of the telecommunications network that is yet to be written, and only time will allow us to judge its influence on that evolution.

1.7 The FCC Computer Inquiries

Starting during the late 1960s and continuing to this day, the FCC, in response to changing transmission technologies and increasing data traffic between computer users, set about determining whether telecommunications regulations should apply to computer and data services. Recall the explosive growth of computer technology and networking during this period—the introduction of time sharing, dial-up modems providing remote access, the IBM PC, the Internet, and the World Wide Web all happened in rapid succession. As a result of these aforementioned innovations, a series of "Computer Inquiries" were initiated by the FCC.

Computer I

Starting in 1966 with a Notice of Inquiry (Computer I), the FCC studied the "interdependence of computer and communications services and facilities." The result of this inquiry was the Computer I Final Decision and Order of 1971. It effectively stated that computers used to facilitate communications belonged to the "basic services" category and everything else was an "enhanced service." Basic services fall under Title II of the Communications Act and are therefore subject to common carrier regulations. In contrast, enhanced-services (Internet services are in this category) are not subject to regulation! During this period the FCC codified the Carterfone decision into Part 68 of the FCC rules. Part 68 allowed the connection of terminal equipment to the PSTN from any source.

Computer II and III

The 1980 Computer Order II dealt with the entrance of the Bell Operating Companies (BOCs) into the enhanced-services market and the further deregulation of customer-provided equipment (CPE). The FCC rulings prompted by Inquiry II required the BOCs to set up separate subsidiary companies to enter the enhanced-services market and to handle the sales of CPE (note that the 1996 Telecommunication Act allows BOCs to reenter this market). The 1986 Computer Order III again dealt with the BOCs' entrance into the unregulated enhanced-services market; however, this time without a structured separation.

In summary, the Computer Inquiries defined the extent that the BOCs could compete in this new area of technology with the intention that they could enter the enhanced-services market subject to certain restrictions designed to prevent discrimination and anticompetitive behavior. Over the course of time, some of these rules have been reconsidered by the FCC, appealed in federal courts, and some are still under open proceedings.

1.8 Divestiture and Deregulation

Although deregulation was designed to increase competition in the telecommunications industry, the government perception was that AT&T continued to employ anticompetitive practices. This led the United States Department of Justice to file a lawsuit against AT&T in 1974. AT&T settled out of court in 1982 with another major compromise. What became known as "The Modified Final Judgment" (MFJ) took effect on January 1, 1984, and replaced the 1956 Consent Decree. AT&T was required to divest itself of all the Bell Operating Companies (BOCs), which would become known as the Regional BOCs, or RBOCs. Additionally, the MFJ deregulated the long-distance market. Again, it must be noted that the 1996 Telecommunications Act allowed AT&T to reenter the local telecommunications-service market!

The MFJ created many new terms and acronyms for the growing number of players in the telecommunications industry. There are presently Local-Exchange Carriers (LECs), formerly the BOCs or independents; Inter-Exchange Carriers (IECs or IXCs), common carriers that provide long-distance service (AT&T, MCI, and Sprint have approximately sixty percent of the market); and incumbent LECs (ILECs) and competitive LECs (CLECs). At the time of divestiture, the twenty-one existing BOCs merged to become seven RBOCs: Ameritech, Bell Atlantic, Bell South, NYNEX, Southwestern Bell, Pacific Telesis, and US West (see Table 1-2). Figure 1-16 on page 30 shows the territories of operation of the RBOCs during the late 1980s.

The original seven RBOCs have taken on an entirely different look as they have merged with other telecommunications businesses. The result of these measures is that there are presently four RBOCs (see Table 1-3 on page 30). Note that US West merged with Qwest, which is not a BOC, and Cincinnati Bell (now owned by Broadwing, Inc.) and Southern New England Telephone (SNET) were large but independent companies when divestiture occurred. They were therefore not considered BOCs. SNET has since been acquired by SBC.

TABLE 1-2 The reduction of the BOCs into RBOCs

RBOC	*Owner of*
Ameritech	Illinois Bell, Indiana Bell, Michigan Bell, Ohio Bell, Wisconsin Telephone
Bell Atlantic	The Chesapeake and Potomac Telephone Company of Maryland, The Chesapeake and Potomac Telephone Company of Virginia, The Chesapeake and Potomac Telephone Company of Washington, D.C., The Chesapeake and Potomac Telephone Company of West Virginia, The Bell Telephone Company of Pennsylvania, The Diamond State Telephone Company, New Jersey Bell Telephone Company
Bell South	South Central Bell, Southern Bell
NYNEX	New England Telephone and Telegraph, The New York Telephone Company
Southwestern Bell	Southwestern Bell
Pacific Telesis	Pacific Telephone, Nevada Bell
US West	Mountain Bell, Northwestern Bell, Pacific Northwest Bell

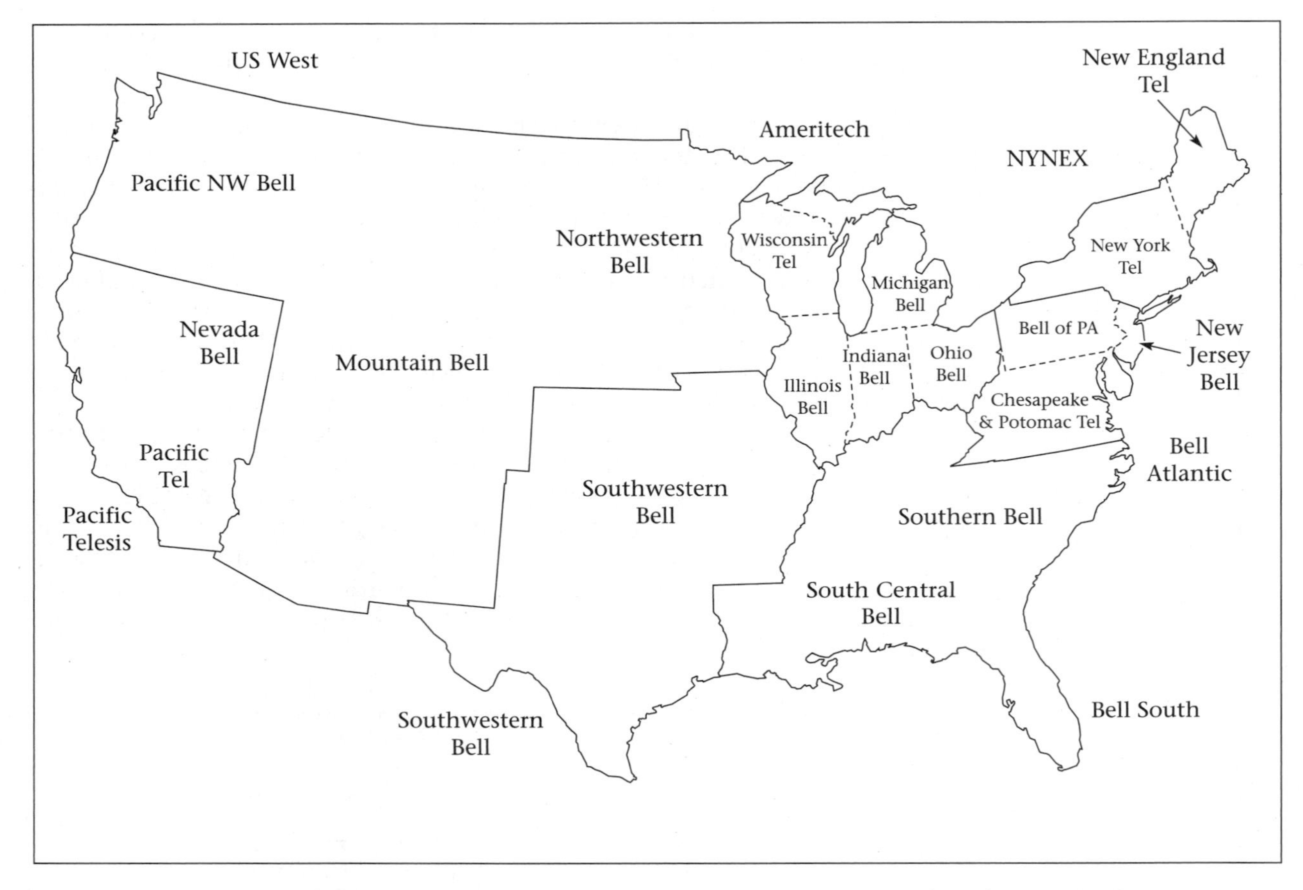

FIGURE 1-16 A map of the Regional Bell Operating Company territories after the MFJ in the late 1980s

TABLE 1-3 RBOCs in the year 2002

RBOC	*Created by Merger of:*
Verizon	Bell Atlantic (created by merger of NYNEX and Bell Atlantic) and GTE
SBC	Southwestern Bell, Ameritech, and Southern New England Telephone
Bell South	South Central Bell, Southeastern Bell, and Southern Bell
Qwest	US West and Qwest

As the industry has evolved, new businesses have formed to take advantage of new technologies and deregulation opportunities. In the telecommunications business there are satellite carriers, specialized common carriers (companies that offer so-called private-line services), and value-added network carriers (companies that operate packet networks). Any company that offers telecommunication services as part of the PSTN is referred to as a common carrier. Common carriers like Sprint can provide service to the general public as part of the PSTN, but they can also provide services for private networks. Furthermore, a private network can be built by leasing circuits from common carriers to construct either all or part of the network.

1.9 The Telecommunications Reform Act of 1996

This law had effectively replaced the MFJ on February 8, 1999. The act provided for further deregulation of both the telecommunications industry and the cable television industry. It removed the prior restrictions on AT&T, the RBOCs, and the IECs. Each is now able to compete in the other's business! The FCC was given the mandate to administer the 1996 Telecommunications Act in such a way as to enhance competition in the industry. In response to the congressional mandate, the FCC established a 14-point checklist for incumbent LECs to meet before they can reenter the interstate long-distance market, a Universal Service Fund to help bring enhanced-services to rural areas, and a fee called the E-rate Fund. There were immediate challenges to the FCC's actions in the courts. The E-rate Fund, which provides access to the Internet for public schools and libraries, became a political football. The FCC's 14-point checklist wound up in court and back in court when the RBOCs claimed they had already met the requirements. In another court ruling, the FCC was deemed to have the authority to set pricing guidelines for the incumbent LECs to charge the competitive LECs for services. In addition, the CLECs claimed the incumbent LECs were dragging their feet in complying with the 1996 Telecommunications Act. All of this has involved more lawyers and court cases.

Merger Mania

Interestingly, after the passage of the 1996 Telecommunications Act, mergers of the seven RBOCs and other telecommunication companies began (see Table 1-3). As an example, NYNEX (created earlier by a merger of New England Telephone and New York Telephone) merged with and became Bell Atlantic and then became Verizon after merging with GTE; Qwest took over

US West; SBC merged with Pacific Telesis and Southern New England Telephone (SNET); AT&T merged with Media One to become AT&T Broadband and then merged with Comcast to become AT&T Comcast Corporation; and on it goes. The telecommunications industry is currently undergoing tremendous change as the result of both deregulation and technological advancement. What the industry will look like in the future is pure speculation, but what will surely happen is even more change with resulting enhanced broadband services available to the consumer.

Some Final Thoughts on Regulation

Even though most of the material in part II of this chapter has been presented as a historical overview of the regulation of the legacy **wireline** telecommunications infrastructure of the United States, it should be noted that the FCC has been continually exercising its authority over the wireless radio services it also oversees. Over the past seven decades, the FCC has been instrumental in:

- development of the FM and television broadcasting bands
- adoption of television standards
- implementation of FM and AM stereo broadcasting
- introduction of the UHF television band
- regulation of satellite broadcasting
- the return of some of the UHF spectrum to the cellular telephone industry
- cellular and PCS licensing and growth
- regulation and expansion of public safety and business radio service
- the adoption of standards for over-the-air High-Definition TV and a phase-in plan to implement the new system.

Even this short list does not do justice to the numerous regulatory decisions that have been made by the FCC in this field.

Refarming by the FCC

The FCC has recently completed a refarming project that involves the frequency bands used by the Public Land Mobile Radio Services. This refarming process, started in the early 1990s, is being done in an effort to develop a workable strategy for the more efficient use of the spectrum needed by these services. Taking advantage of technological innovations, this rule-making by the FCC will provide additional use of the frequencies presently available to users of these radio services by narrowing present channel spacings and introducing channel sharing with UHF-TV on a geographical basis. The future is likely to see more refarming efforts like this one as the need for more wireless bandwidth (spectrum) drives regulation.

1.10 Telecommunications Standards

Let us turn our attention to telecommunications standards. In his *Telecommunications Primer*, Bryan Carne defines **standards** as

> ... documented agreements containing precise criteria to be used as rules, guidelines, or definitions of characteristics. Their consistent application ensures that materials, products, processes, and services are fit for their purpose.

Since the MFJ of 1984, and more recently with the 1996 Telecommunications Act, competition in the United States between telecommunication providers has increased the complexity, speed, and capacity of available facilities. Indeed, this is a worldwide phenomenon. This fact, coupled with rapid technological advances, has produced increased activity of standards bodies. Standardization is usually considered to be necessary for low-cost implementation and speed in bringing services to the market. Furthermore, with the present global nature of the telecommunications industry, standards are necessary to ensure interoperability of equipment from different vendors on a worldwide basis. Standards organizations usually consist of manufacturers, service providers, and users working together to promote physical characteristics for the anticipated telecommunications requirements of the future. With standards in place, users can develop applications that build upon the standards. There are many levels of standards organizations, with their sphere of influence depending upon their makeup. Standards bodies are sponsored at the regional, national, and international levels.

International Standards Organizations

Since the advent of the telecommunications industry in the middle of the 1800s, nations have been involved in the standardization of both operating procedures and equipment performance. There are currently three standards organizations at the global level, which are:

International Standards Organization (ISO) (http:\\www.iso.ch)

International Electrotechnical Commission (IEC) (http:\\www.iec.ch)

The International Telecommunications Union (http:\\www.itu.org)

Originally founded in 1865 as the International Telegraph Union, the ITU has since become a specialized agency of the United Nations (1947). In 1987, the ISO and IEC formed a joint committee (JTC1) for standards for the information industry. In 1993 the ITU was reorganized into the Radiocommunications Sector (ITU-R), the Telecommunications Standardization Sector (ITU-T), and the Telecommunications Development Sector (ITU-D).

Regional Standards Organizations

Regional standards organizations include:

European Telecommunications Standards Institute (ETSI)
Telecommunication Technology Committee (TTC) (Japan)
Committee T1—Telecommunications (ANSI-T1) (United States)

National Standards Organizations

United States standards organizations include:

The American National Standards Institute (ANSI) http:\\www.ansi.org

ANSI facilitates standards development and is the sole United States representative to the ISO and the IEC. Some of the accredited standards committees that ANSI performs its work through are:

The International Committee for Information Technology Standards (INCITS) http:\\www.t10.org (formerly the NCITS Committee and prior to that the X3 Committee)

The Accredited Standards Committee (ASC) X12 http:\\www.atis.org (concerned with Electronic Data Interchange standards)

Committee T1 http:\\www.t1.org (concerned with Telecommunications standards)

The Electronic Industries Alliance (EIA) http:\\www.eia.org

The Institute of Electrical and Electronics Engineers (IEEE) http:\\www.ieee.org

The IEEE and the 802.x Subcommittees

The IEEE, a professional organization, developed the standards for local-area networks. The Institute of Electrical and Electronics Engineers (IEEE) 802 committee of the Technical Activities Board consists of groups from industry and academia. These "working groups" decide on the final standards for the different technologies under their jurisdiction. The following list details the various 802 subcommittees presently in place:

802.1 High-Level Interface
802.2 Logical Link Control
802.3 CSMA/CD Networks (Ethernet)
802.4 Token Bus Networks
802.5 Token Ring Networks
802.6 Metropolitan Area Networks
802.7 Broadband Technical Advisory Group

802.8	Fiber-Optic Technical Advisory Group
802.9	Integrated Data and Voice Networks
802.10	Network Security
802.11	Wireless Networks
802.12	Demand Priority Networks
802.14	Residential Networks
802.15	Wireless Personal-Area Networks
802.16	Broadband Wireless-Access Networks

The IEEE standards Web site is extremely informative and can be found at http:\\www.standards.ieee.org.

Another United States standards organization is the National Institute for Standards and Technology (NIST) at http:\\www.nist.gov. Formerly the National Bureau of Standards, the NIST is an agency of the United States Department of Commerce. In addition to these standards organizations, there are also many industry "forums." These organizations are usually involved in efforts to promote new or emerging technologies and to harmonize the efforts of different developers in the application of these new technologies to the creation of products.

Industry Forums

Folowing is a brief sampling of some of the industry forums in existence:

ATM Forum http:\\www.atm.org

Frame Relay Forum http:\\www.frforum.com

ADSL Forum http:\\www.dslforum.org

Telecommunications Information Networking Architecture Consortium (TINA-C) http:\\www.tinac.com

UMTS Forum http:\\www.umts-forum.org

Cable Modem Forum http:\\www.cablemodem.com

Summary of Part II

If the future is anything like the past, we can expect periods of both deregulation and regulation to occur in the telecommunications industry as time moves forward. Rapid technological change will drive a great number of the regulatory issues of the future. Furthermore, telecommunications standards are necessary for the successful interoperability of equipment in a global environment of rapid technological change.

Questions and Problems

Part I

1. Perform an Internet search on one of the early pioneers of telecommunications and record your findings. Name one other technical contribution by this person other than what they are remembered for.
2. Make a short list of three or four Internet Web sites that are devoted to the history of either the telephone or the telegraph. Write a short paragraph on some event that helped to shape the telecommunications industry.
3. What technological innovation was necessary for the successful implementation of a transcontinental telephone line?
4. What affect did technology have on the pony express?
5. Define the *golden era of radio.*
6. Define *geosynchronous satellite.*
7. Assuming that Moore's Law holds for all components in a PC, determine what the capacity of a PC hard drive will be in the years 2008 and 2012.
8. Take a position on the need for additional fiber-optic cable and summarize your reasons.
9. Some people believe that the PSTN will not survive. What do you think, and why?
10. Can the growth rate of the Internet continue unabated? Explain.

Part II

1. Do an Internet search on one of the various "Acts" that affected the telecommunications industry. Write a short paragraph describing your understanding of why it was enacted.
2. Who was J.P. Morgan and what was his legacy?
3. Do an Internet search on your state's Public Utilities Commission. What do they regulate?
4. Search the FCC's Web site for information about how it operates. Write a short paragraph about a recent telecommunications ruling.
5. Do an Internet search of http:\\www.ieee.org and the standards activities of the 802 subcommittees. Determine what the present hot topic in this arena is.
6. Do an Internet search on one of the global standards organizations. Find out what the hot topics of the day are.
7. Do an Internet search on one of the listed forums. What does its Web site have to offer?
8. What is the WARC (telecommunications organization) and when will they meet next?
9. Pick a country on another continent and determine what organization regulates their telecommunications.
10. Go to the Web site http:\\www.nctt.org and describe what this organization does.

Basic System Elements in Telecommunications

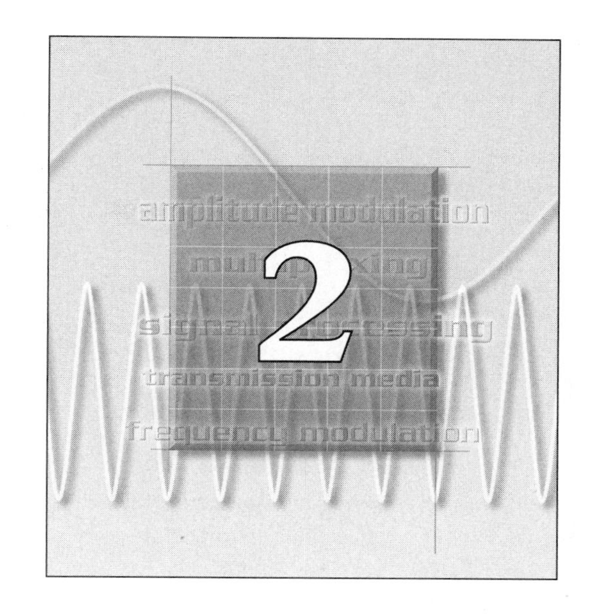

Objectives Upon completion of this chapter, the student should be able to:

- Describe the basic elements of a telecommunications system and explain their functions.
- Explain the concept of modulation and why it is required.
- Explain the concepts of bandwidth, frequency spectra, and the difference between analog and digital signals.
- Describe how signals are represented in both the time and frequency domain.
- Discuss the use of Fourier analysis in the evaluation of signal composition.
- Describe the operation of electronic filters and their effect on signal components.
- Understand and be able to work with dBs and dBms.
- Discuss sources of noise, how noise affects communications, and various industry-standard noise measurements.
- Understand the basic operation of a spectrum analyzer and the types of measurements it is used for.
- Be familiar with the type of measurement capability that is provided by network analyzers.

Outline

2.1 Introduction to Telecommunications Systems
2.2 Key Terms and Definitions
2.3 The Telecommunications System Block Diagram
2.4 Review of Telecommunications System Components
2.5 Introduction to the Tutorials

Key Terms

analog
audio frequency
bandwidth
baseband
Bode Plotter
channel
dB
decade
demodulation
demultiplexing
digital modulation
duty cycle
electronic filters
Fourier analysis
frequency
frequency domain
frequency response
frequency span
markers
modulation
multiplexing
network analyzer
noise
noise figure
noise temperature
octave
packet switching
pulse repetition rate
Q
radio frequency
receiver
resolution
roll-off rate
signal
shape factor
signal-to-noise ratio
spectra
spectrum analyzer
switching fabric
symbol
system dropout
TCP/IP
time domain
transmitter
video
wavelength

Introduction

There are five basic elements to any telecommunications system: the source, a transmitter, the channel, a receiver, and the destination. In this chapter, the student will learn about the operation of these basic elements, the need for modulation, types of signals, signal bandwidth, and modulation types. Other concepts introduced will be signal wavelength and multiplexing. An introduction to the various types of channels/transmission media will be presented. The concept of a switching fabric embedded in the channel will be examined and examples of several typical telecommunications systems discussed.

With the fundamental concepts and definitions introduced, the student is led through five brief tutorials. The topics covered in these short sections are:

- a review of the language of telecommunications: dBs and dBms
- basic electronic filters

- telecommunications test and measurement equipment
- noise
- Fourier analysis

The purpose of these five tutorials is to provide the student with the tools needed to understand the more complex topics to be presented in later chapters. There is a great deal of emphasis placed on the concepts and the practical matters of how signals are represented in the time and frequency domain. This material is presented along with an overview of the digital oscilloscope, the spectrum analyzer, and the network analyzer. These test and measurement tools are invaluable in their ability to provide insight into the characteristics of signals and systems and the correct operation of these systems. A review of dB and basic filter concepts is also presented, as these topics may have slipped out of the student's memory or been given little emphasis in previous study. The limiting effect of noise is explained and some of the various industry-standard noise measurements presented. Lastly, in an effort to tie everything together, the student is presented with some examples of Fourier analysis of common signals and a comparison is made between this mathematical model and real measurements made of these same common signals.

2.1 Introduction to Telecommunications Systems

This chapter starts with an overview of a basic, generic telecommunications system. As can be seen in the block diagram in Figure 2-1, the typical telecommunications system consists of a source of the information to be sent; a hardware subsystem called a **transmitter**, the **channel**, or means by which the signal travels; another hardware subsystem called a **receiver**; and

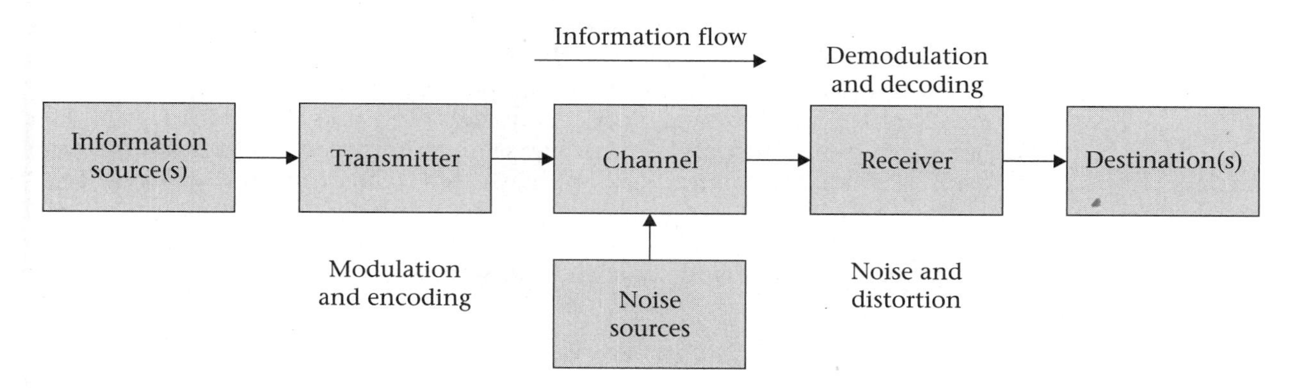

FIGURE 2-1 Typical telecommunications system

the final destination for the information. The channel may also contain a **switching fabric** that directs the signal to its final destination. Another block labeled noise is shown in the diagram. Although this block does not represent hardware, it does represent any sources of interference that might impair the transmission of the information through the transmission medium in a successful fashion. For those familiar with the open-systems interconnection (OSI) reference model, the primary emphasis of this book is on the physical layer (layer 1).

2.2 Key Terms and Definitions

Before describing the functions of the various blocks in the basic system, several terms need to be defined. In the telecommunications world, we often refer to the transmission of signals and symbols. A **signal** describes some physical phenomena. The first telecommunications signals were the flow of electrical charge (current) through a telegraph cable. The duration of the electrical current defined a **symbol**—the dot or dash of Morse code.

Analog Signals

The early telephone only became possible when inventors like A.G. Bell and others discovered how to convert sound waves into an electrical signal that could be applied to a wire for conduction to another location. The resulting electrical signal that contained the speech information would propagate along the wire to the receiving end, where the telephone receiver would convert the electrical impulses back into sound waves. These types of signals, of which Figure 2-2 is an example, are known as **analog** signals because they can take on any value within a given range.

Marconi's wireless telegraph used a spark-gap transmitter that produced an electromagnetic signal that propagated through free space. The signal's duration again encoded a dot or a dash of Morse code. The very crude coherer detector used to receive the signal would convert the signal into a form recognizable by a human operator. Wireless radio was rapidly adapted to voice communications using amplitude modulation, again an analog system. The past century has seen various technological innovations that have led to extremely sophisticated telecommunications systems for the worldwide transmission of voice, music, video data, and digital data through the use of complex modulation and multiplexing schemes employed to utilize and share the transmission medium.

Digital Signals

The invention of the computer, with its representation of information as 0s and 1s has brought us back to the transmission of symbols that represent

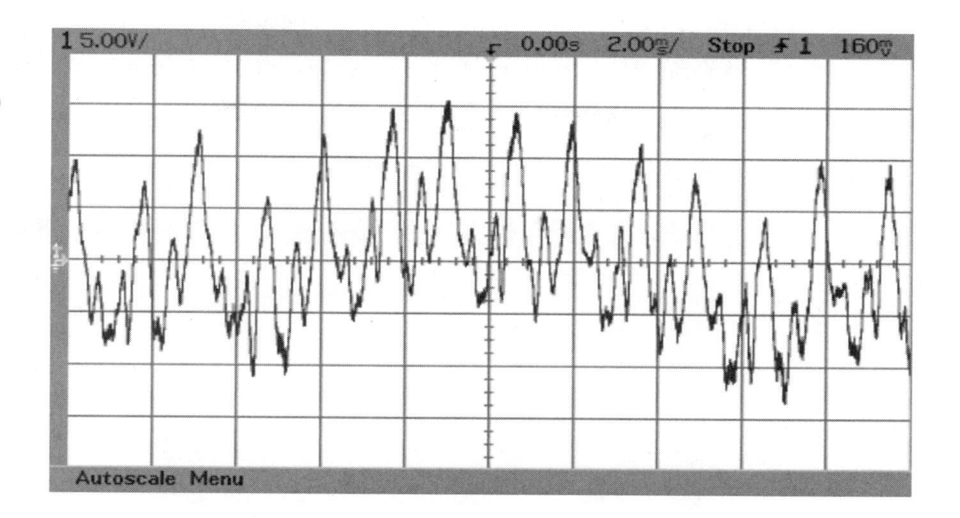

FIGURE 2-2 Digital oscilloscope display of an analog (music) signal

the desired information to be transmitted. Telecommunications systems that use **digital modulation** techniques have a finite number or set of symbols (or signals that represent the symbols) that are transmitted. The receiver's function is to decide which of the symbols in the set has been received. With today's technology, all information, including rich multimedia content, is being transmitted by symbols at increasingly faster rates. This high-speed transmission of data is being accomplished with progressively more complex and bandwidth-intensive modulation schemes over wireline, wireless, or fiber-optic transmission media. The PSTN (with the exception of the local loop) and the high-speed long-distance links of the network have all become digital systems.

Signal Characteristics

Let us examine some of the terminology used with signals. Because a signal represents some particular physical phenomena, it can contain many individual frequencies. Also, as previously described, an analog signal can take on a continuous range of values, whereas a digital signal only has a limited number of discrete values.

Frequency

The **frequency** of a signal is defined as the number of cycles of a periodic waveform that occur in one second. The unit of measurement is the Hertz (Hz). One Hertz equals one cycle per second. A single audio tone would be identified as having a frequency of a certain number of Hertz. For a periodic pulse waveform, the typical identifying measurements are the **pulse repetition rate**, given in pulses per second (pps), and the **duty cycle** (the percentage of "on" time). There is a direct relationship between the repetition rate and duty cycle and the frequency content of the pulse signal, which will be explored later in the chapter. Shown in Figure 2-3 is a 100-kHz sine-wave signal and a 20% duty cycle, 100-Kpps pulse signal displayed on an oscilloscope.

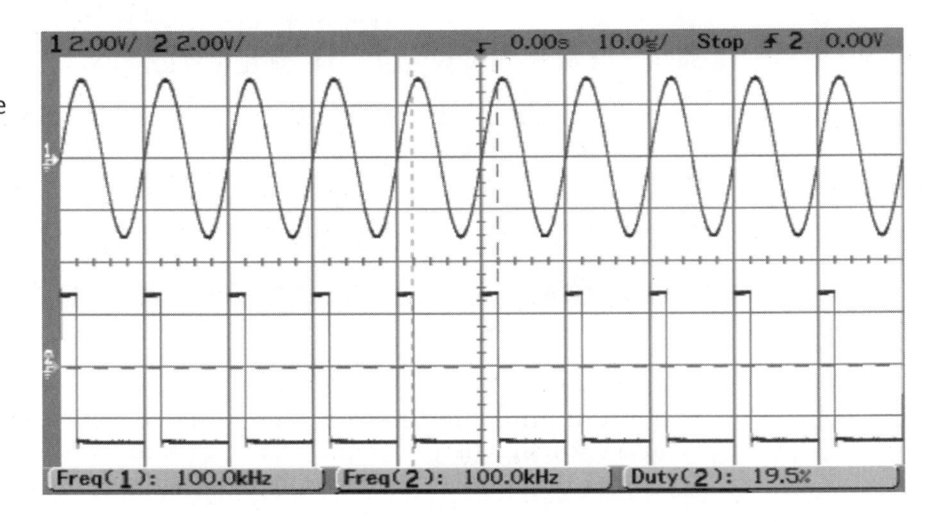

FIGURE 2-3 A 100-kHz sine wave and a 20% duty-cycle pulse waveform

Signal Designations

Depending upon their frequency content, various labels are given to signals. An **audio-frequency** (AF) signal contains frequencies in the typical range of human hearing—approximately 20 to 20,000 Hz (for a young person). The PSTN usually restricts a transmission channel for a voice signal to the range of 300 to 3,400 Hz (known as a voice-grade circuit). However, a Compact Disc (CD) player will reproduce the entire range of audio frequencies; therefore, its audio operation is termed high fidelity or HiFi. A **video** signal is one that extends from close to dc (0 Hz) to several MHz (the typical analog TV signal). A **radio-frequency** (RF) signal refers to a signal with no frequency components close to 0 Hz. The output signal of a cellular or cordless telephone is an RF signal. A baseband signal is usually considered to be the low frequency modulating signal in a communications system. Typically, it would have frequency components from near dc (0 Hz) upward. Voice or music signals would be considered **baseband** signals, as would certain types of data-transmission systems.

Signal-Frequency Spectra and Bandwidth

The frequency spectra of a signal consists of the signal's component frequencies. Frequency spectra is related to bandwidth, but there is a difference. The **bandwidth** of a signal is simply a measurement of the range of frequencies (given in hertz) in the signal that contains the majority of the signal power (there are actually many different definitions in use for *signal bandwidth* that will be discussed at a later time). The frequency **spectra** of a signal, on the other hand, is that which can be displayed on a **spectrum analyzer**, a specialized piece of RF test equipment. The spectrum analyzer has a very large dynamic range and is able to display important information about a signal's distortion, harmonic content, or spurious frequency content. Typically, some of this information is not contained within the frequency range specified as the signal's bandwidth. Bandwidth can also refer to the capacity of the telecommunications channel. This concept will be discussed further when the channel is covered in detail later in the chapter.

Signal Wavelength

With many types of telecommunications systems it is necessary to deal with a signal's wavelength. Wireless systems in particular utilize this term when dealing with the system hardware. Furthermore, the operating characteristics of fiber-optic cable systems and components are given in terms of wavelength, rather than in terms of their frequency. **Wavelength** is defined as the distance an electromagnetic wave travels in one period or cycle of the signal. This distance is transmission-media dependent. For an electromagnetic wave in free space, the wavelength, λ, is equal to the velocity of light, c, divided by the frequency of the signal, as shown in Equation 2.1.

$$\lambda = \frac{c}{f} \tag{2.1}$$

EXAMPLE 2.1

Determine the free-space wavelength of a 1200-MHz signal.

Solution Free-space wavelength is found from the following equation:

$$\lambda = \frac{300}{f}$$

where f is given in Megahertz and λ is given in meters.

Therefore,

$$\lambda = \frac{300}{1200} = .25 \text{ m or } 25 \text{ cm}$$

2.3 The Telecommunications System Block Diagram

Let us return to the basic block diagram of a telecommunications system and examine the middle four blocks (the hardware subsystems) in greater detail, as shown in Figure 2-4. The transmitter and receiver (T/R) combination is used to convert the information to be transmitted into a signal that is amenable to the transmission medium (channel), and then to convert that signal back into its original form. This is done through a process known as **modulation**, or encoding, at the transmitter and **demodulation**, or decoding, at the receiver, corresponding to either analog or digital signals respectively. Additionally, if the channel contains a switching fabric, the transmitter must also embed addressing or routing instructions into the signal being transmitted. The following section defines these concepts and describes the processes in more detail.

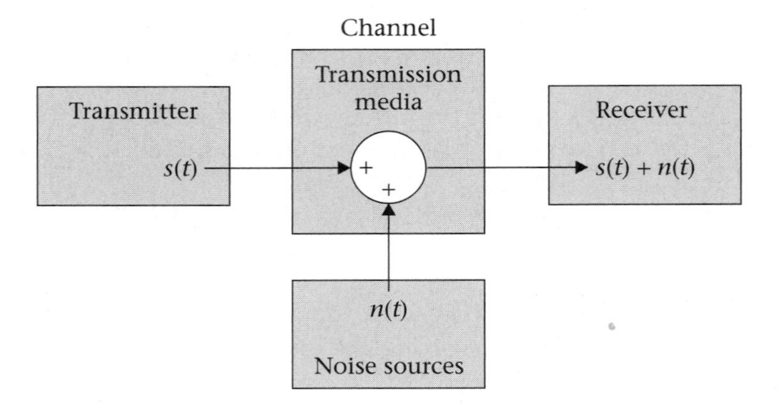

FIGURE 2-4
Telecommunications system block diagram

The Need for Modulation

Clearly, all transmission media exhibit various properties due to their physical nature. For example, a long run of a pair of copper wires acts like a lossy electrical low-pass filter and can therefore be useful for the transmission of low-frequency voice signals over a limited distance. A transmitter using this type of channel need only convert a voice signal from sound waves to an electrical signal and then apply it to the copper pair at a sufficient amplitude to overcome the losses in the system. The receiver of this same channel converts the electrical signal variations back to sound waves with some form of electromechanical transducer.

A fiber-optic cable does not conduct electricity. Therefore, for it to be able to do something like transmit someone's voice, the voice signal must undergo an additional change from an electrical signal to some form of optical signal that can propagate down the cable and be detected at the receiving end.

Electromagnetic waves propagate through free-space and dielectric materials. However, some characteristic of the wave (amplitude, frequency, or phase) must be varied in accordance with the message signal for any useful transmission of information to take place. In each example described here, the transmitter has changed or modulated the signal in such a way that the signal can be delivered by the channel to the receiver, at which point the signal is demodulated back to its original form. Modulation techniques will be discussed in more detail in chapters 3 through 6.

In wireline systems the transmitter-receiver pair is commonly referred to as a *modem* (modulator/demodulator); in wireless systems the terms *transmitter* and *receiver* continue to be used; and for fiber-optic cable systems the terms *electrical-to-optical* (E/O) and *optical-to-electrical* (O/E) *interfaces* are used to describe the transmitter-receiver pair.

Multiplexing and Demultiplexing

The transmitters and receivers in telecommunications systems also perform the functions of multiplexing and demultiplexing. **Multiplexing** is the process of combining together many separate signals to send them over the same transmission media. This process might be a sharing of frequency, time, or space, or a combination of these methods. Some of the more common schemes are frequency and time-division multiplexing. All forms of over-the-air broadcasting (AM, FM, and TV) use frequency-division multiplexing (FDM) and space-division multiplexing (SDM) when the same modulation techniques are being used. There are no geographical locations in the United States where there are either co-channel (same frequency) or adjacent channel (next closest channel) stations assigned. Substantiation of this fact is found in the design of the common Video Cassette Recorder (VCR). The VCR is designed for an interference-free output on either TV channel 3 or 4, as both channels will not be assigned in any given location. On the other hand, High-Definition TV (HDTV), which uses a different type of modulation system, can be assigned an adjacent-channel spot on the TV band because the two forms of modulation don't interfere with one another. **Demultiplexing** reverses the multiplexing process at the receiver. Multiplexing and demultiplexing will be covered in greater detail in chapter 7.

In summary, the T/R hardware configuration is channel dependent. Whether the channel is wireless, wireline, or fiber-optic will dictate the correct hardware and modulation technique needed to deliver the signal from its source to the destination. Also, within the broad definitions of wireless, wireline, and fiber-optic channels are many different families of technologies suitable for transmission applications at different frequencies or wavelengths and data rates.

It should be noted that in the many public radio-service systems that do not employ a switching fabric in the channel, another possible function of the transmitter-receiver pair is to choose a clear frequency channel (many cordless telephones do this automatically). Furthermore, in the reception of broadcasting services, tuning is employed so that the receiver can select a single signal from the many available. All consumer electronics-type radios and televisions can perform this selection/tuning function.

The Channel

A long-standing definition of the telecommunications channel has been: "the channel is the means by which the signal propagates." This definition would tend to emphasize the transmission media—either wireless, wireline, or some other guided-wave media like the coaxial cable, fiber-optic cables, wave guides, or other types of specialized transmission lines. All transmission media have different characteristics that dictate how much bandwidth is available, how much signal attenuation occurs with distance, and how much noise and distortion might impair the signal as it propagates. These characteristics dictate the required transmission and reception hardware, and eventually limit channel bandwidth or capacity.

Today this definition is inadequate. The exponential growth in computer-to-computer data transmission has changed our concept of the channel. Now the channel is much more complicated than ever before, as it also includes the switching fabric that directs the signal to the correct destination. Furthermore, the switching-fabric technology is almost totally digital and implements higher layer functions in our open-systems interconnection (OSI) model. Even a local telephone call that is transmitted though the local loop as an analog signal (using a copper wire pair) will be converted to a digital signal at the PSTN central office as it is passed through the switch on its way to its destination. This occurs even if the destination is the next-door neighbor.

Since the middle of the 1990s an accelerating, radical change has been taking place in the switching fabric of the world's telecommunications infrastructure. With the advent of the Internet and the World Wide Web came a shift away from connection-oriented circuit-switching to connectionless circuit-switching technology. The adoption of **TCP/IP** (Transmission Control Protocol/Internet Protocol) as the de facto networking protocol standard has been a major driving force in this change. The legacy network switch used in the telecommunications channel physically ties up a circuit connection, hence using up its limited resources. This type of switching is usually done by an AT&T 5ESS or a Nortel DMS-100 in a PSTN central office. The new switching paradigm is **packet switching**. Packet-switching technology sets up virtual or connectionless circuits by using TCP/IP and routers. In this technology, packets of data are forwarded from router to router by the use of an Internet Protocol (IP) address that is part of the message. When a data packet reaches a router, the router searches its memory to determine the correct output link to send the data out on. The packet eventually reaches the correct IP address assigned to a particular computer on a network somewhere in the world. In this scenario, even though the switching fabric is part of the channel there are no resources tied up for long durations.

Noise

Although noise is not a part of the telecommunications system, it is introduced here because it plays an integral role in the system's operation. **Noise** is defined as anything that interferes with the transmitted signal or symbol. Electrical noise is present everywhere and is produced by many different sources. Noise can add to a signal and cause the signal to be "masked." An analog signal can become unintelligible or a digital signal can contain too many errors to be usable due to the addition of noise. The block diagram of the telecommunications system in Figure 2-4 showed the noise being introduced into the system as the signal traverses the channel. This occurs because the signal is usually at its weakest level in the channel. However, it should be pointed out that in addition to the distortion introduced in the channel, noise can also be introduced by both the transmitter and the receiver of the system. Signals can also suffer distortion due to the modulation/encoding and demodulation/decoding process. Analog signals are much more susceptible to noise than digital signals, a topic which will be discussed at greater length at a later point.

2.4 Review of Telecommunications System Components

At this point, let us do a quick review of the telecommunications block diagram. In general, a telecommunications system consists of the following subsystems:

- A source of information
- A transmitter to modulate or encode the information
- A channel or transmission medium to deliver the information from the transmitter to the receiver (noise enters the system here)
- A receiver to demodulate or decode the information
- A destination for the information

The Source

The information source supplies a signal to the transmitter. The signal might be audio, video, data, or a combination of all three (multimedia), as depicted in Figure 2-5 on page 48.

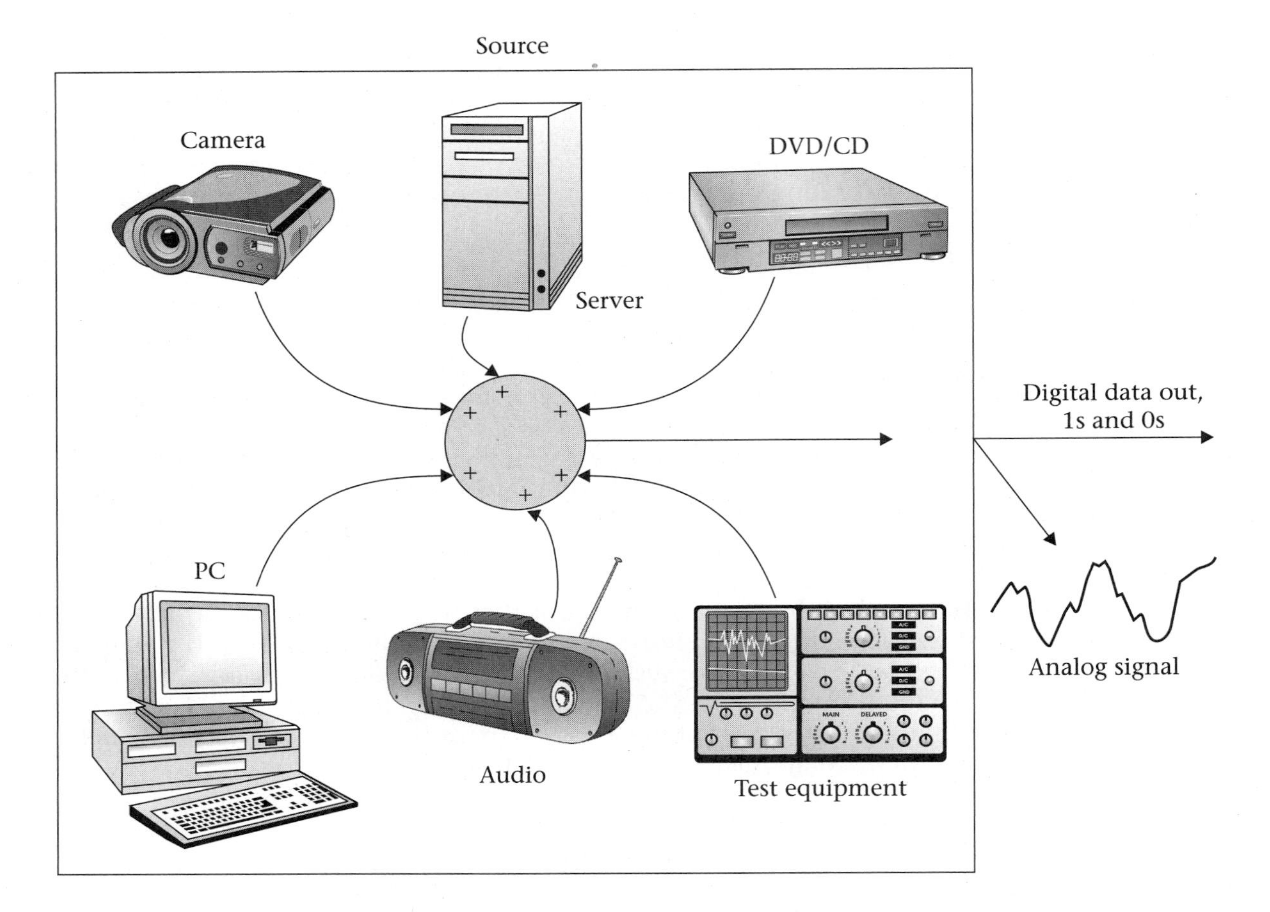

FIGURE 2-5 The source supplies a message or data to the transmitter

The Transmitter

The transmitter makes the signal compatible with the channel by modulation or encoding as seen in Figure 2-6. The transmitter may also convert the signal from analog to digital (ADC) or from electrical to optical (E/O) and perform some form of multiplexing and possibly supply addressing or routing information.

The Channel

The channel is the means by which the signal propagates: free space (the air interface), unshielded twisted pairs (UTP), CAT-5 LAN cable, coaxial or fiber-optic cable, and so forth. The channel will also include any switching fabric that directs the signal to its final destination (see Figure 2-7). Noise

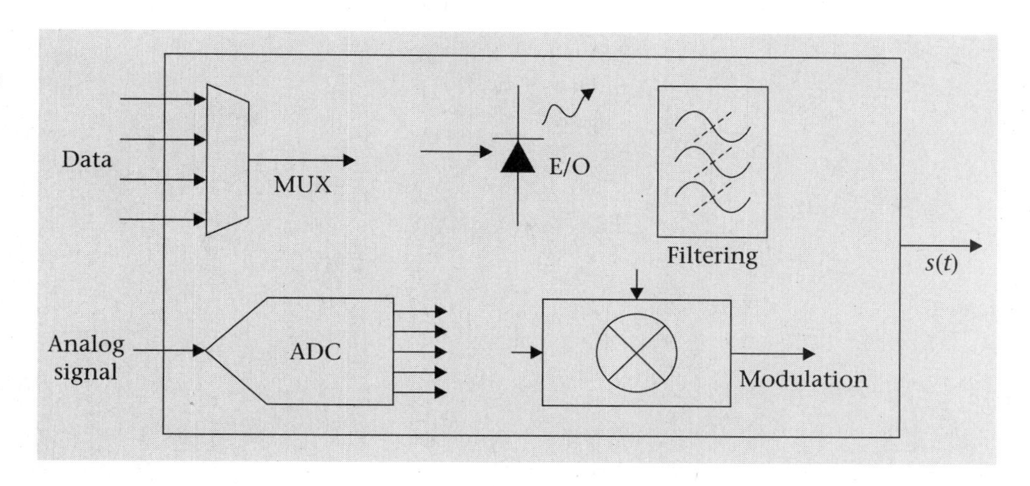

FIGURE 2-6 The transmitter sends the signal through the channel

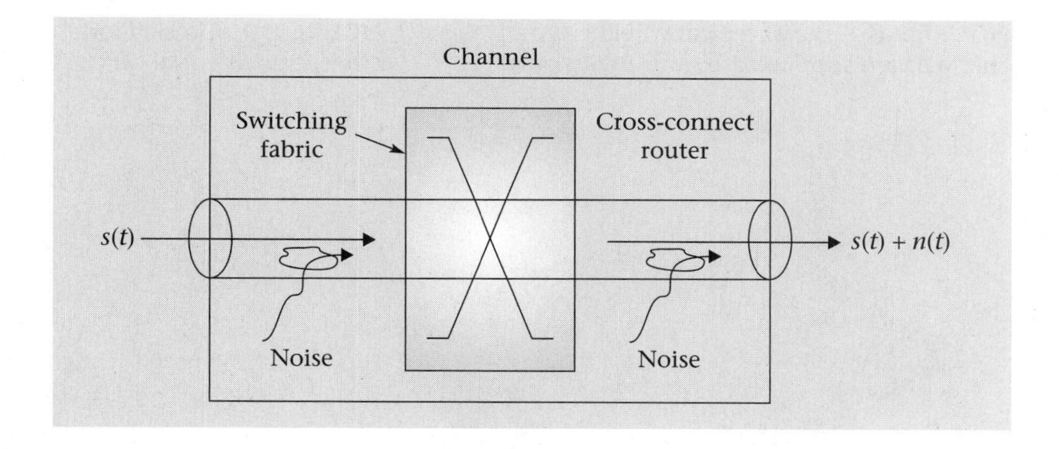

FIGURE 2-7 The channel

adds to the signal in the channel and causes distortion or errors to occur at the destination. Figures 2-8 through 2-11 show various parts of physical channels and switching fabrics. Figure 2-8 on page 50 shows a typical utility pole with a coaxial cable, fiber-optic cable, and a multiple-pair copper wire telephone-feeder cable. The wireless free-space channel is shown in the background.

The next three figures show three different examples of switching fabric. Figure 2-9 shows a classic central office, stored-program control telephone switch. This type of switch is rapidly falling out of favor in the Internet age. Figure 2-10 shows an optical switch that operates with micro-mirrors using micro-electromechanical system (MEMS) technology. Finally, Figure 2-11 shows spatial switching. Each side of the triangular platform contains antennas that direct signals into 120-degree segments of the cell site.

FIGURE 2-8 Several typical wireline-type channels are supported by this utility pole

FIGURE 2-9 Classic Nortel circuit switch used in a telephone central office (*Courtesy of Nortel*)

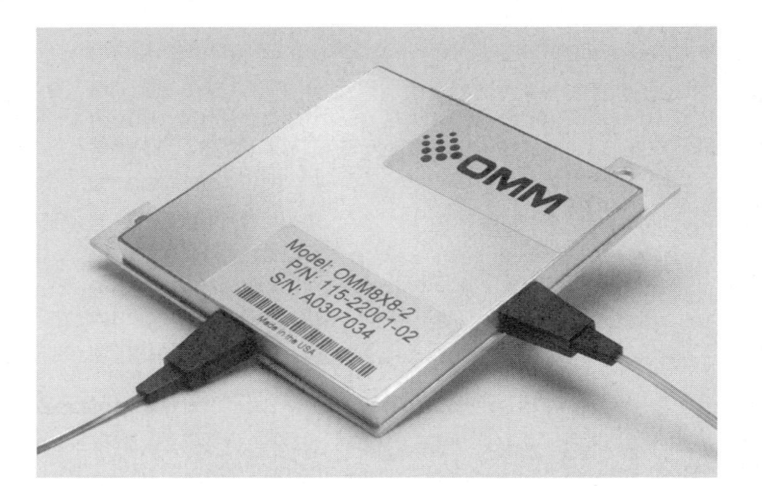

FIGURE 2-10 Optical switch using micro-electromechanical systems (MEMS) (*Courtesy of OMM, Inc.* © 2000 OMM, Inc.)

FIGURE 2-11 Spatial switching is employed in the cellular wireless channel

The Receiver

The receiver, depicted in Figure 2-12, selects the signal and demodulates or decodes it back to its original form. The receiver will also add noise to the signal, causing distortion or errors to occur. The receiver might also need to convert the signal from digital to analog (DAC) or from optical to

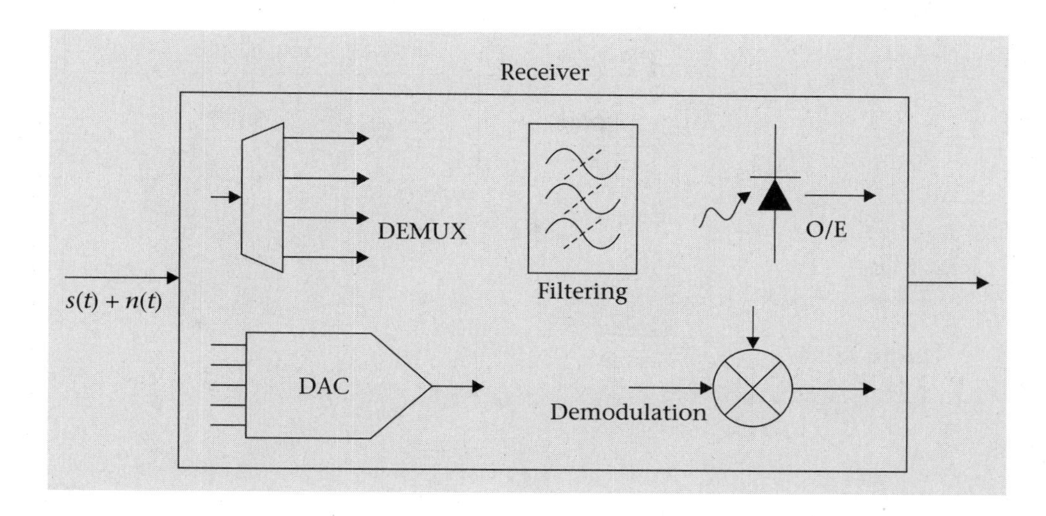

FIGURE 2-12 The receiver

electric (O/E) and perform demultiplexing. In the computer world the receiver will have a unique medium-access control (MAC) address that identifies it to the network and allows only it to receive a message sent to it. In the cellular-telephone world each mobile unit has a unique identification number that is used in this receiver selection process.

The Destination

The destination of the information being sent can be as widely varied as the source and type of information. However, the growing use of the Internet is increasingly causing more and more telecommunications traffic to be from one computer to another and with increasing amounts of multimedia content. These facts, coupled with the advantages of digital modulation, are driving the changes in the telecommunications networks of the world. Figure 2-13 on page 52 shows some of the possible destinations of information sent over a telecommunications network.

2.5 Introduction to the Tutorials

The five tutorials in this section of the chapter are practical in nature and are extremely important to one's understanding of telecommunications theory, as the other chapters in this book will build upon the concepts. These short tutorials on topics pertinent to telecommunications are going to be presented with numerous examples and illustrations. The first will be a review of the language of signal levels and telecommunications hardware performance—the dB and dBm. Then, and presented in the following order, will be: a summary of electronic filters and their characteristics, an

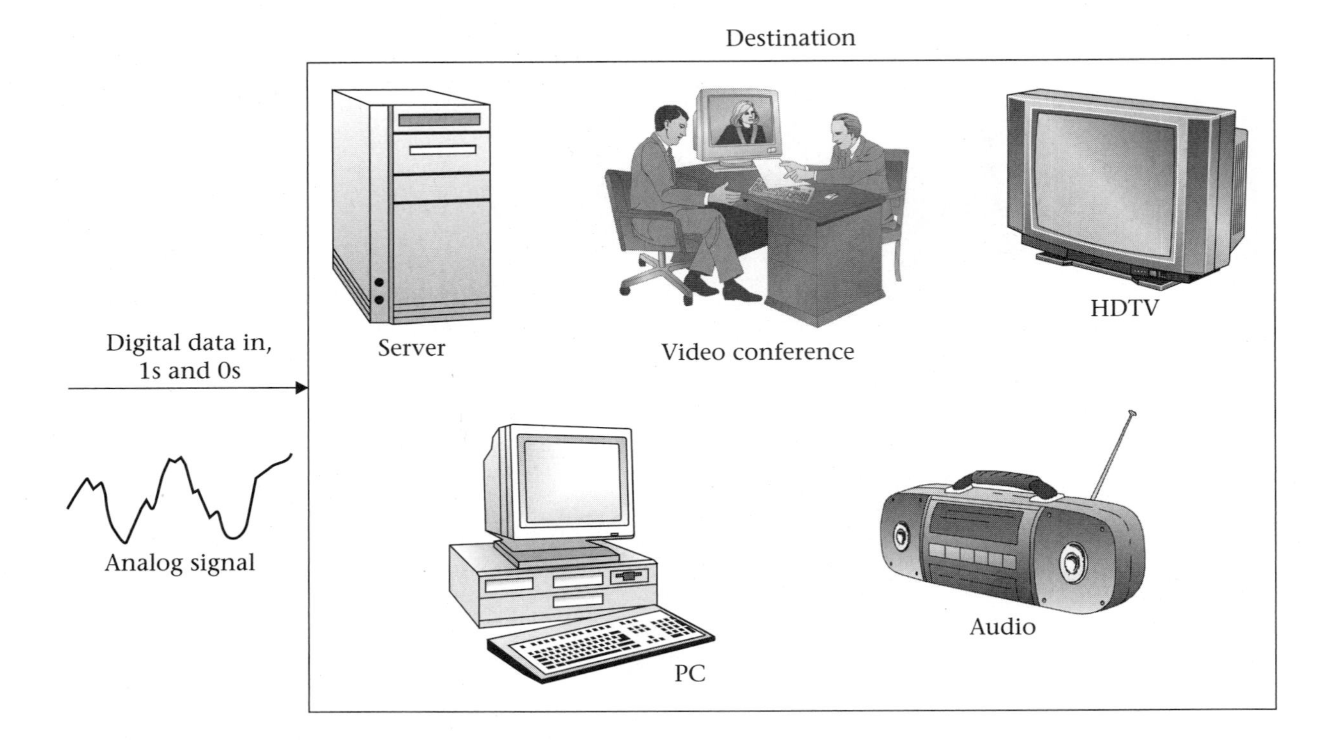

FIGURE 2-13 The destination

overview of modern electronic test and measurement equipment used in the telecommunications field, a short coverage of noise measurements and calculations, and an introduction to Fourier analysis and the frequency spectra of common waveforms.

To reiterate, these tutorials are *extremely* important in the preparation of the reader for the material to be presented in later chapters. One should have a good grasp of the concept of what a signal consists of, what its bandwidth is, and how to examine it in both the time and frequency domain. With these concepts well-understood, the various forms of modulation and signal processing described in succeeding chapters should not be as difficult to understand as they would be without this material.

2.6 Tutorial 1: The Language of Telecommunications

The field of telecommunications involves the use of many acronyms, with new ones being added on a frequent basis. However, the **dB** is an integral part of the language of telecommunications, and without an understanding of the dB one is quickly lost when discussing common measurements of signal strength or the attenuation characteristics of transmission media. The dB is used extensively to represent ratios of power and voltage. Figure 2-14 shows an electronic network or system with a signal applied to the input. If the electronic system functions as an amplifier, one can talk about a power or voltage gain given in dB. If the system is lossy (i.e., a coaxial cable), one can talk about a loss or attenuation in dB. Note that because the dB represents a ratio, it is a dimensionless quantity.

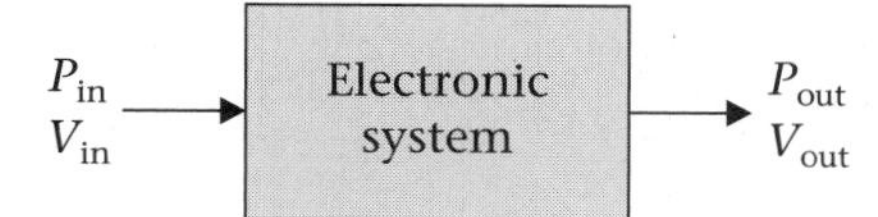

FIGURE 2-14 Block diagram of a typical electronic system

Mathematical Definition

Mathematically, the equations for gain or loss in dB are

$$\text{Power Gain or Loss in dB} = 10\log\left(\frac{P_{out}}{P_{in}}\right) \tag{2.2}$$

and

$$\text{Voltage Gain or Loss in dB} = 20\log\left(\frac{V_{out}}{V_{in}}\right) \tag{2.3}$$

The equations for gain or loss are the same; however, a power or voltage gain yields a positive dB value, while a loss yields a negative dB value, as seen in the following examples:

EXAMPLE 2.2

What is the dB gain (or loss) for the following situation?

For a certain system, the output power is 100,000 times the input power.

Solution In this case,

$P_{out} = 100{,}000P_{in}$

Therefore,

Gain (or Loss) = $10\log_{10}(100{,}000) = 10\log_{10}(10^5) = 50$ dB

Note that the log of 100,000 = 5. That is the same power one needs to raise 10 to in order to get 100,000.

Some common values of power gain for powers of 2 are shown in Table 2-1.

TABLE 2-1 Common dB values of powers of 2

Gain of	*Equals*
2	3 dB
4	6 dB
8	9 dB
16	12 dB
32	15 dB
64	18 dB

To go from a given dB value to the ratio that the number of dBs represents, one needs a calculator to perform the following operation:

$$\text{Power Gain or Loss Ratio} = 10^{(dB\ Value/10)} \qquad \textbf{2.4}$$

and

$$\text{Voltage Gain or Loss} = 10^{(dB\ Value/20)} \qquad \textbf{2.5}$$

EXAMPLE 2.3

Convert a power gain of 64 dB into a ratio.

Solution

$$64 \text{ dB} = 10^{(64/10)} = 10^{(6.4)} = 2.5212 \times 10^{6}$$

Now let us look at another example where a loss is involved.

EXAMPLE 2.4

What is the dB gain or loss for a certain system when the power out of a certain system is .01 times the power in?

Solution In this case

$$P_{out} = .01P_{in}$$

Therefore,

$$\text{Power Gain in dB} = 10\log_{10}(.01) = 10\log_{10}(10^{-2}) = -20 \text{ dB}$$

Note that the answer is negative, indicating a system loss or an attenuation of the input signal.

The beauty of using dBs is that if we have cascaded systems, which is commonplace in communications and telecommunications systems, we simply algebraically add the dBs from each subsystem to get an overall gain or loss for the system.

EXAMPLE 2.5

What is the overall gain of the system in Figure 2-15?

FIGURE 2-15 Example 2.5

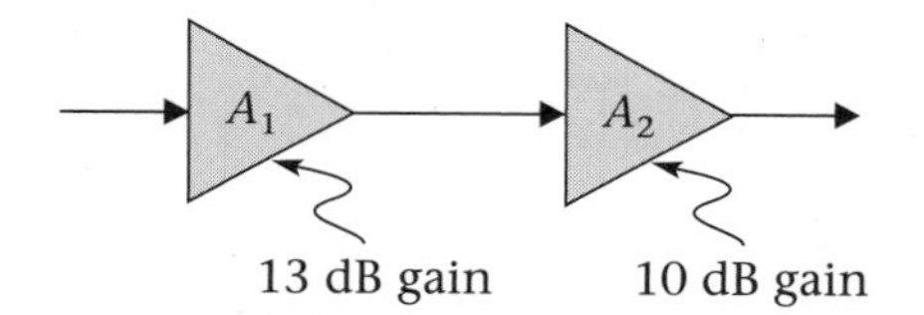

Solution In this case, simply add the dB gains of each stage together:

Total dB Gain = gain of stage 1 + gain of stage 2

Total Gain = 13 dB + 10 dB = 23 dB

To check the answer, first change the dB values to ratios

$13 \text{ dB} = 10^{1.3} = 20$ and $10 \text{ dB} = 10^{1.0} = 10$

Total gain is therefore $20 \times 10 = 200$, and because $23 \text{ dB} = 10^{2.3} = 200$ we are certain of our answer. Recall that we only need to add the dB values together to determine the overall system gain in dB. Note that this technique also works if one of the system stages has a loss (a negative dB value).

Referenced dB Values

It is commonplace to reference dBs to power or voltage levels. Some of the more commonly referenced values are represented in Table 2-2 on page 56.

In all cases, a reference or standard impedance value must be specified. Some common nominal values are:

- Audio signals 600Ω
- Video signals 75Ω
- RF signals 50Ω
- Twisted pairs 100Ω

These values would be used when doing any power calculations.

TABLE 2-2 Table of commonly referenced dB values

Value	*dBs referenced to*
dBm	milliwatts
dBW	watts
dBkW	kilowatts
dBμV	microvolts
dBV	volts

When dBs are referenced to power or voltage levels, the dB equation becomes

$$\text{Power in referenced dB} = 10\log_{10}\left(\frac{\text{Power}}{\text{Reference Power}}\right) \qquad \textbf{2.6}$$

with the reference value in the denominator of the equation.

EXAMPLE 2.6

What is 5 W equal to in dBm?

Solution Simply plugging values into the equation yields

$$\text{Power in referenced dB} = 10\log_{10}\left(\frac{5\text{ W}}{1\text{ mW}}\right) = 10\log_{10}(5000)$$

$$= 10\log_{10}(5000) = 10(3.699) = 36.99\text{ dBm}$$

So another way of indicating a power level of 5 W is to use +36.99 dBm (commonly rounded off to +37 dBm). The reason for the plus sign is to differentiate between levels over 1 mW and those under 1 mW (negative values of dBm). These levels are commonly encountered in many areas of telecommunications.

EXAMPLE 2.7

What is 500 μV in dBμV?

Solution The equation is as follows:

$$\text{Voltage in referenced dB} = 20\log_{10}\left(\frac{500\ \mu\text{V}}{1\ \mu\text{V}}\right) = 20\log_{10}(500)$$

$$= 20(2.69897) = 53.98\text{ dB}\mu\text{V}$$

So another method of representing 500 μV is to assign it a value of +53.98 dBμV.

Additionally, one can use referenced dB values with values of system gain or loss to determine overall system operational characteristics. Recall that dBs are dimensionless; therefore, dBs added to dBms will yield dBms and vice versa.

EXAMPLE 2.8

For the system shown in Figure 2-16, if the input power is –12 dBm, what is the power out?

FIGURE 2-16 Example 2.8

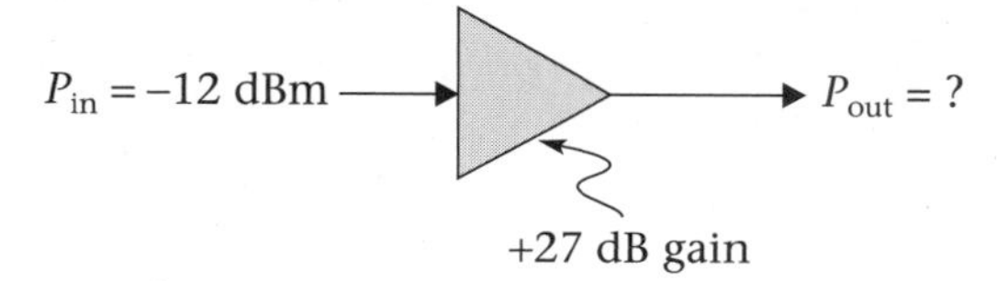

Solution

$$P_{in} = -12 \text{ dBm}, \quad \text{Power Gain} = +27 \text{ dB}, \quad P_{out} = ?$$

One simply adds the values together:

$$P_{out} = -12 \text{ dBm} + 27 \text{ dB} = +15 \text{ dBm}$$

Note again how the signs and the mixed labels are dealt with.

EXAMPLE 2.9

For the system shown in Figure 2-17, the input voltage is +20 dBμV. What is the output voltage?

FIGURE 2-17 Example 2.9

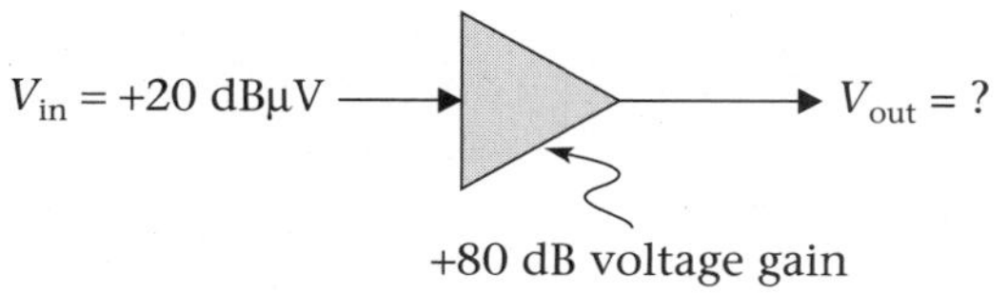

V_{in} = +20 dBμV → V_{out} = ?

+80 dB voltage gain

Solution

$$V_{in} = +20 \text{ dB}\mu\text{V}, \quad \text{Amplifier Voltage Gain} = +80 \text{ dB}, \quad V_{out} = ?$$

Again, by simply adding the dB values we get

$$V_{out} = +20\ \text{dB}\mu\text{V} + 80\ \text{dB} = 100\ \text{dB}\mu\text{V} = 100{,}000\ \mu\text{V}$$

Let us check our results.

$$V_{in} = +20\ \text{dB}\mu\text{V} = 10\ \mu\text{V},\ 80\ \text{dB Gain} = 10{,}000x$$

and therefore,

$$V_{out} = 10\ \mu\text{V} \times 10{,}000 = 100{,}000\ \mu\text{V}$$

Everything works. Recall that adding dBs is equivalent to multiplying numbers.

2.7 Tutorial 2: Review of Electronic Filters

The characteristics of almost every electronic circuit or network vary with frequency. In some electronic systems this is an undesired effect, and steps are taken to prevent it from happening. However, many other electronic systems use this effect to an advantage in the processing of signals and the performance of a host of other useful functions. **Electronic filters** are a class of circuit that exhibits a selective frequency response and hence is extremely useful in the field of telecommunications. Recall that capacitors and inductors have frequency-dependent characteristics. Therefore *RC*, *RL*, and *RLC* circuits will exhibit behavior that is frequency dependent. This being the case, one can build analog filter circuits from these components. The most common types of electronic filters are the low-pass filter (LPF), the high-pass filter (HPF), the bandpass filter (BPF), and the bandstop filter (BSF). The BSF is sometimes referred to as a notch filter.

Digital filters can be implemented with digital-signal processors (DSPs). The DSP is a special type of microprocessor designed for signal processing. Using digital filters requires that the signal be converted from analog into digital format. This can be accomplished through the use of an analog-to-digital converter (ADC). After any digital filtering, the signal can be converted back to analog form by using a digital-to-analog converter (DAC) and the appropriate analog filter to eliminate any unwanted spurious frequencies.

Ideal Filters versus Real Filters

The ideal filter circuit would have an extremely sharp cutoff frequency as shown in Figure 2-18. Here one can see the four most common types of filters—the LPF, HPF, BPF, and BSF. Real analog filters, depending upon their configuration, usually do not possess these sharp cutoff characteristics, nor do they possess a flat gain across the passband portion of their response. Shown in Figure 2-19 is the typical frequency response of a simple *RC* low-pass filter. Figure 2-20 shows the circuit of the same *RC* LPF.

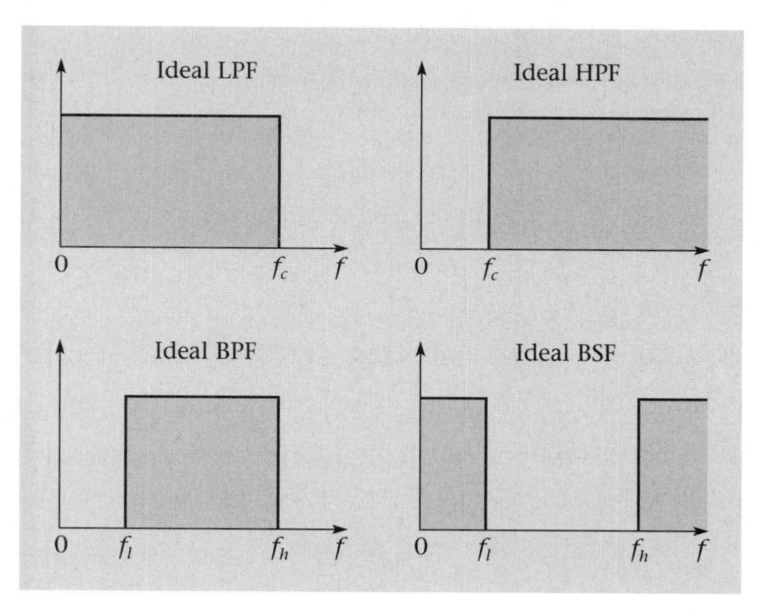

FIGURE 2-18 Ideal filters: LPF, HPF, BPF, and BSF

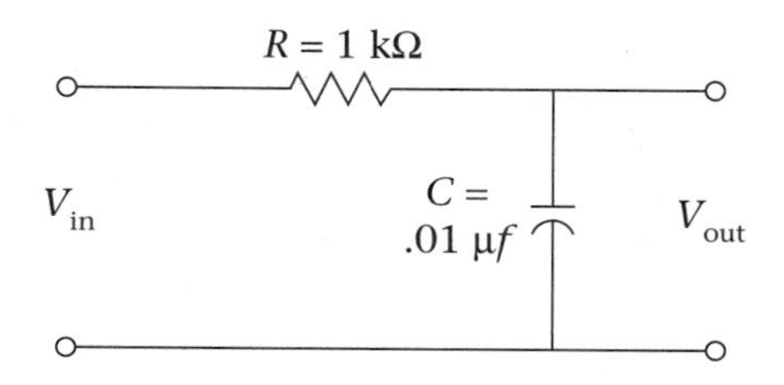

FIGURE 2-20 Simple *RC* LPF circuit

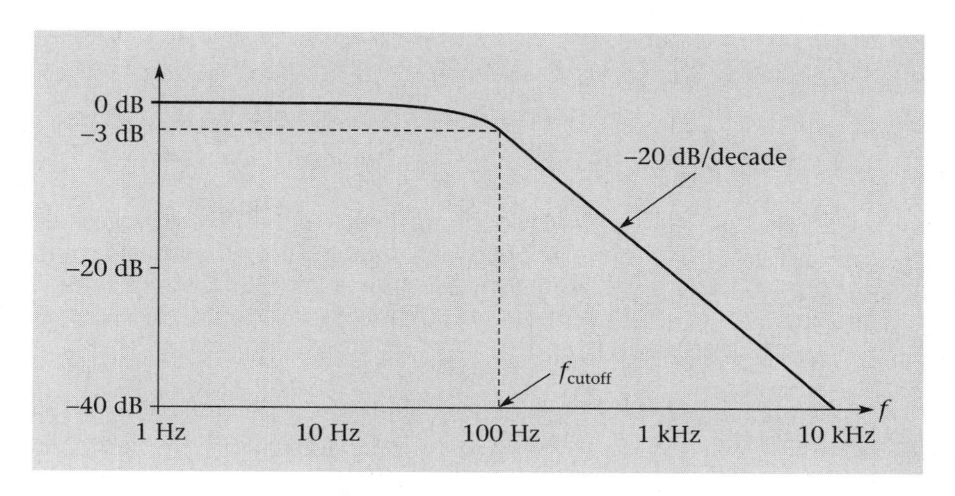

FIGURE 2-19 Simple *RC* LPF response curve or Bode Plot

Characteristics of Filters

Because the purpose of an electronic filter is to be frequency selective, several measures of frequency response have evolved. Today, a plot of the filter's frequency response, either generated by simulation software or measured with a network analyzer, is often used to evaluate a filter's performance. There are several legacy measures of the sharpness of a filter's response related to classical mathematical analysis. Additionally, filters will have a characteristic impedance that usually is designed to match the system that they will be incorporated into.

Roll-Off Rates

LPF and HPF filters have various roll-off rates. The **roll-off rate** (ROR) is given in dB per decade (dB/dec) or dB per octave (dB/oct) where a **decade** is a factor of ten in frequency and an **octave** is a factor of two in frequency. The ROR in dB/dec or dB/oct can be related as follows:

$$\#\ \text{dB/dec} = (20/6) \times \#\ \text{dB/oct} \tag{2.7}$$

Q or Shape Factor

BPF and BSF are types of resonant circuits that have either a selectivity, **Q**, or a **shape factor** (SF) associated with their frequency response curves.

$$Q = \frac{X_L}{R} \text{ or } \frac{X_C}{R}$$

or

$$Q = \frac{\text{3-dB Bandwidth}}{f_0} \tag{2.8}$$

and

$$\text{Shape Factor} = \frac{\text{60-dB Bandwidth}}{\text{6-dB Bandwidth}} \tag{2.9}$$

In both equations, the bandwidth measures relate to the range of frequencies between predetermined signal-level points. Figure 2-21 shows the frequency response of a typical BPF. As shown here, the center frequency, f_0 is at 100 kHz and the bandwidth is defined as the range of frequencies between f_l and f_h. These two points have been defined to be where the output of the filter falls off to 0.707 of the maximum output (this is a commonly used standard).

It is possible to analyze a simple *RC* low-pass filter by writing a voltage-divider equation for the output voltage as a function of frequency and then dividing by the input voltage.

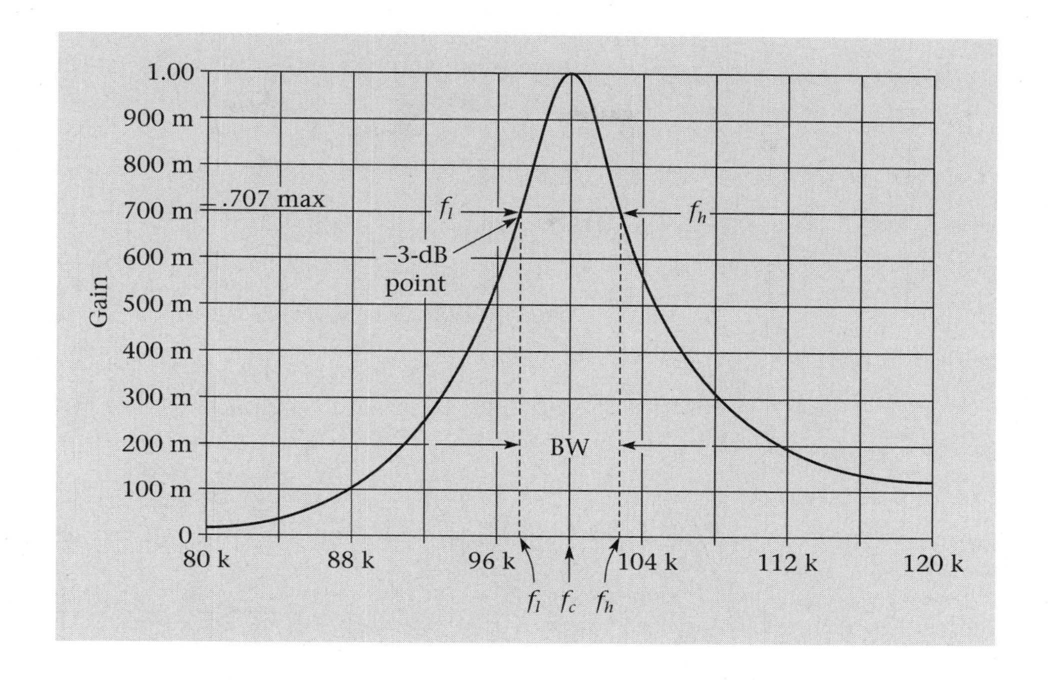

FIGURE 2-21 BPF frequency response

•—EXAMPLE 2.10

For a simple low-pass *RC* filter like the one shown in Figure 2-22, determine the transfer equation of its operation. The circuit gain (transfer equation) is given by the following equation:

$$\frac{V_{\text{out}}}{V_{\text{in}}} = \left(\frac{X_C}{\sqrt{R^2 + X_C^2}} \right) = \text{Filter Gain} \qquad \textbf{2.10}$$

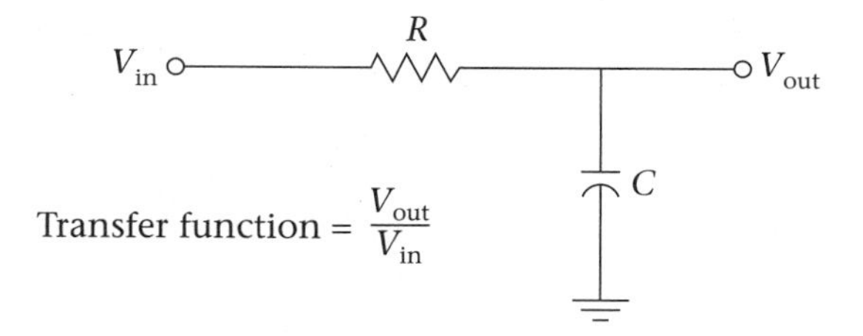

FIGURE 2-22 Simple *RC* low-pass filter circuit

•—Solution Let us examine the results given by this equation for three different frequencies of operation. At a frequency where $X_C = R$, the output is .707 of the input voltage (or the circuit output power is ½ of the input power). This frequency is called the –3-dB point or cutoff frequency, f_{cutoff}.

This frequency can be shown, to be given by, $f_{cutoff} = 1/(2\pi RC)$. Now look at the two extreme frequency points. At dc, or 0 Hz, the output is equal to the input or the circuit gain is 1.0. At a very high frequency ($f = \infty$) the output drops to zero (an LPF response).

Software Simulation of Filters

This type of circuit analysis is extremely easy to perform with a software simulation package like MultiSIM™, formally known as "Electronic Work-Bench." Additionally, MultiSIM's circuit-evaluation suite has a virtual instrument called a **Bode Plotter** that will graph the frequency response of the circuit and allow detailed measurements of bandwidth and gain through the use of cursors and markers. The results are shown in Figure 2-23.

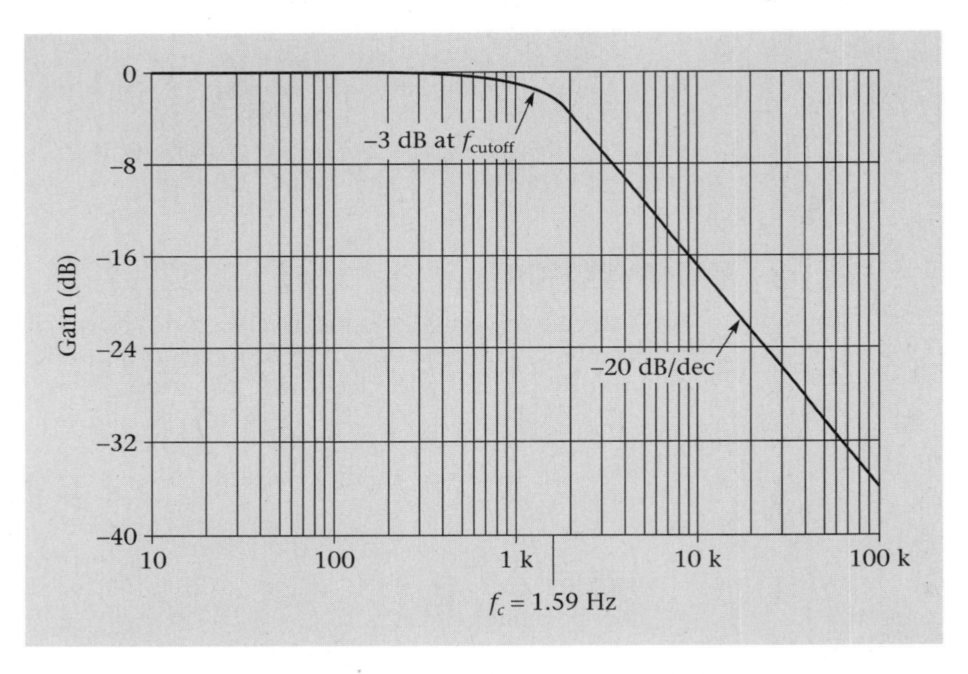

FIGURE 2-23 Typical LPF Bode Plot generated with MultiSIM

As can be seen, the MultiSIM circuit analysis yields the *ideal* mathematical frequency response for the simple *RC* low-pass filter. One can measure the ROR as –20 dB/dec or –6 dB/oct if we are past the cutoff frequency, f_{cutoff}. The nature of simple *RC* filters is to have RORs that are multiples of –20 dB/dec or –6 dB/oct. This is because mathematical analyses of these LPFs tend to yield equations of the following type:

$$\text{Output } \alpha \frac{1}{f^n}$$

where n = # of poles. Each capacitor in a filter can be equated to a pole. So, if one cascaded two simple low-pass filters, then $n = 2$ and the ROR would be –40 dB/dec or –12 dB/oct. The higher the ROR, the sharper the cutoff of the filter.

For BPFs and BSFs, the value of circuit Q (magnification or quality factor) determines the filter's selectivity. The larger the Q, the sharper the frequency response. If it is necessary to have a larger bandwidth filter, one can cascade several BPFs with staggered center frequencies. MultiSIM can eliminate many tedious calculations involving analog filters, easily plotting frequency response and providing bandwidth measurements. See Figure 2-24 for an example.

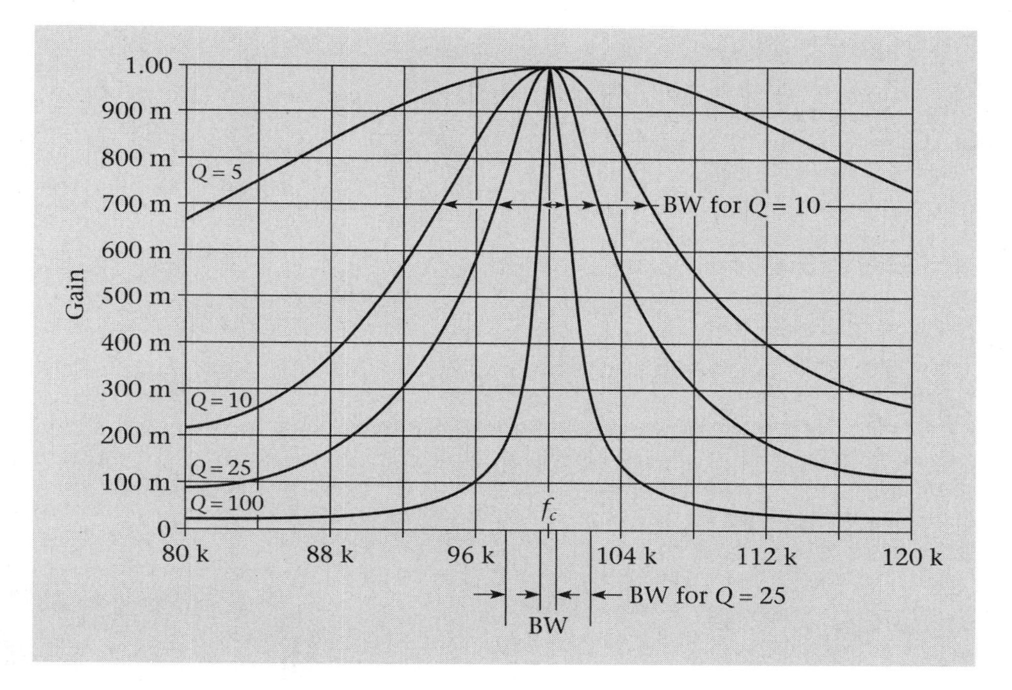

FIGURE 2-24 This figure shows several different values of circuit Q and the bandwidth associated with them

Digital Filtering

The digital filters introduced earlier in the chapter are another story. Although complex and usually implemented using hardware and software, their frequency response is repeatable. Depending upon the level of complexity, they can almost approximate an ideal "brick-wall" or "boxcar" filter. As we move toward the all-digital network and the conversion of signals into a digital format whenever possible, system designers are implementing digital filtering whenever it is cost-effective or necessary.

2.8 Tutorial 3: Telecommunications Test Equipment

Because telecommunications deals with the transmission of signals, it is important to be able to observe and measure the parameters of these signals as they are processed through a telecommunications system and to be able to measure the frequency response of the various subsystems and transmission networks that make up the entire system. In addition to using test equipment to monitor system performance and troubleshoot faults, modern sophisticated test and measurement equipment is necessary in the manufacturing environment to perform the final testing of critical product parameters and to document these results for quality-control purposes. The telecommunications technician should be familiar with typical general-purpose test and measurement equipment designed to aid in these processes. The basic testing and display tools of telecommunications are the digital oscilloscope and spectrum and network analyzers. Additionally, specialized equipment is available to monitor and evaluate the LAN and WAN network environment; digital transmission systems using SONET; T-carrier; ATM; and the transmission of data over microwave, CATV, cellular, and lightwave transmission facilities. This specialized test equipment is used to measure system conformance to worldwide standards and to evaluate system operation in terms of bit error rate (BER) measurements, timing jitter, and other system-specific impairments.

Signal Displays in the Time Domain

The most common method used to display a signal is by using an oscilloscope, of which an example is pictured in Figure 2-25. Because the oscilloscope shows how the signal amplitude changes with time, the display is said to show the signal in the **time domain**.

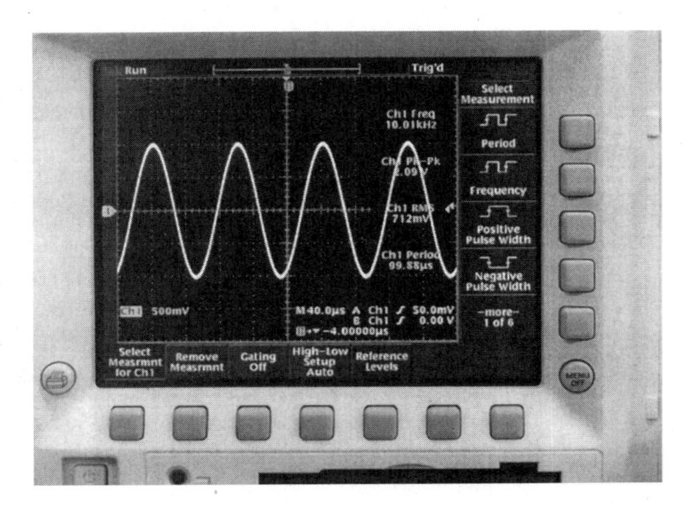

FIGURE 2-25 Typical digital oscilloscope time-domain display of a 10-kHz sine wave with automatic measurements

The time display of a signal can be very instructive because it shows the basic signal parameters of amplitude, frequency or period, and shape. Modern digital oscilloscopes usually provide automatic measurement of signal amplitude and frequency; however, the accuracy of these measurements is limited. The limited dynamic range (sensitivity) of an oscilloscope and its limited frequency response usually reduces its effectiveness as a measurement tool. Use of a dual-trace oscilloscope, shown in Figure 2-26, usually provides a technician or engineer the ability to perform a first-approximation type of analysis of signals and system operation. An additional feature of this instrument is the ability to mathematically calculate the fast-Fourier transform (FFT) of a standard time-domain signal into its frequency components, thus providing limited spectrum-analysis capabilities (albeit through mathematical calculation). The oscilloscope is a standard virtual instrument (VI) in the test-equipment suite of MultiSIM. It is assumed that the reader has some knowledge of oscilloscope operation, therefore little detail will be presented here. However, it is important to remember that the main instrument controls are the vertical gain and horizontal sweep time. The vertical gain is usually specified in volts/division while the horizontal scale is calibrated in seconds/division.

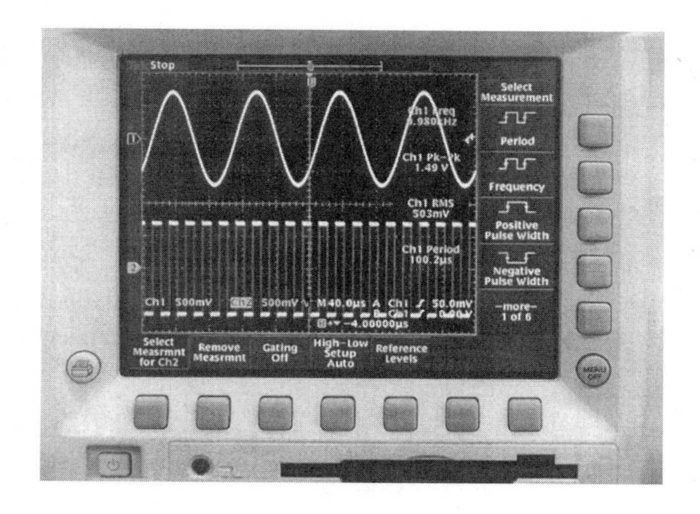

FIGURE 2-26 Dual-trace oscilloscope displaying two signals at the same time

Signal Displays in the Frequency Domain

The spectrum analyzer (SA) is an extremely sensitive, specialized, wideband radio receiver that displays the amplitude of a signal versus its frequency. This display is depicted in Figure 2-27 on page 66. The display of a signal in this manner is known as a **frequency-domain** display. The spectrum analyzer holds a definite advantage over the oscilloscope due to its large dynamic range and large bandwidth. One is able to see signal imperfections

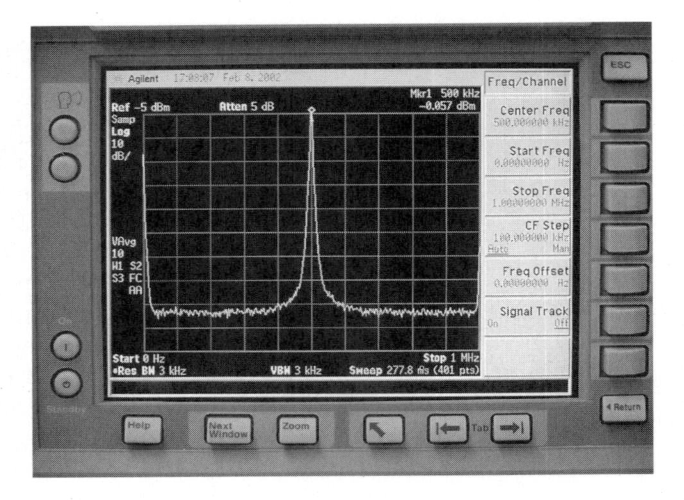

FIGURE 2-27 Typical spectrum-analyzer frequency-domain display of a 500-kHz sine wave

and detect system nonlinear operation through the observation of signal harmonics and spurious frequencies that are not obvious on the time display of an oscilloscope.

The logarithmic vertical scale of the SA usually displays the power of the signal and is calibrated in dB/division, while the linear horizontal scale is the **frequency span**, or range that is being displayed. As with an oscilloscope, the sensitivity of the SA can be adjusted and the range of frequencies displayed can be altered. The amplitude display of the SA is different from that of an oscilloscope in that the top of the display is set as the reference-power level (commonly 0 dBm) for the rest of the display. Observed signal power therefore decreases as one approaches the bottom of the display. For an oscilloscope, the zero reference level usually is set to the middle line of an oscilloscope screen and signal amplitude increases with the distance from the center line. Another interesting capability of the spectrum analyzer is that one is able to select the ranges of frequencies to be displayed. One might be interested in looking at a large span of frequencies, an entire service band, or alternately a narrow band of frequencies with a higher frequency resolution. Frequency span is calibrated in Hz/division. The **resolution** of an SA is its ability to display or separate signals that are closely spaced in the frequency domain. Some additional features of the SA are the ability to modify the vertical-scale display from logarithmic to linear, to change the top reference level, and to perform automatic power, frequency, and bandwidth measurements through the use of markers. **Markers** are small cursors that can be automatically deployed by the instrument or positioned by the user.

Spectrum Analyzer Specifications

The most important specifications of the SA are those of sensitivity, dynamic range, resolution, and frequency range. The low-cost SA shown in

Figure 2-27 is capable of detecting signals as small as –120 dBm (1 fW) and displaying a range of 100 dB of signal-power difference. The resolution of this SA is 1 Hz, and it can display any frequency range from approximately 0 Hz to 3.0 GHz. It must also be noted that single frequency signals usually appear as shown in Figure 2-28 due to the shape factor or resolution of the filters used in the SA. Note that the SA has been introduced as a VI in the latest editions of MultiSIM. Figures 2-28 and 2-29 illustrate the typical display of an SA and the SA's ability to show a large dynamic range. Figure 2-28 shows a single sine wave and Figure 2-29 shows a signal and its harmonic content. In Figure 2-29 on page 68, a "Marker" table is included in the display. One can see that the third harmonic (Marker 2) is –54.24 dB down from the fundamental signal.

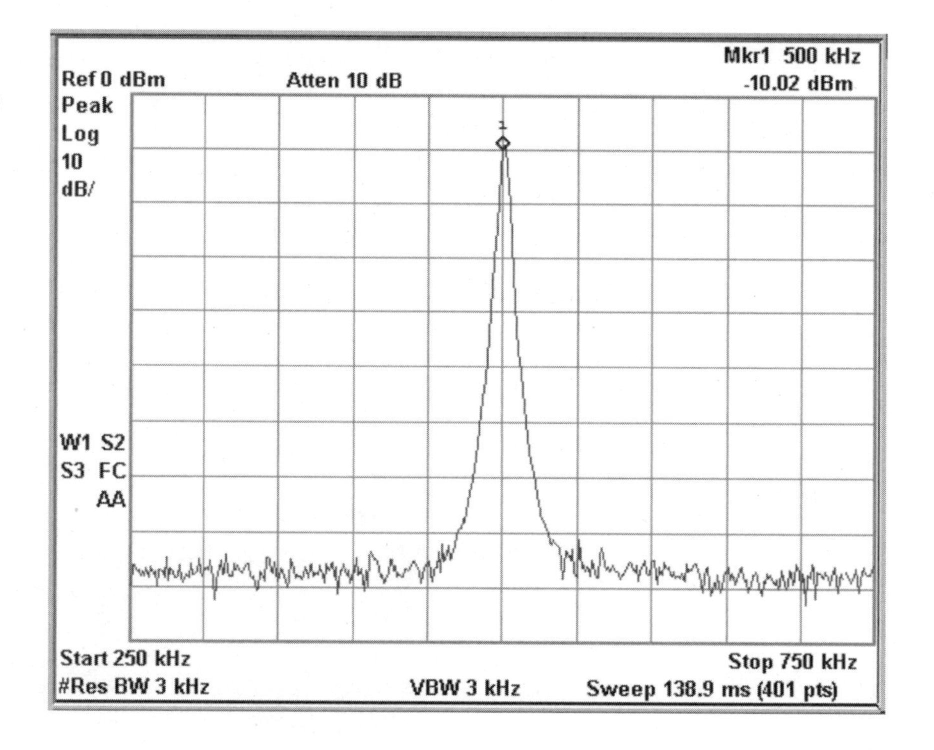

FIGURE 2-28 A spectrum-analyzer display of a 100-kHz sine wave

Measuring Network Frequency Response

Another important type of measurement is the **frequency response** (otherwise known as the *system transfer function* in the engineering world) of the various subsystems of a telecommunications system. A knowledge of the frequency response of the channel or transmission media allows one to determine system bandwidth, and hence channel capacity. It also allows for an understanding of transmission problems and limitations caused

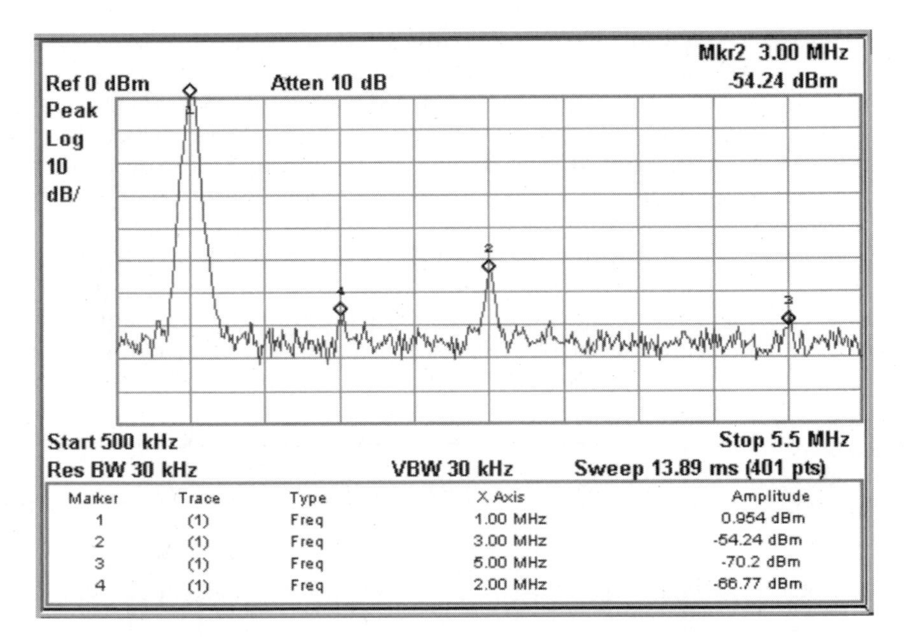

FIGURE 2-29 A spectrum-analyzer display of a sine wave and its harmonics

by **system dropout**, a high loss at a certain frequency or frequencies. A **network analyzer** (NA) is commonly used in the telecommunications manufacturing environment to evaluate the T/R pair or subsystems of the T/R systems and to align these systems to meet manufacturing specifications. The NA is also used to document the specifications of various systems at their time of manufacture. A low-cost scalar and vector network analyzer are shown together in Figure 2-30.

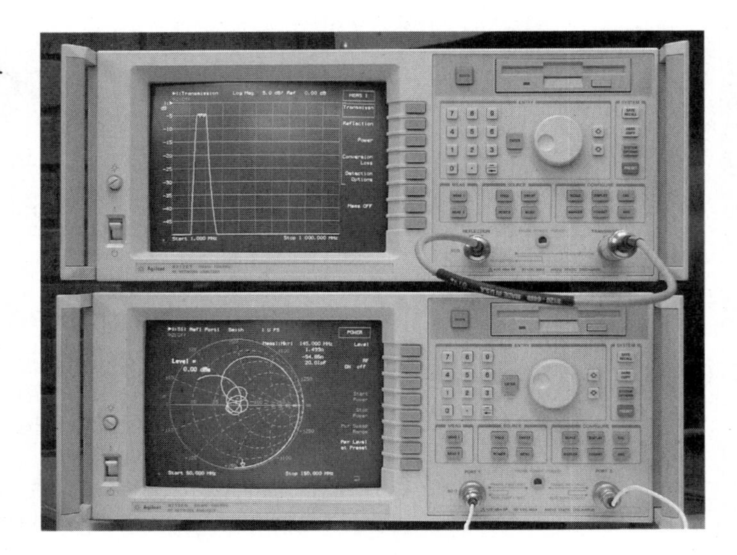

FIGURE 2-30 Low-cost scalar and vector network analyzers

The Network Analyzer

The network-analyzer display is extremely similar to that of the SA. The basic NA consists of two sections: a transmitter and a receiver. A device or system under test (DUT or SUT, respectively) is placed in line between the transmitter output section and the analyzer's receiver input section; thus, the network analyzer is analogous to our model of a basic telecommunications system (transmitter, channel, receiver). The analyzer's transmitter outputs a constant amplitude RF signal that is swept across a range of frequencies. The display is usually calibrated in so many dB/division in the vertical direction (with the reference at the top line like the SA) and in frequency span in the horizontal direction. The display indicates the device or system's frequency response. The scalar NA can display the forward power-transfer characteristic (defined as S_{11}) and reflected power or reflection coefficient, Γ (see chapter 9 for more details), while the more comprehensive vector network analyzer can determine all of the S-parameters for a two-port device or system. Recall the virtual instrumentation Bode Plotter in MultiSIM—it is a virtual network analyzer with limited functionality. Markers with automatic measurement capability are usually provided as a feature of the network analyzer.

Figures 2-31 and 2-32 show typical scalar and vector network-analyzer screen displays. Figure 2-32 on page 70 displays the frequency response of a BPF plotted on a Smith chart. Note the marker (arrow) at the end of the trace. The marker indicates an impedance measurement made at a certain frequency. This value is annotated on the top of the vector network-analyzer screen.

FIGURE 2-31 A scalar network analyzer display of the frequency response of a BPF

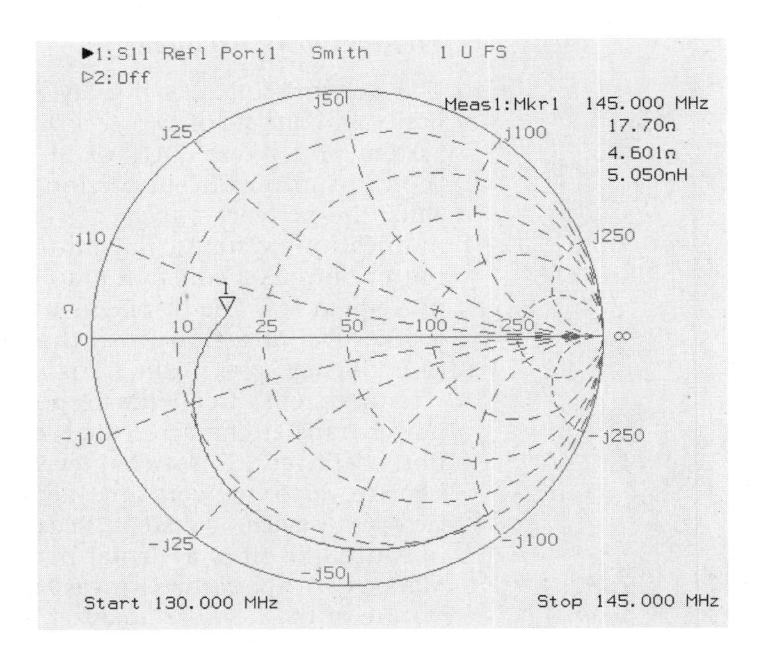

FIGURE 2-32 A vector network analyzer display of the frequency response of a BPF

Other Basic Telecommunications Test Equipment

Although available as multifunctional stand-alone units, the basic functions of a frequency counter or RF power meter are built into both the spectrum and network analyzers. As mentioned before, there is a vast selection of specialized test equipment to monitor and evaluate the numerous types of networks and transmission protocols, systems, and technologies currently in use. Unfortunately, this type of application-specific telecommunications test equipment tends to be extremely expensive, industry specific, and not readily accessible to most educational institutions and their students.

Virtual Instruments and Software Simulation

Simulation software packages are becoming more realistic and accurate with each new software release and they are becoming more mainstream, increasingly found in both the designer's and the educator's hands. Designers can simulate the operation of their evolving plans for new telecommunications systems and then adjust them according to feedback from the simulation program. The educator can use complex system simulations to reinforce concepts and theories discussed in class. Performing a "case study" of a telecommunications system by means of simulation usually allows for the changing of parameters, which can aid in learning about the nuances of system behavior.

2.9 Tutorial 4: Noise and Noise Measurements

Where does noise come from? Noise can be either external or internal in nature, originating in both the channel and the telecommunications equipment (transmitter and receiver). In general, noise is anything that tends to impair the received signal, thus being the ultimate limiting factor in all telecommunications systems. For analog systems, one talks about the signal-to-noise ratio and the system-noise figure as measures of performance. For digital systems, one talks about the bit error rate to indicate performance characteristics. A common term for noise is *electromagnetic interference* (EMI).

External Noise

External noise is introduced during signal transmission and can be:

- Man-made
- Atmospheric
- Extraterrestrial

Man-made noise is generated by the activities of civilizations using electricity to power industrial processes (manufacturing, transportation, lighting, and so forth). The industrialized nations of the world produce electrical pollution in the form of noise.

Atmospheric noise is produced by active weather patterns all over the world. Watch the Weather Channel or go to http:\\www.weather.com to access a map of a given day's lightning strikes in the United States. Constant electrical activity in the atmosphere (estimated to be one hundred lightning strikes per second worldwide) produces background noise. This type of noise tends to be predominately in the 0- to 30-MHz range and its intensity is inversely proportional to its frequency.

Extraterrestrial noise has many sources. Most of it is produced by our own sun, but one can include noise from everything else in the cosmos in this category. Note that what is being labeled noise here is what a radio-astronomy person considers a signal. External noise is often characterized as having a certain power per unit of bandwidth (e.g., so many μW/MHz or pW/Hz).

Internal Noise

Internal noise is introduced by the system and is produced by small, random fluctuations of electrical current through either passive or active devices. Known as *thermal noise* for passive devices like resistors and *equivalent resistance* for capacitors and inductors, this noise can be calculated.

Noise produced by most active devices, on the other hand, is known by various names such as *shot*, *partition*, and *transit-time noise*, and is not easily calculated.

Noise Calculations

The noise voltage produced by a resistor and the equivalent resistance of a passive device can be calculated by using the following equations:

$$P_n = kT\Delta f \quad \textbf{2.11}$$

and

$$e_n = \sqrt{4kT\Delta f R} \quad \textbf{2.12}$$

where $k = 1.38 \times 10^{-23}$ eV, T is the temperature in degrees Kelvin (K), and Δf is the bandwidth appropriate to the particular situation.

Noise generated by a diode (shot noise) can be calculated and is given by the following equation:

$$i_{\text{rms}} = \sqrt{2ei_{\text{diode}}\Delta f} \quad \textbf{2.13}$$

where e is the charge on an electron, 1.6×10^{-19} C, and i_{diode} is the diode current.

Noise generated by active devices like transistors and ICs is difficult to calculate, so an equivalent noise resistance is assigned to these devices. Let us take a look at noise displayed on an oscilloscope (Figure 2-33) or a spectrum analyzer (Figure 2-34).

The time-domain display of noise shown in Figure 2-33 indicates the random nature of noise. Noise that occurs equally at all frequencies is known as "white noise," while noise that demonstrates some particular

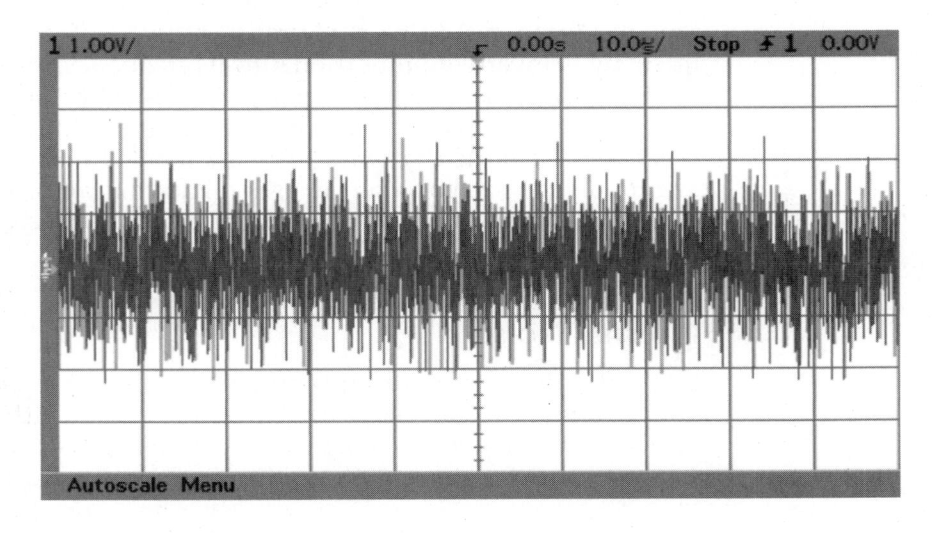

FIGURE 2-33 This time display of noise shows its random nature

frequency characteristics is known as "colored noise." Figure 2-34 shows a spectrum-analyzer display of noise voltage. Note the reference level; full scale is 150 μV and the frequency span is from 0 Hz to 3 GHz.

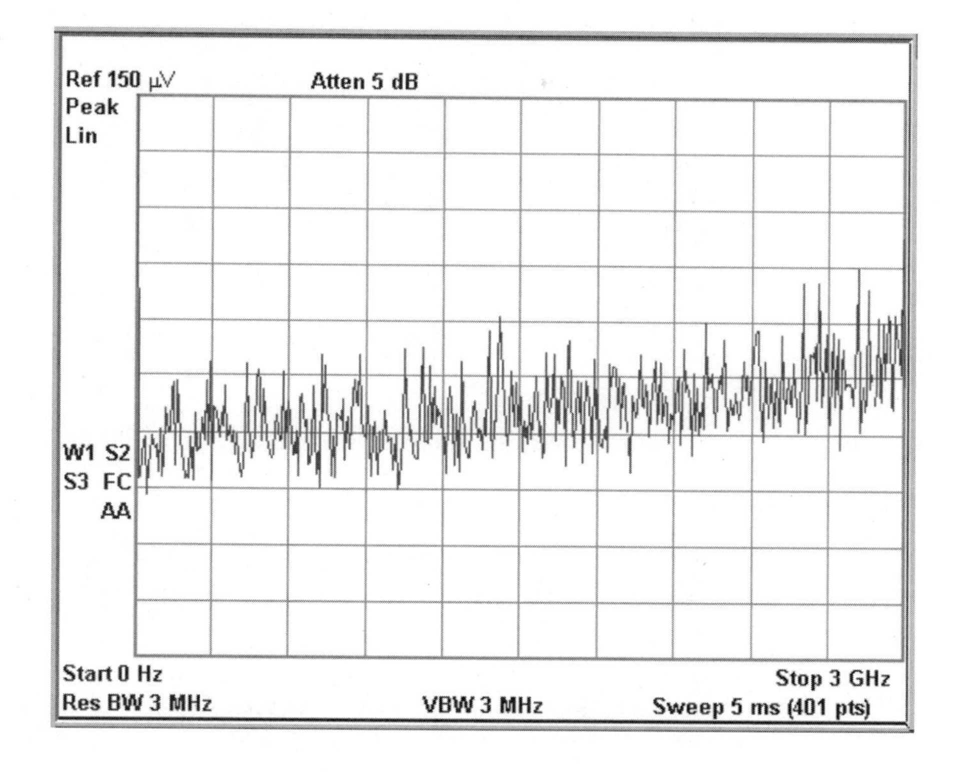

FIGURE 2-34 Noise as displayed on a spectrum analyzer

Effects of Noise on Signals

Now take a look at the effect of noise on two different types of signals, shown in Figures 2-35 and 2-36 on page 74. The sine wave in Figure 2-35 has been affected by the addition of noise to its amplitude, as has the pulse waveform. However, if information has been encoded in terms of 1s and 0s, the pulse waveform will be more immune to the effects of the noise. Why is this so?

Due to the noise, any device receiving the noisy sine wave will have a difficult time resolving the correct amplitude of the signal as it is received. If, for instance, the amplitude of the signal contains transmitted information, the receiver will invariably make errors in its output of the signal as compared to the original one. However, for the noisy pulse signal the receiver only has to determine the presence or absence of a pulse. This general determination is much easier to make than determining the precise voltage of the noisy sine wave.

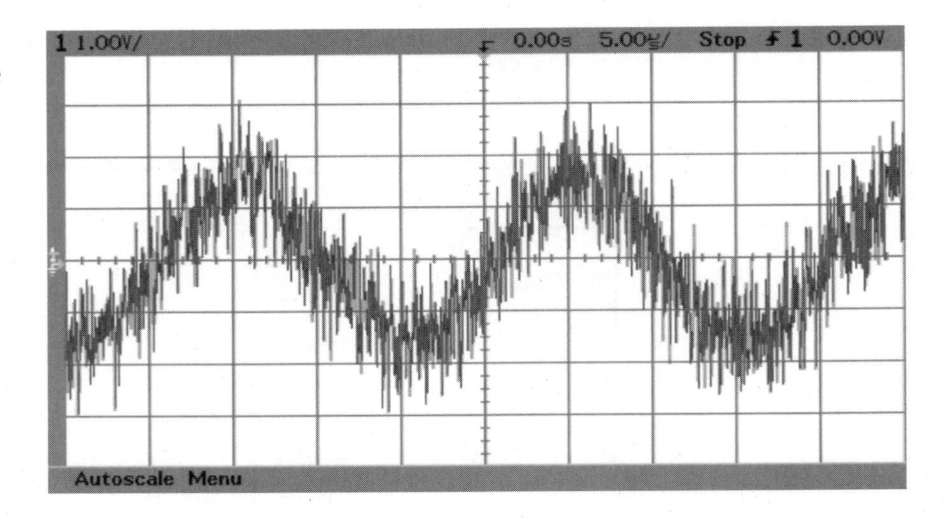

FIGURE 2-35 Noise effect on a sine wave

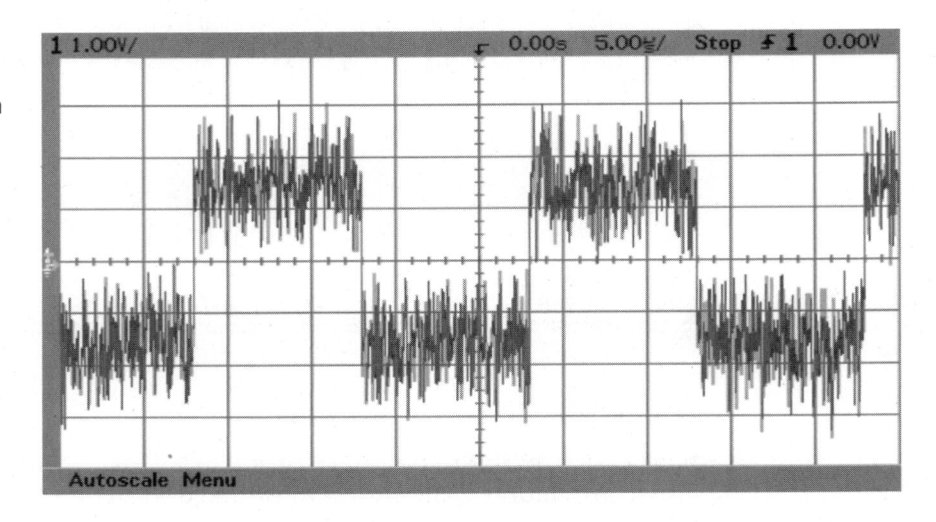

FIGURE 2-36 Noise effect on a square-wave pulse waveform

Noise Measurements: SNR

For telecommunications systems we use several different noise measurements. One of the most widely known is **signal-to-noise ratio** (SNR). This is a measurement (usually at the input to the receiver portion of the telecommunications system) of the quality of the signal as it is received. Mathematically, this is represented by the following equations:

$$SNR = \frac{\text{Signal Power}}{\text{Noise Power}} = \frac{P_S}{P_N} \tag{2.14}$$

or

$$SNR = 10\log\left(\frac{P_S}{P_N}\right) = 20\log\left(\frac{E_S}{E_N}\right) \qquad \textbf{2.15}$$

What constitutes a good SNR? Video-information (analog) signals are considered to be of perfect quality when the SNR is 60 dB (signal-power to noise-power ratio is 10^6). Digital signals often experience a very low bit error rate (BER) for an SNR of 10 dB. A CD player usually has an SNR of greater than 90 dB (a perfect output for practical purposes). Look back at Figures 2-35 and 2-36 to see if you agree that digital signals are more immune to noise.

Other Variations on SNR

SINAD is used by the wireless industry to rate receivers. SINAD is given by the following equation, where D represents distortion:

$$SINAD = 10\log\left(\frac{S+N+D}{N+D}\right) \qquad \textbf{2.16}$$

Usually, the wireless receiver's sensitivity is specified for a certain value of SINAD. For example, a particular radio's sensitivity is 12 fW for a SINAD of 12 dB.

Noise Figure

Noise figure (or *noise ratio*) is a measurement of the quality of the receiver in a telecommunications system and is given by:

$$NF = 10\log\left(\frac{SNR_{\text{in}}}{SNR_{\text{out}}}\right) \qquad \textbf{2.17}$$

This measurement indicates how much noise has been added to the signal by the receiver. What constitutes a good noise figure? A noise ratio (NR) of 1 or a noise figure of 0 dB indicates a perfect system.

Noise Temperature

Noise temperature is another way of specifying the noise performance of a device. It is becoming a more common measurement and is given by:

$$T_{eq} = T_0(NR - 1) \qquad \textbf{2.18}$$

What is a typical noise temperature? A direct-broadcast satellite (DSB) receiver might have a low-noise amplifier (LNA) with a noise temperature of 60 K. As a final comment on noise figure and noise temperature, it must be pointed out that the first stage of a receiver or system usually sets the noise figure or noise temperature for the rest of the system. This is the reason why a more costly (better quality) low-noise amplifier (LNA) is used in the first stage of most sensitive or high-frequency equipment.

Bit Error Rate

When dealing with the transmission of digital data, the system's bit error rate (BER) is typically used in conjunction with the signal-to-noise ratio to specify system performance. Chapter 6 will present examples of this for various digital modulation schemes.

2.10 Tutorial 5: Fourier Analysis and Signals

This tutorial will give a brief introduction to the application of **Fourier analysis** to signals. Liberal use of time and frequency displays will be used to convey the mathematics of Fourier analysis.

Many years ago a great French physicist and mathematician by the name of Jean Fourier developed a mathematical procedure that allows one to represent or model any periodic signal by an infinite series of harmonically related sine and/or cosine waves. The amplitude and frequency of the various components can be calculated mathematically. It has already been mentioned that many modern oscilloscopes have the ability to calculate and display what is known as the fast-Fourier transform (FFT) of a signal. Let us start our discussion of Fourier analysis by looking at a square wave.

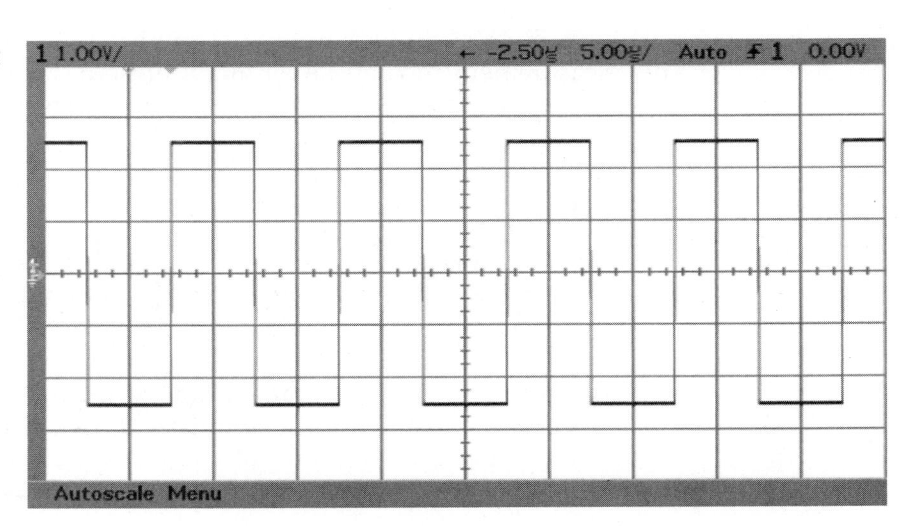

FIGURE 2-37 100-kHz square wave

Using Fourier analysis, it can be shown that for a 100-kHz square wave like that shown in Figure 2-37, the signal can be represented by an infinite series of odd harmonics as shown by the following equation:

$$v(t) = \frac{4V}{\pi}\left(\cos \omega t - \frac{1}{3}\cos 3\omega t + \frac{1}{5}\cos 5\omega t - \ ...\right) \quad \textbf{2.19}$$

From the Fourier-series equation, it is seen that the amplitude of the harmonics decreases by $\frac{1}{n}$, where n is the harmonic number.

Therefore, the power of the nth harmonic is given by:

$$\text{Power Harmonic} = \text{Power Fundamental} \times \left(\frac{1}{n^2}\right) \quad \textbf{2.20}$$

The power of the third harmonic should be $\frac{1}{9}$ of the fundamental, or –9.54 dB down from the fundamental frequency component. Similarly,

The fifth harmonic is	$\frac{1}{25}$, or –13.98 dB down
The seventh ...	$\frac{1}{49}$, or –16.90 dB down
The ninth ...	$\frac{1}{81}$, or –19.08 dB down
The eleventh ...	$\frac{1}{121}$, or –20.83 dB down

Now let us look at this square-wave signal with the SA, as illustrated in Figure 2-38 on page 78. Look at the display and note the frequency components that are shown. The Marker table shows values of dB down from the fundamental frequency for the 3rd, 5th, 7th, and 9th harmonic. The predicted values compare quite faithfully with the measured values.

What is the meaning of what we have seen? All periodic signals consist of an infinite number of harmonically related sine or cosine waves. The vector (amplitude and phase) addition of these frequency components will yield the time display of the signal.

Are there no such things as square waves? The answer to that question is both yes *and* no. A square wave is only what we observe in the time domain. We are comfortable with time-domain representations of signals because we relate to how things change with the passage of time. In the time-domain display, the signal appears to alternate back and forth between two signal levels, and hence looks like what is known as a square wave. However, the electronic process that is causing this periodic change in voltage level is actually producing a signal that is really a large number of harmonically related sine waves, as shown to us in the frequency domain by the spectrum-analyzer display, or as predicted by the appropriate Fourier series. Do not be confused by the mathematics. We use it to model the process; it does not cause the results to happen.

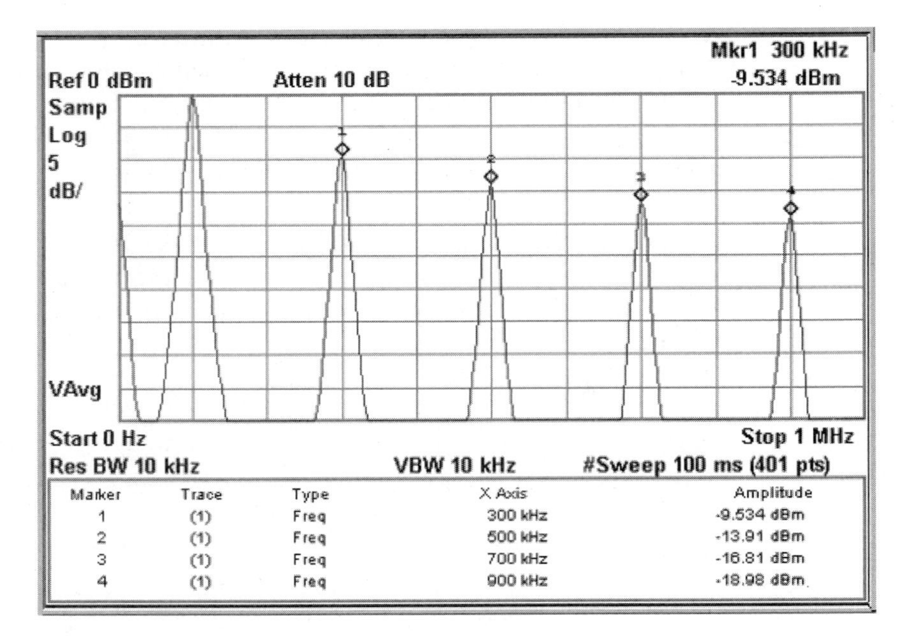

FIGURE 2-38
100-kHz square-wave spectrum

For another example, let us look at a triangular wave, as shown in Figure 2-39.

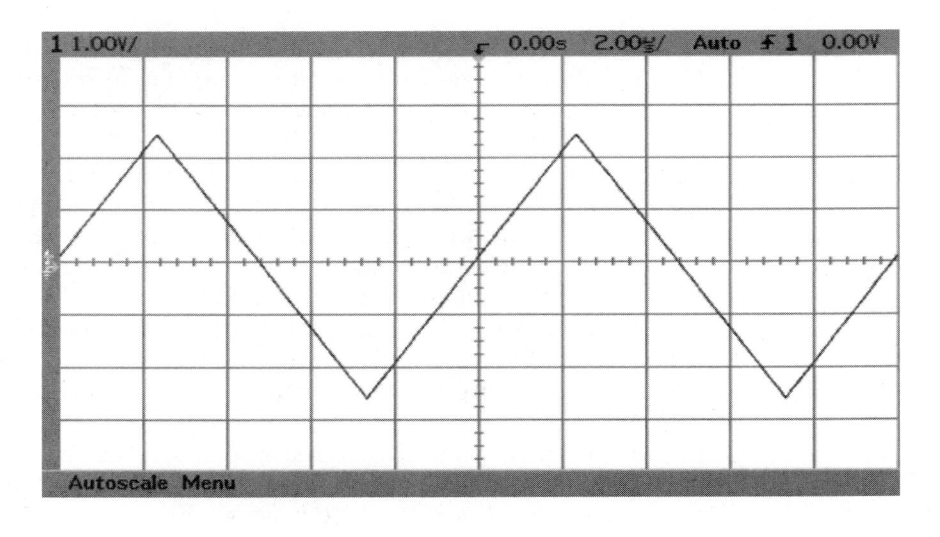

FIGURE 2-39
100-kHz triangular wave

For this waveform, the Fourier series is given by:

$$v(t) = \frac{8V}{\pi^2}\left(\cos\omega t - \frac{1}{3^2}\cos 3\omega t + \frac{1}{5^2}\cos 5\omega t - \ \ldots\right) \qquad \textbf{2.21}$$

Again, notice how the Fourier-series equation indicates that the triangular wave is made up of only odd harmonics, which in this case decrease in intensity as the square of the harmonic number, n. Therefore, the 3rd harmonic is –19.08 dB down, the 5th harmonic is –27.96 dB down, the 7th harmonic is –33.80 dB down, the 9th harmonic is –38.17 dB down, and so forth. Now look at the SA display of this signal, shown in Figure 2-40. Compare Figure 2-38 (the square-wave spectra) with Figure 2-40 (the triangular-wave spectra). The amplitudes of the triangular-wave components are seen to decrease more rapidly, than those of the square wave, as predicted by the mathematical model given by the Fourier series. It should be pointed out that the electronic process that is producing this periodic triangular waveform is also producing the harmonic components that make up the signal. Again, the mathematics only models the process.

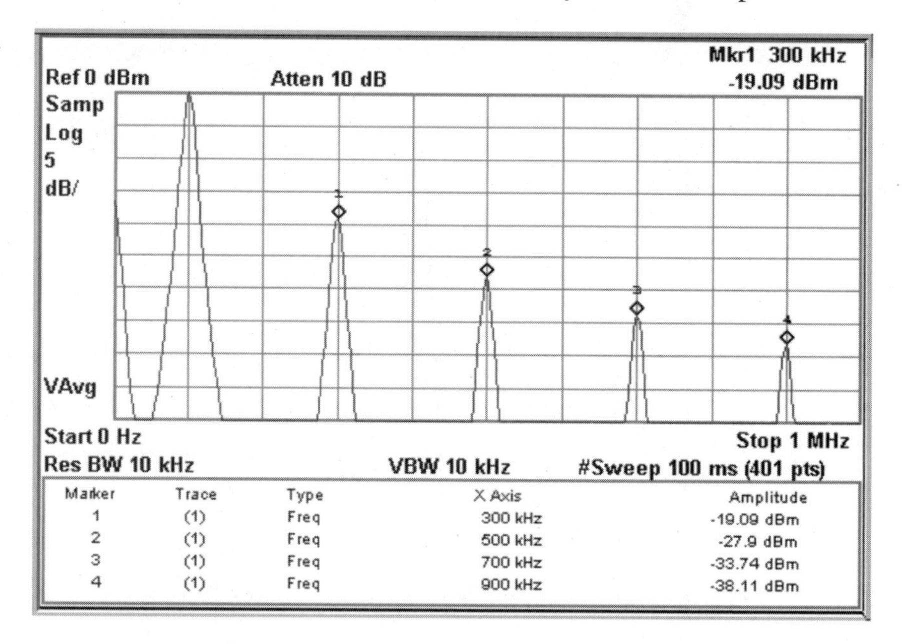

FIGURE 2-40 The SA display of a 100-kHz triangular wave

Another example is the ramp waveform shown in Figure 2-41 on page 80. In this case, the SA display of the ramp waveform (Figure 2-42) shows that both even and odd harmonics constitute the waveform. At this point we stop and ask, How important is our understanding of this concept of signal composition? It is extremely important! Anyone studying telecommunications should be aware of the fact that any periodic or nonperiodic signal consists of a spectrum of frequencies. Analog signals tend to consist of a continuous range of frequencies starting near dc and gradually diminishing in amplitude. Digital signals, which are increasingly being used for the transmission of data, tend to consist of harmonically related components that gradually diminish in amplitude.

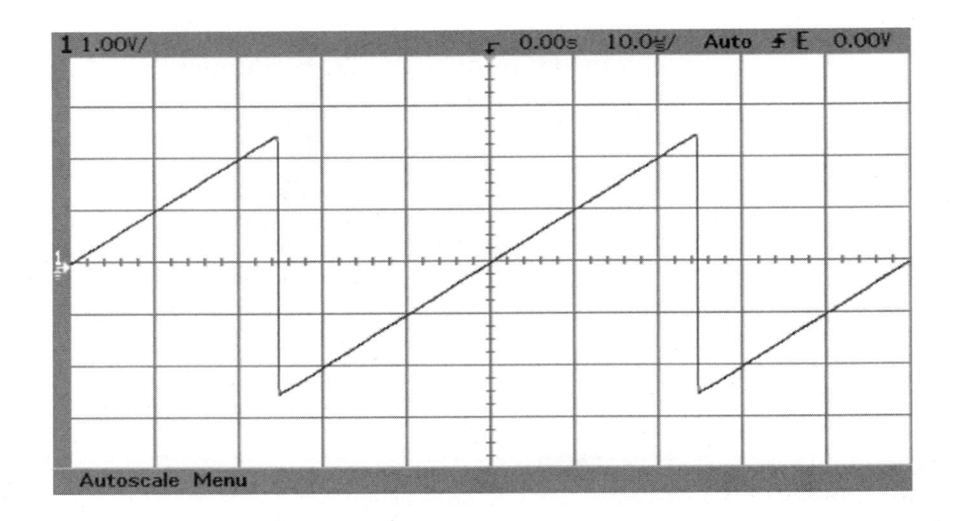

FIGURE 2-41 100-kHz ramp waveform

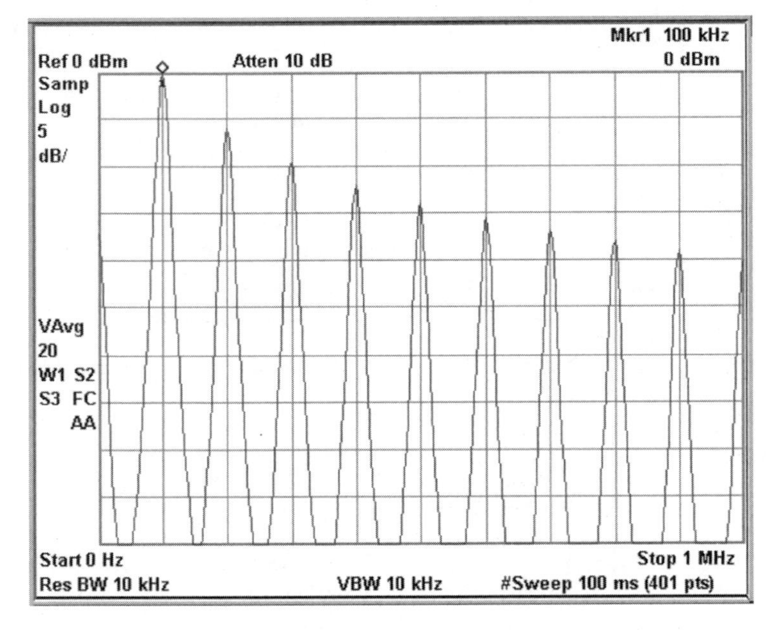

FIGURE 2-42 SA display of a 100-kHz ramp waveform

The concept of signal bandwidth has been introduced before. As can be seen by the SA displays presented here, digital signals tend to have large bandwidths. In all cases of signal transmission, sufficient bandwidth is necessary to achieve distortion-free or error-free transmission of a signal. If insufficient bandwidth exists in the channel, the rate at which information can be transmitted will be limited. Additionally, digital signals with their large bandwidths will tend to interfere with signals in other frequency channels if steps are not taken to reduce signal bandwidth. As one can see, limited bandwidth is at odds with our desire for high-speed transmission of data. We will look at the transmission of digital data in more detail in chapter 6.

Summary

In this chapter, the material necessary for the student's successful understanding of the chapters that will follow has been presented with numerous examples and illustrations. Fundamental telecommunications definitions and concepts have been introduced and the basic telecommunications system block diagram has been explored in detail. Examples of present-day physical channel infrastructure were presented. Chapter 8, which discusses transmission media, will further explore this topic. The chapter concludes with a series of five short tutorials on topics extremely relevant to the student's future grasp, perception, and comprehension of the more complex topics in this field.

The tutorials first review dBs and referenced values of dBms, the understanding of which is an absolute necessity when one is discussing telecommunications system hardware. The second tutorial reviews the theory of electrical filters. These frequency-selective circuits are part of the telecommunications-system designer's basic building blocks. They are used for numerous signal-processing tasks in the implementation of practical telecommunications systems. The third tutorial looks at modern test and measurement equipment. Technical personnel employed by the telecommunications industry use this equipment to test and verify proper operation of newly manufactured systems or those systems already deployed in the field. Also introduced briefly in this tutorial and the previous one is the topic of virtual instrumentation and simulation software. Software simulation of telecommunications systems and the use of virtual test and measurement equipment is becoming extremely popular as a convergence of test equipment and the PC continues. The utility of such methods becomes greater with each succeeding software release and with each increase in PC functionality. The fourth tutorial deals with noise—the ultimate limiting variable of our telecommunications systems—and noise measurements. The last tutorial ties several topics together with the discussion and illustration of signal-frequency spectra with both Fourier analysis and modern test equipment. With examples shown in both the time and frequency domain, this final section of the chapter is designed to solidify the reader's understanding of just what an electronic signal is.

Questions and Problems

Sections 2.1 to 2.4

1. Describe the difference between analog and digital signals.
2. What is meant by the term *signal*?
3. Describe the acoustical channel.
4. Determine the wavelength of an 18-GHz signal.
5. Determine the frequency of a light signal with a wavelength of 1420 nm.
6. Describe the difference between circuit-switched and connectionless circuit-switching technology.
7. Explain the operation of SONAR in terms of the telecommunications block diagram.
8. Explain the operation of RADAR in terms of the telecommunications block diagram.
9. Do an Internet search on "microelectromechanical systems." Write a short paragraph about what is involved in this technology.
10. If Global-Positioning-System (GPS) technology becomes embedded in the receivers of wireless telecommunications systems, what are the potential benefits to the system?

Section 2.6

11. A certain system has three stages with power gains of 15, 65, and 52. Determine the power gain in dB of each stage and the overall system gain.
12. A certain transmission line has a loss of 6.7 dB. Determine the relative signal strength at the output compared to the input.
13. A certain amplifier has a voltage gain of 64 dB. Express the gain as a numeric ratio.
14. Use dBm to indicate the value of a 24-mW signal.
15. If a –35-dBm signal undergoes amplification of +42 dB, determine its resulting level in both dBm and watts.
16. Determine the amplitude of a –86-dBm signal.
17. A microwave oven's magnetron has a power output of 800 W. Reference this value to dBW.
18. The signal level from the local cable-TV company is –40 dBm. Determine its voltage value. Recall that cable TV operates with a 75-Ω standard.

Section 2.7

19. A certain filter has a roll-off rate of 18 dB/oct. What does this mean?
20. What determines the *Q* of a filter?
21. What is the bandwidth of a filter with a *Q* of 75 and a center frequency of 375 MHz?
22. Describe what occurs if an LPF and an HPF are cascaded? Give an example.
23. Using a software simulation program (if available), plot the response curve of a resonant circuit constructed from a 1-μH inductance and a 2-μF capacitor. Assume a circuit resistance of 12 Ω. Determine the resonant frequency, *Q*, and bandwidth.
24. If the –6-dB bandwidth of a filter is 3.1 kHz, determine the –60-dB bandwidth if the filter's shape factor is 1.45.
25. A certain low-pass filter has a roll-off rate of –40 dB/dec. If an input signal that is out of the passband has its frequency raised by a factor of 4, what happens to its output level? Assume the input level is constant.

Section 2.8

26. Determine the signal frequency and amplitude of the oscilloscope display shown in

Figure 2-43. Give the amplitude of the signal in three different ways.

27. Determine the signal frequencies and amplitudes of the spectrum-analyzer displays in Figures 2-44a and 2-44b.

28. For the spectrum-analyzer display in Figure 2-45 on page 84, determine the number of dB down from the larger amplitude signal the smaller signal is.

29. What is meant by the spectrum analyzer specification of *frequency resolution*?

30. What does the *span* setting of the spectrum analyzer refer to?

31. Compare the dynamic range of the spectrum analyzer with that of the oscilloscope.

32. What is an optical spectrum analyzer (OSA)? Where is it used?

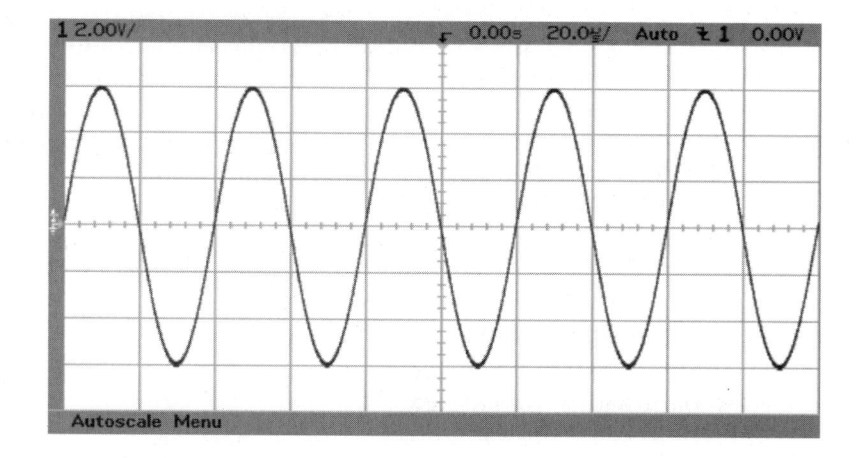

FIGURE 2-43
Question 26

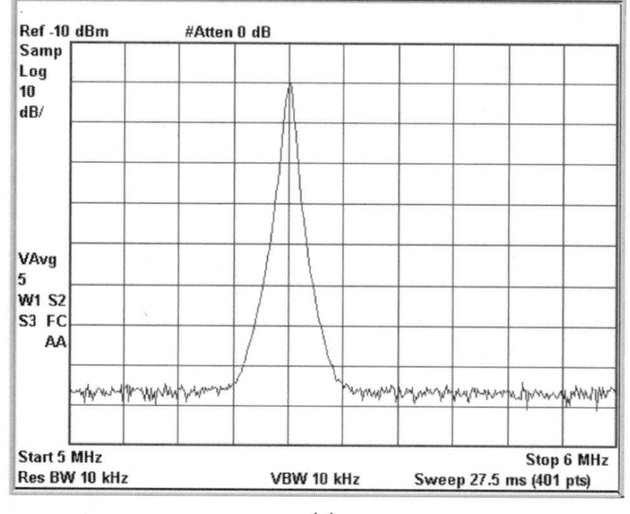

(a)

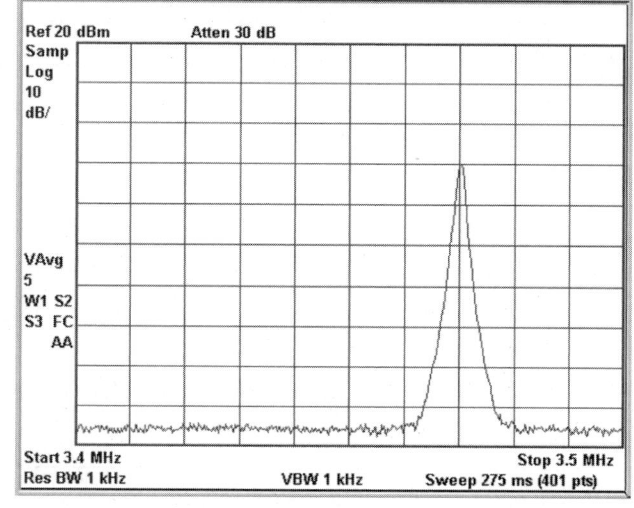

(b)

FIGURE 2-44 Question 27

33. Describe the basic operation of a network analyzer.

34. Do an Internet search to find out the meaning of the term *reflection coefficient* as it refers to the transmission of a signal.

Section 2.9

35. Determine the noise voltage produced by the 75-Ω input resistance to a TV receiver at a temperature of 295 K. Use the Internet to determine the bandwidth of a TV-channel tuner.

36. Determine the signal-to-noise ratio for the following conditions: input signal of 25 μV, input noise of 150 nV.

37. For a certain system, the input SNR is 53 dB and the output is 51.8 dB. Determine the system noise figure and noise temperature. Assume the ambient temperature is 27°C.

38. Do an Internet search to find the sensitivity of a typical mobile radio (walkie-talkie).

39. Deep-space telecommunications requires extremely narrow bandwidth signals. Explain why.

Section 2.10

40. Determine the type (shape) of the signal in the spectrum-analyzer display in Figure 2-46.

41. Determine the bandwidth of the signal in question 40 if all harmonics less than 40 dB down are to be included.

42. What does the output signal consist of, if the signal in question 40 is passed through a BPF centered at a frequency of 300 kHz? The BPF has a bandwidth of 50 kHz.

43. What occurs if the signal in question 40 is passed through an LPF with a cutoff frequency of 400 kHz?

44. Determine what is meant by the FFT of a signal. Use the Internet to help determine your answer.

45. If a periodic signal consists of harmonic components, try to determine the frequency content of an impulse (e.g., a spark or lightning bolt).

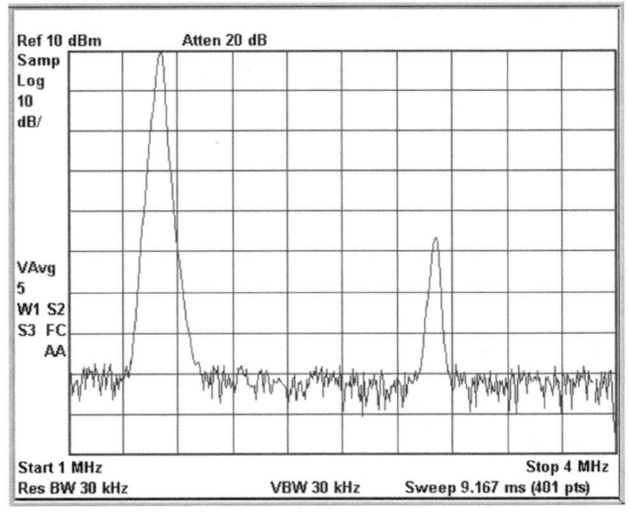

FIGURE 2-45 Question 28

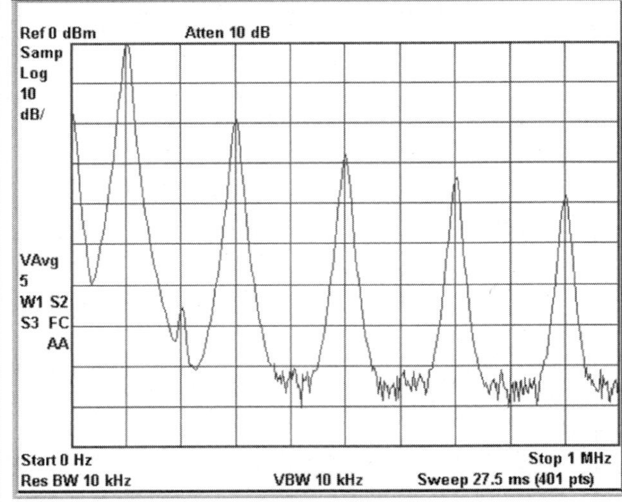

FIGURE 2-46 Question 40

Chapter Equations

$$\lambda = \frac{c}{f} \qquad 2.1$$

$$\text{Power Gain or Loss in dB} = 10\log\left(\frac{P_{\text{out}}}{P_{\text{in}}}\right) \qquad 2.2$$

$$\text{Voltage Gain or Loss in dB} = 20\log\left(\frac{V_{\text{out}}}{V_{\text{in}}}\right) \qquad 2.3$$

$$\text{Power Gain or Loss Ratio} = 10^{(db\ Value/10)} \qquad 2.4$$

$$\text{Voltage Gain or Loss} = 10^{(db\ Value/20)} \qquad 2.5$$

$$\text{Power in referenced dB} = 10\log_{10}\left(\frac{\text{Power}}{\text{Reference Power}}\right) \qquad 2.6$$

$$\#\ \text{dB/dec} = (20/6) \times \#\ \text{dB/oct} \qquad 2.7$$

$$Q = \frac{\text{3-dB Bandwidth}}{f_0} \qquad 2.8$$

$$\text{Shape Factor} = \frac{\text{60-dB Bandwidth}}{\text{6-dB Bandwidth}} \qquad 2.9$$

$$\frac{V_{\text{out}}}{V_{\text{in}}} = \left(\frac{X_C}{\sqrt{R^2 + X_C^2}}\right) = \text{Filter Gain} \qquad 2.10$$

$$P_n = kT\Delta f \qquad 2.11$$

$$e_n = \sqrt{4kT\Delta f R} \qquad 2.12$$

$$i_{rms} = \sqrt{2ei_{diode}\Delta f} \tag{2.13}$$

$$SNR = \frac{\text{Signal Power}}{\text{Noise Power}} = \frac{P_S}{P_N} \tag{2.14}$$

$$SNR = 10\log\left(\frac{P_S}{P_N}\right) = 20\log\left(\frac{E_S}{E_N}\right) \tag{2.15}$$

$$SINAD = 10\log\left(\frac{S+N+D}{N+D}\right) \tag{2.16}$$

$$NF = 10\ \log\ (\text{Noise Ratio}) = 10\log\left(\frac{SNR_{\text{in}}}{SNR_{\text{out}}}\right) \tag{2.17}$$

$$T_{eq} = T_0(NR-1) \tag{2.18}$$

$$v(t) = \frac{4V}{\pi}\left(\cos\omega t - \frac{1}{3}\cos 3\omega t + \frac{1}{5}\cos 5\omega t - \ \ldots\right) \tag{2.19}$$

$$\text{Power Harmonic} = \text{Power Fundamental} \times \left(\frac{1}{n^2}\right) \tag{2.20}$$

$$v(t) = \frac{8V}{\pi^2}\left(\cos\omega t - \frac{1}{3^2}\cos 3\omega t + \frac{1}{5^2}\cos 5\omega t - \ \ldots\right) \tag{2.21}$$

Amplitude Modulation

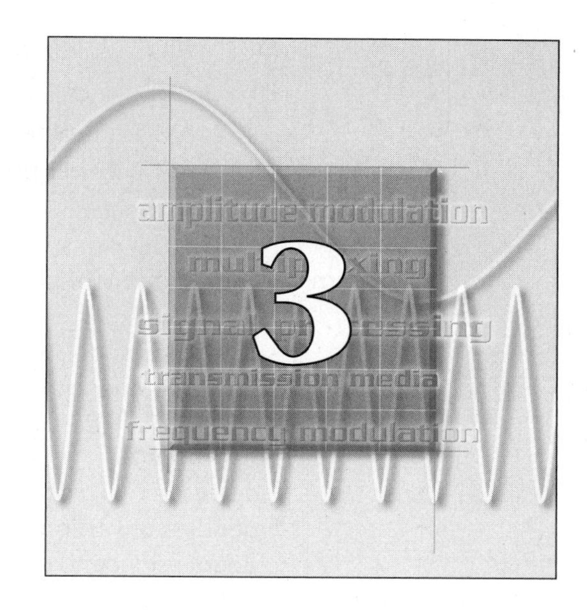

Objectives Upon completion of this chapter, the student should be able to:

- Discuss the evolution of amplitude-modulation (AM) technology.
- Describe an AM signal in both the time and frequency domain.
- Understand the difference between the addition, mixing, and modulation of two distinct signals.
- Define the frequency components of an AM wave and determine its bandwidth.
- Define the meaning and significance of the AM modulation index and be able to determine its value from measurement.
- Calculate the power relations of an AM signal and sketch its frequency spectra.
- Draw the block diagrams of typical low- and high-power AM transmitters.
- Describe basic mixer and balanced-modulator functions.
- Draw the block diagram of the basic superheterodyne receiver, describe the functions of its subsystems, and be aware of typical receiver specifications.
- Describe the software digital radio.
- Describe the operation of alternative forms of AM.

Outline

Key Terms

amplitude modulation
audio mixer
balanced modulator
carrier frequency
cross-modulation products
detector
double conversion
double-sideband full-carrier AM
double-sideband suppressed-carrier AM
dynamic range
envelope
image frequency
image rejection
image signal
intermediate frequency
large-carrier AM
local oscillator
mixer
modulation index
pilot subcarrier
product detector
RF amplifier
selectivity
sidebands
single-sideband suppressed-carrier AM
software radio
superheterodyne
tuned radio-frequency receiver
vestigial sideband AM

Introduction

The first form of modulation used for voice communications was **amplitude modulation** (AM). This modulation scheme has advantages in its simplicity, but disadvantages in its relatively poor performance in the presence of noise and its inefficient use of power. Today, the basic technique of AM can be modified and combined with other modulation techniques to realize a variety of forms of digital modulation and to achieve bandwidth efficiency. It is therefore essential that one understand the process of AM in some detail.

A fairly rigorous treatment of AM will serve the student as a foundation for the understanding of other, more modern complex modulation schemes and their signal characteristics. The student will be introduced to practical examples of signal bandwidth and frequency content through numerous examples of time- and frequency-domain signal representations. Topics covered in this chapter include:

- a short history of the evolution of AM
- the theory of AM operation
- mathematical analysis of AM
- the modulation index

- signal bandwidth and signal power relations
- transmission and receiving hardware
- the superheterodyne technique
- mixer and balanced modulator theory
- practical superheterodyne receiver characteristics
- an introduction to digital radios
- an overview of alternative versions of standard amplitude modulation and their hardware implementation.

3.1 The History and Evolution of AM

We will start our discussion of the various analog modulation techniques available with amplitude modulation (AM). AM is one of the earliest forms of modulation to be used in both telegraph and wireless communication. These early information-transmission schemes used both short and long bursts of either a voltage/current or electromagnetic energy to represent the dots and dashes of Morse code. The telegraph evolved to use several different mechanically produced frequencies for signaling. These various "multiplex" formats used an early form of amplitude-shift keying (ASK). The first wireless transmitters used a form of AM known as on-off keying (OOK).

Early AM Wireless Systems

A typical early wireless transmitter is shown in Figure 3-1 on page 90. Note the inductance and capacitance used to tune the output frequency of the spark gap. The resonant frequency of these two components tended to maximize output power at a particular frequency. Due to the nature of the spark-gap emission, this typically occurred at a low frequency with its corresponding long wavelength.

Although little was known about antenna theory at the time, it was discovered early on that for a conductor to effectively radiate long wavelength signals it had to be oriented vertical to the earth and physically be some appreciable fraction of a wavelength. It was common for early wireless experiments to use balloons and kites to support long lengths of wire that served as the antenna. Also, it was thought that transmitting distance would be limited by the curvature of the earth, which became an additional rationale for tall antennas.

The wireless transmitter shown in Figure 3-1 would emit a signal of either long or short duration depending on the length of time the telegraph key was closed. The signal was the electromagnetic noise produced by a spark-gap discharge that propagated through the air to a receiver located at

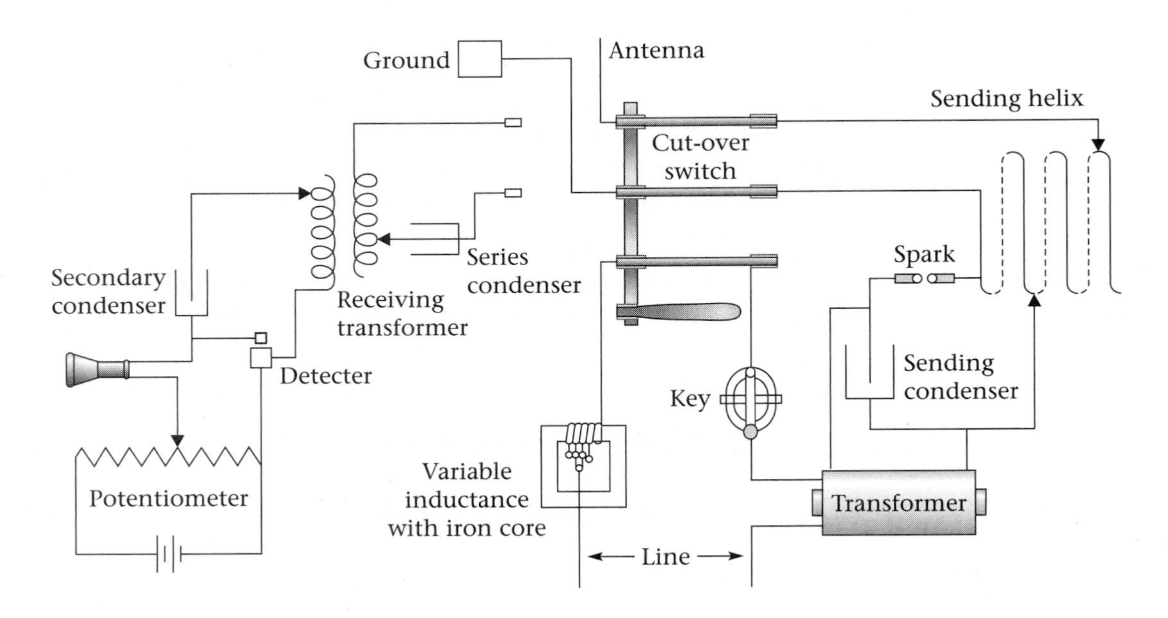

FIGURE 3-1 An early wireless spark-gap telegraph transmitter

a distance. At the receiver the detected signal was interpreted as a dot or a dash depending on its duration. Using Morse code, combinations of dots and dashes stood for various alphanumeric characters.

The next generation of wireless transmitters used more stable radio-frequency alternators or high-powered Poulsen spark-gap transmitters for their signal sources. The RF alternators were used to transmit another form of AM that is sometimes referred to as binary amplitude-shift keying (BASK), which was essentially the same as OOK. The Poulsen transmitters used a form of frequency-shift keying (FSK) to transmit a signal that was received and interpreted as a BASK signal by the receivers.

The First Broadcast

Beginning in 1905, Reginald Fessenden conducted experiments in wireless transmission at Brant Rock, Massachusetts, using 50-kHz high-frequency alternators built by General Electric. The output of this type of generator was much more stable than that of a spark-gap transmitter, allowing him to experiment with a continuous form of amplitude modulation. Fessenden was eventually able to broadcast voice using AM. His experiments culminated on Christmas Eve of 1906, when he is credited with transmitting the first ever radio broadcast. This broadcast was repeated on New Year's Eve. Refer back to Figures 1-7a and 1-7b, which show the inside of the Brant Rock Station and its over 400-foot-tall transmitting antenna.

Modern AM

Amplitude modulation is now used for low-frequency legacy radio broadcasting, which had its corporate beginnings after World War I (the present AM band is 530–1700 kHz); short-wave broadcasting; low-definition (NTSC) television video-signal transmission; amateur and CB radio; and various other services. Newer uses of AM also include quadrature amplitude modulation (QAM or n-QAM, where n is a power of 2). QAM is a hybrid form of amplitude and phase modulation (PM) used for high-speed data transmission.

Go to **http:\\www.ntia.com** for a United States frequency chart providing details of frequency assignments for various radio services, including AM broadcasting.

3.2 The Classical Definition of AM

The classical definition of *amplitude modulation* is: "The amplitude of a high-frequency (RF) sine wave is varied in accordance with the instantaneous value of the modulating signal." As can be seen from the oscilloscope display shown in Figure 3-2, the characteristics of the modulating signal (its amplitude and frequency) are superimposed upon the high-frequency sine wave. The frequency of the modulating signal is represented by the number of times per second that the modulated wave's amplitude varies. The amplitude of the modulating wave is represented by the relative amount of variation in the modulated wave's amplitude.

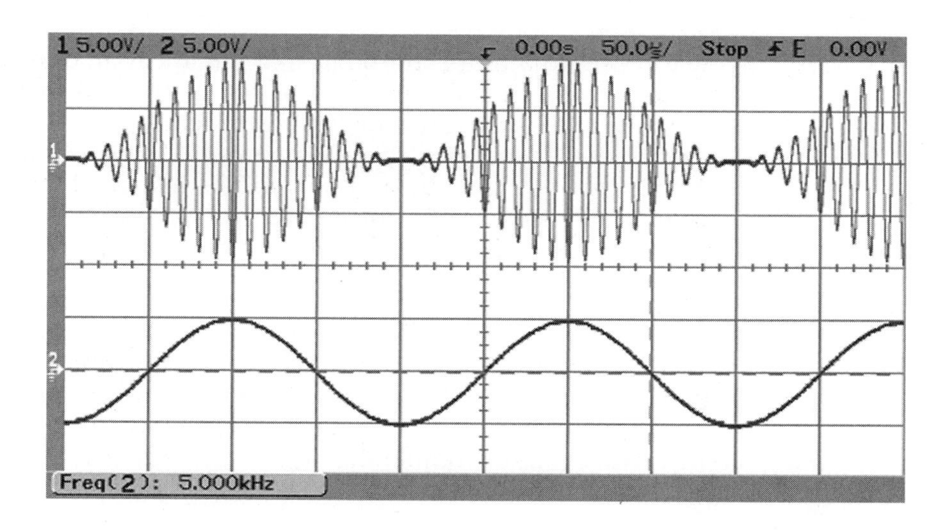

FIGURE 3-2 A time display of the modulating signal at 5 kHz and the resulting AM signal

The Mathematics of Modulation

A sine wave can be represented in equation form, as shown in equation 3.1. For amplitude modulation, the value of E_P (which represents the high-frequency sine wave's amplitude) is varied by the modulating signal.

$$e(t) = E_P \sin(\omega t + \phi) \qquad \textbf{3.1}$$

Recall the need for modulation; it is required in order to make the signal compatible with the channel. For early wireless transmissions of voice or music, a high-frequency carrier wave was needed. Without it the length of the vertical antenna necessary to launch an audio-frequency (AF) signal would be physically unrealizable due to the long wavelength of the signal. However, an additional difficulty would have existed even if the dimensions of the antenna were not a problem. Without the high-frequency carrier to locate or place the information in its own little niche in the radio-frequency spectrum, all voice or music transmissions would share the same baseband frequency spectrum. This would create mass confusion in most wireless systems unless there were some special means of coding the signals. See chapter 6 for an explanation of how today's wireless systems are able to share the same passband frequencies through the use of complex access technologies like time-division multiple access (TDMA) and code-division multiple access (CDMA).

Several of the early telegraph multiplex schemes also recognized the fact that a basic form of frequency-division multiplexing combined with amplitude-shift keying (ASK) would allow for the transmission of several messages simultaneously on the same wire.

3.3 The Process of AM

Before we do a more thorough mathematical analysis of an AM signal, let us make sure that we understand the distinction between several similar looking and commonly occurring processes involving electrical signals. Two of these processes, the addition of two signals and the amplitude modulation of one signal by another, will be discussed at this time. Later in the chapter the theory of another similar process, signal mixing, will also be considered.

The Concept of Signal Addition

We will compare an AM signal to the sum (addition) of two distinct signals by observing the two processes in both the time and frequency domain.

Figure 3-3 shows a time display of two distinct sine waves. The upper sine wave has a frequency of 10 kHz and the lower sine wave has a frequency of 100 kHz.

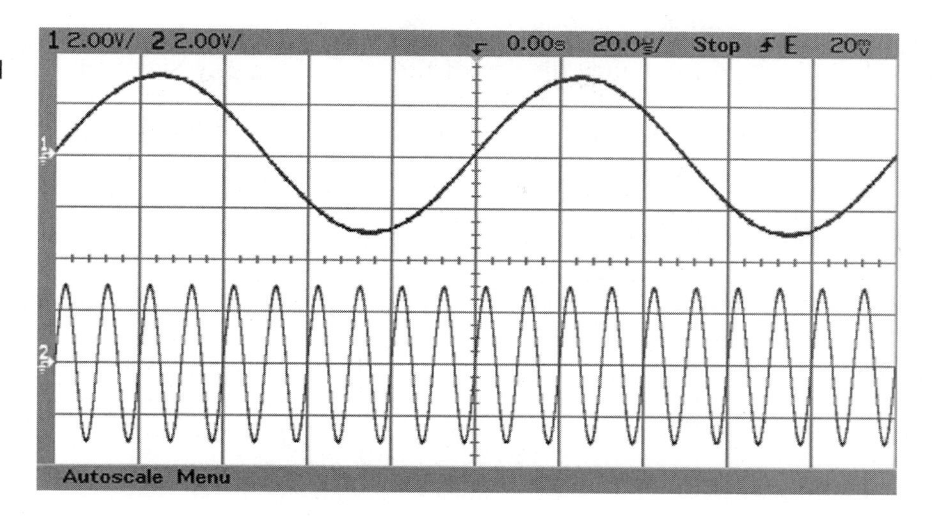

FIGURE 3-3 A 10-kHz sine wave and a 100-kHz sine wave

Figure 3-4 shows a time display of the resulting waveform of the sum of the two different frequency signals shown in Figure 3-3. A simple adder circuit could be used to obtain this signal; the **audio mixer** will perform the same summing function. An audio mixer is typically used by musicians to combine several microphone outputs into a single signal to be applied to a speaker amplifier.

One can see the two distinct frequencies in the composite waveform. There is a wave with a slow variation of 100 μsec per cycle (10 kHz) and a wave with faster variations of 10 μsec (100 kHz). We might ask the question, How does this look in the frequency domain? Figure 3-5 on page 94 shows a frequency domain display of the 10-kHz wave and Figure 3-6

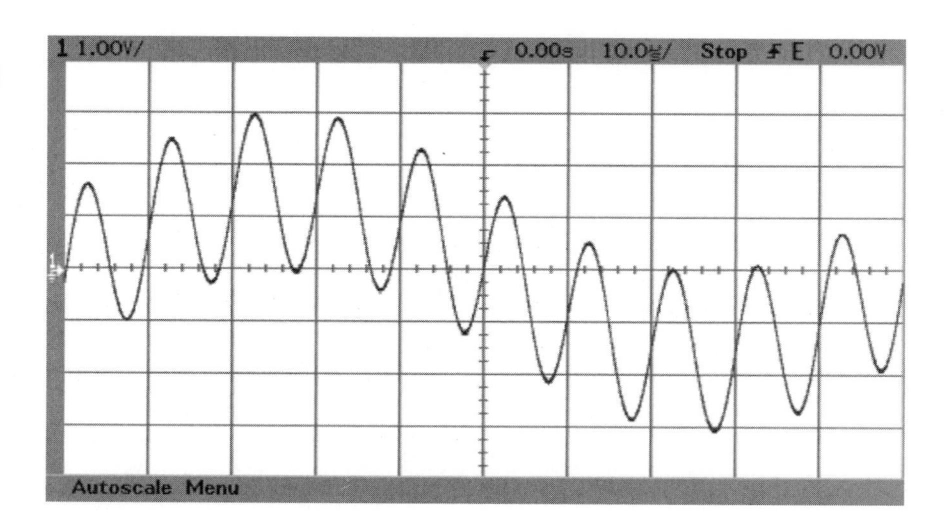

FIGURE 3-4 The sum of a 10-kHz and a 100-kHz sine wave

shows the 100-kHz wave. Both displays are prior to signal addition. The frequency display of the sum of the two signals is shown in Figure 3-7.

Note that the addition of the two signals results in only the two signals. This process can be described as the algebraic or linear addition of two signals. This process does not yield any new frequency components.

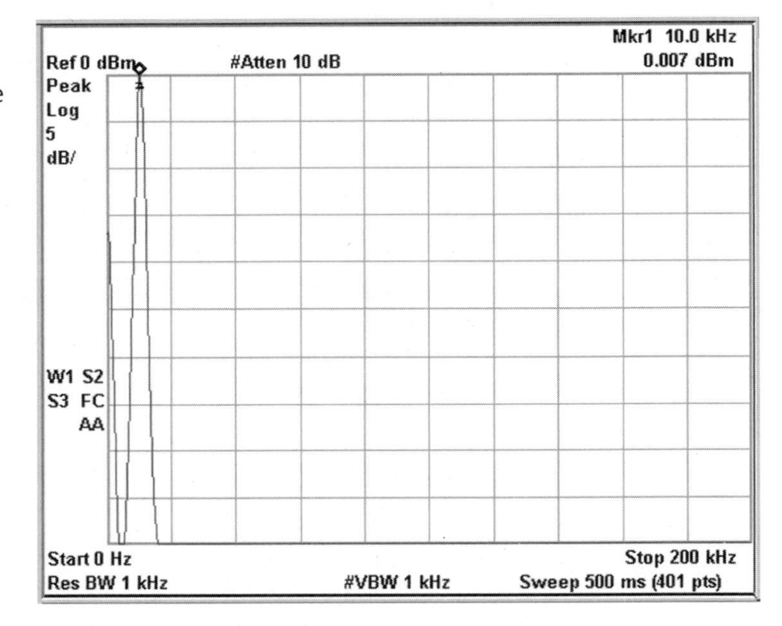

FIGURE 3-5
SA display of the 10-kHz sine wave

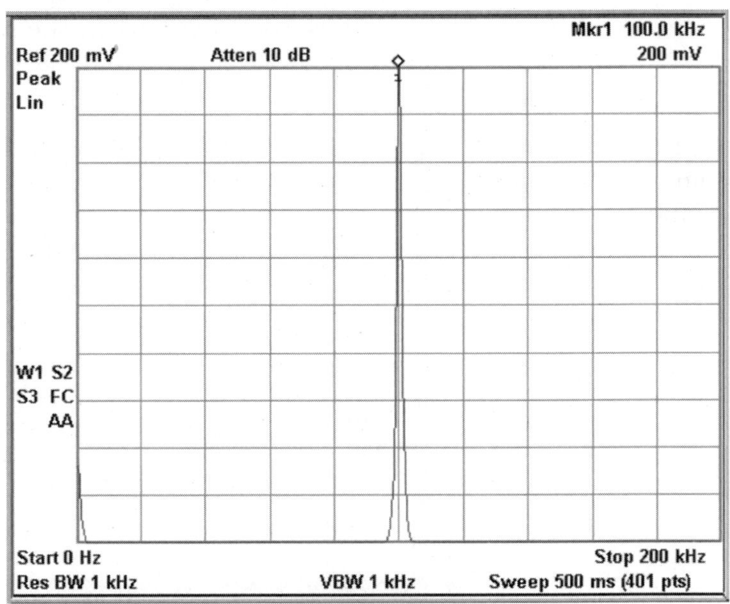

FIGURE 3-6
SA display of the 100-kHz sine wave

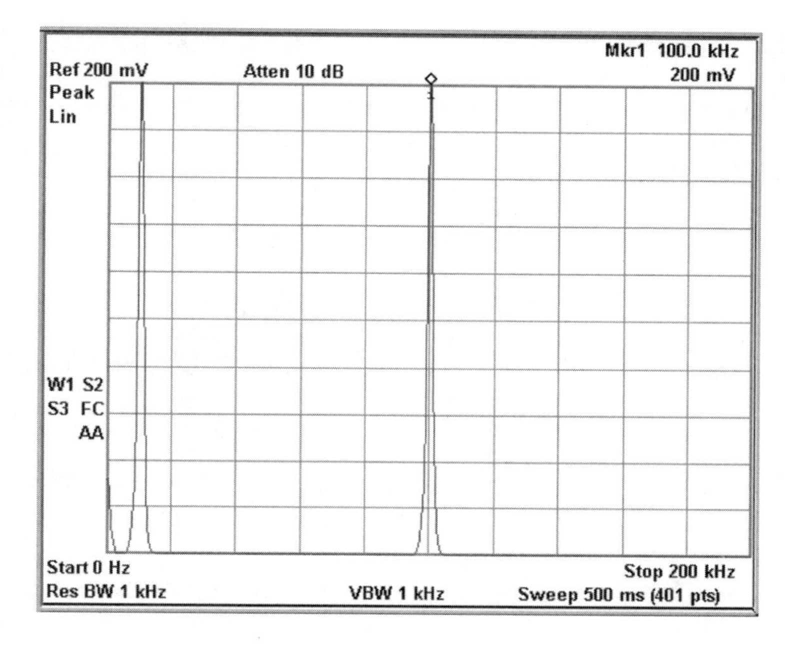

FIGURE 3-7 SA display of the sum of two sine waves

The Concept of AM

Let us return our attention to the AM process to see what happens when a low-frequency sine-wave amplitude modulates a high-frequency sine wave. Figure 3-8 shows the two signals before modulation. The audio signal is at 5 kHz and the high-frequency signal is at 100 kHz.

Now take a look at Figure 3-9 on page 96, which shows the result of the lower-frequency signal amplitude modulating the higher-frequency signal,

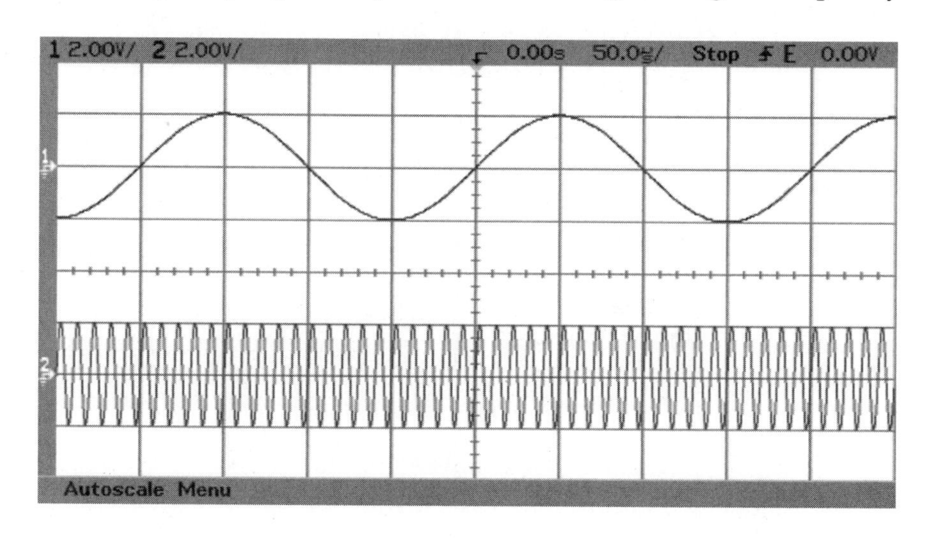

FIGURE 3-8 Time display of low- and high-frequency sine waves before modulation

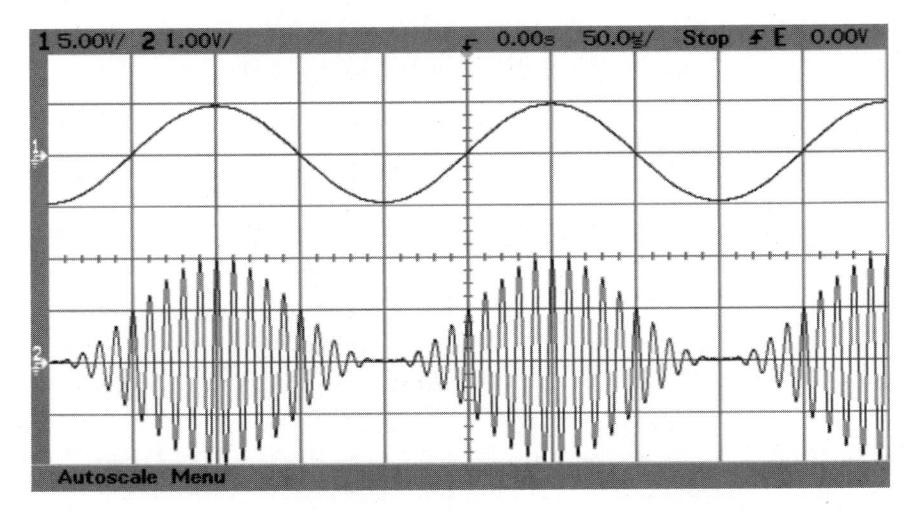

FIGURE 3-9 Time-domain display of an AM wave and the modulating wave

resulting in a composite AM signal displayed in the time domain (the process that accomplishes this will be discussed later). The modulating wave has been included to show the correlation between the modulated waveform and the modulating wave. The oscilloscope time display shows the slow audio variations in the amplitude of the high-frequency sine wave.

Figure 3-10 is a spectrum-analyzer display of an AM signal, shown in the frequency domain. As can be seen on the display, the composite AM signal consists of one of the original frequencies and some new frequencies. Specifically, these three frequencies are the **carrier frequency**, the

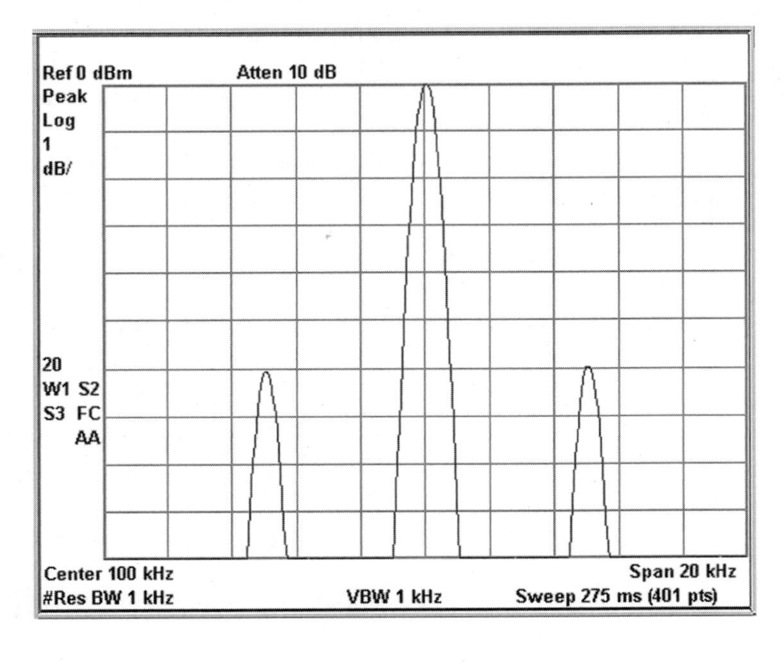

FIGURE 3-10 Frequency-domain (spectrum-analyzer) display of an AM wave. The carrier frequency is 100 kHz, the sidebands are at 95 and 105 kHz, and the modulating signal is 5 kHz

higher-frequency sine wave of the two original sine waves, and two new frequencies called **sidebands**, which are symmetrically spaced around the carrier frequency. One might wonder what has happened to the other low-frequency sine wave—the modulating wave. There are several reasons why it might not be visible. One reason is that the typical spectrum analyzer (SA) cannot display very low audio frequencies. Another reason is that the process that produced the AM signal usually filters out low-frequency (AF) sine waves. Finally, another reason that applies in this particular case is that the span of the SA does not cover a large enough frequency range to include the modulating signal.

3.4 Mathematical Analysis of AM

At this time it will be instructive to perform a mathematical analysis of the AM process. We can represent the modulating signal, e_{mod}, and the carrier signal, e_{carrier}, as

$$e_{\text{mod}} = E_M \sin\omega_M t$$

and

$$e_{\text{carrier}} = E_C \sin\omega_C t$$

where

$$\omega_M = 2\pi f_M$$

and

$$\omega_C = 2\pi f_C$$

Note that these equations use the Greek symbol ω as the argument of the sine wave. Because it is mathematically impossible to "take" the sine of a frequency, we use the relationship of $\omega t = 2\pi ft$ to make the mathematics work out (note that ωt is an angle). The resultant AM wave is shown in Figure 3-11 on page 98. The modulating wave has been superimposed on the envelope of the modulated waveform to show the correlation between the two signals.

The Mathematical Model of AM

It is noted that the amplitude of the waveform, its **envelope**, is equal to:

$$\text{Amplitude} = E_C + e_{\text{mod}} = E_C + E_M \sin \omega_M t \qquad \textbf{3.2}$$

and the instantaneous voltage of the amplitude modulated wave is given by:

$$e(t) = (E_C + E_M \sin \omega_M t) \times \sin \omega_C t \qquad \textbf{3.3}$$

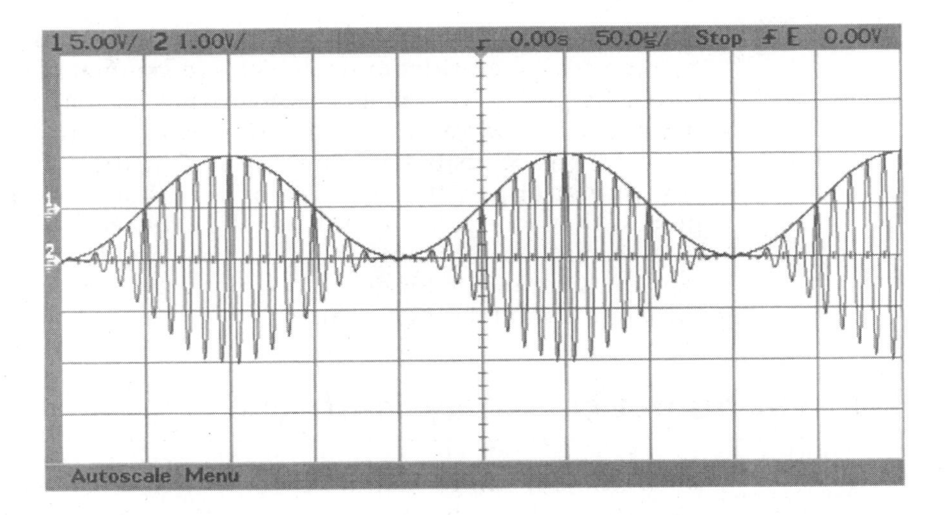

FIGURE 3-11 The envelope of an AM-modulated wave

Through mathematical manipulation and the use of a common trigonometric identity, the following equation results:

$$e_{AM}(t) = E_C \sin(\omega_C t) + \frac{mE_C}{2}\cos(\omega_C - \omega_M)t - \frac{mE_C}{2}\cos(\omega_C + \omega_M)t \quad \textbf{3.4}$$

Again, note the difference between the addition of two sine waves and the process of amplitude modulation. The resultant AM signal consists of the original carrier frequency, f_C, and two new frequencies, $f_C - f_M$ and $f_C + f_M$. The first frequency, $f_C - f_M$, is called the lower sideband (LSB) and the second frequency, $f_C + f_M$, is called the upper sideband (USB).

The Modulation Index

Equation 3.4 also contains a new variable, m, which is the **modulation index**. The modulation index is defined as follows:

$$m = \frac{E_M}{E_C} \quad \textbf{3.5}$$

The allowed value of m is between 0 and 1.

Alternatively, we can express m as a percentage by using the following equation:

$$m = \frac{E_M}{E_C} \times 100\% \quad \textbf{3.6}$$

Now m is between 0 and 100%. When talking about AM, we sometimes use the term *depth of modulation* to denote the value of m. A small value of m

is considered *shallow modulation,* whereas a value of m approaching 1 (or 100%) is termed *deep modulation*. Figures 3-12 and 3-13 indicate the reason for the choice of these terms.

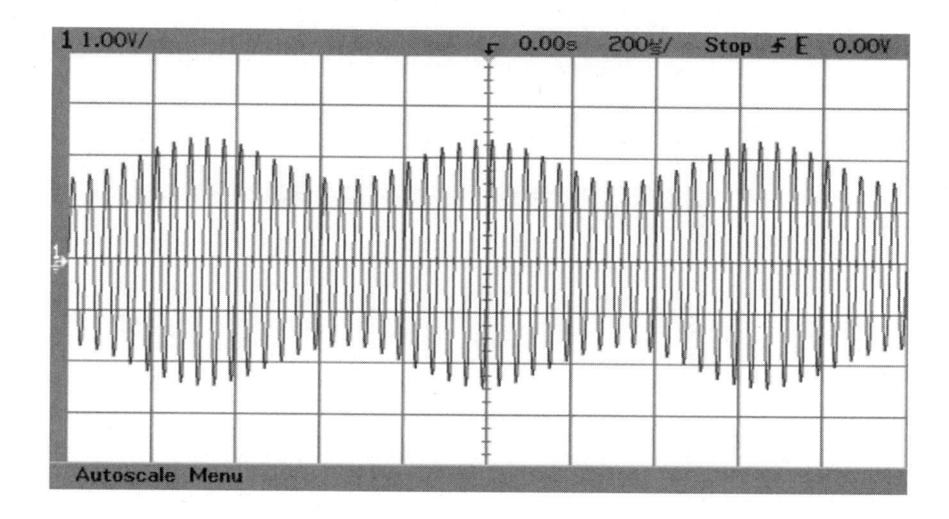

FIGURE 3-12 An AM wave with a modulation index of $m = 20\%$

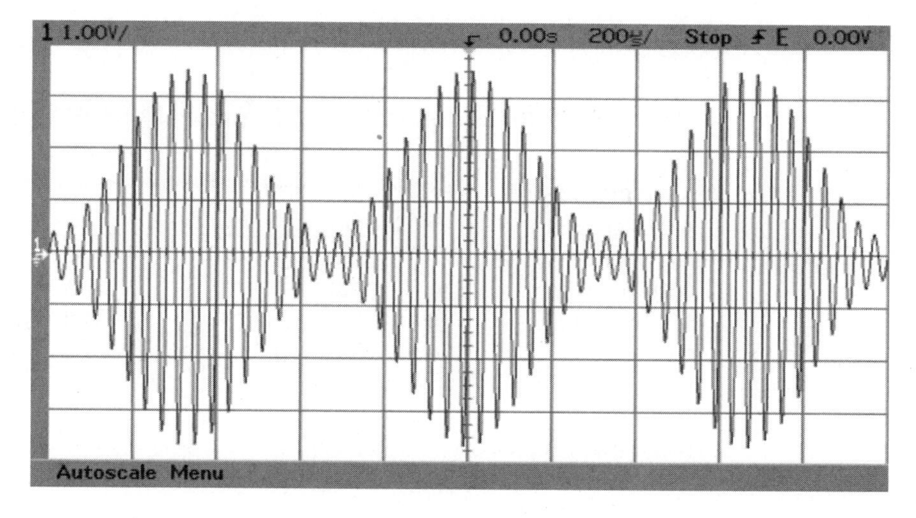

FIGURE 3-13 An AM wave with a modulation index of $m = 80\%$

Over-Modulation

What about values of m greater than 1 (100%)? This is called over-modulation. Under these circumstances distortion will occur through the creation of new frequencies (additional sidebands) in the modulated wave.

Figure 3-14 shows the time display of an over-modulated ($m > 1$) AM wave. As can be seen, the modulated wave does not appear as it should. What is happening is that new sidebands are being created that represent frequencies that did not exist prior to the process of over-modulation. The FCC has strict regulations about over-modulation because it almost always causes interference to other adjacent (in the frequency spectrum) communication links and services. If one looks at the frequency display of an over-modulated signal (see Figure 3-15) one will see the newly formed sidebands that have been caused by the over-modulation. Note that these new sidebands represent a signal equal to twice the frequency of the modulating wave (i.e., its second harmonic).

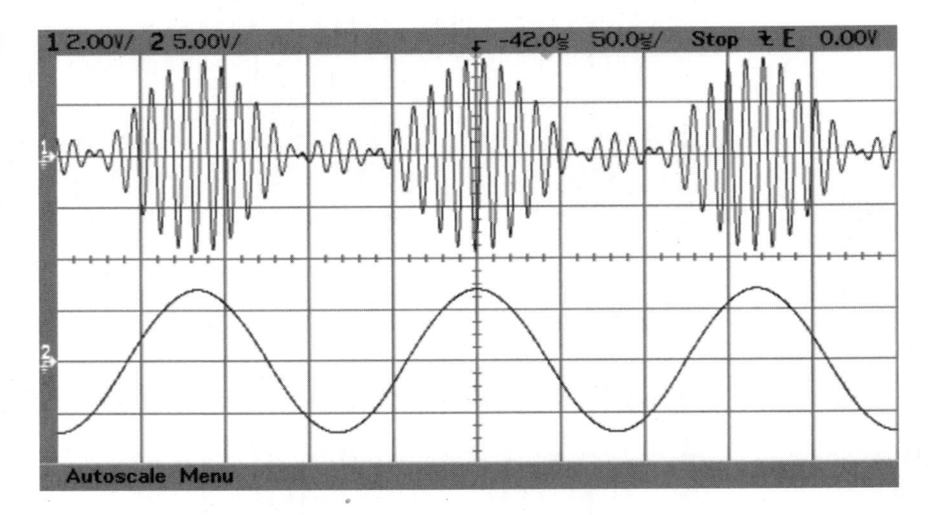

FIGURE 3-14 An AM wave with *m* greater than 100%

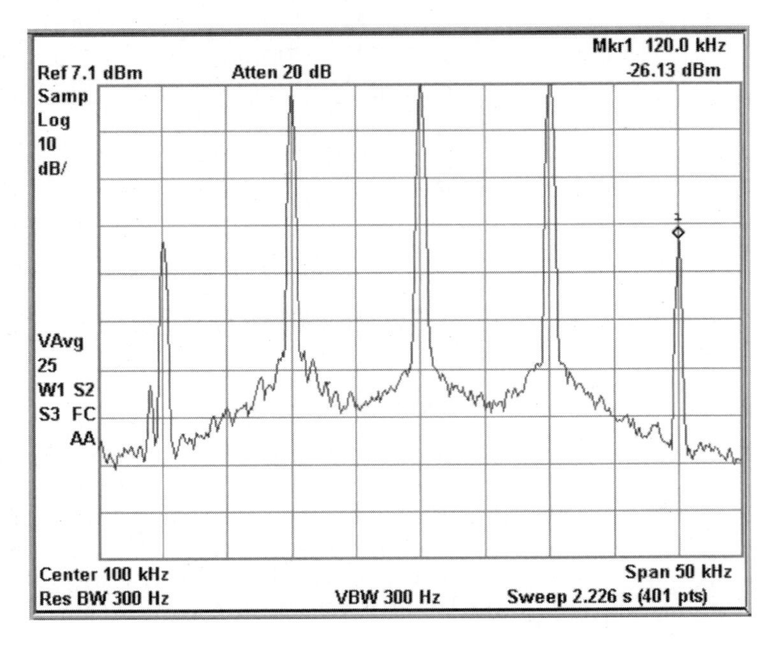

FIGURE 3-15 An over-modulated signal with a modulating frequency of 10 kHz. Note the second pair of sidebands. A marker indicates their relative amplitude.

Measurement of the Index of Modulation

It is possible to determine the index of modulation, m, from a time display of an AM waveform if it is the simple case of a single modulating frequency or tone. One can calculate the modulation index from the signal waveform by using equation 3.7.

$$m = \frac{E_{max} - E_{min}}{E_{max} + E_{min}} \quad \textbf{3.7}$$

where E_{max} and E_{min} are the peak-to-peak maximum and minimum values of the envelope of the modulated AM wave.

EXAMPLE 3.1

Figure 3-16 shows a time display of an AM wave. What is the value of m?

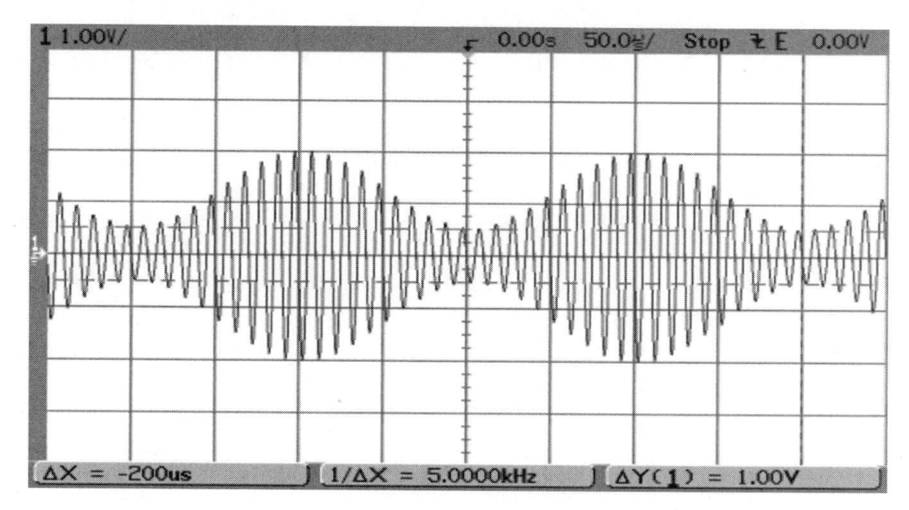

FIGURE 3-16 An AM wave. The oscilloscope cursors (dashed horizontal lines) show a 1 $V_{p\text{-}p}$ value of E_{min}.

Solution From Figure 3-16, the value of E_{max} can be seen to be 4 $V_{p\text{-}p}$.

According to the oscilloscope measurement, $E_{min} = 1\ V_{p\text{-}p}$.

Therefore, using

$$m = \frac{E_{max} - E_{min}}{E_{max} + E_{min}}$$

substitution yields,

$$m = \frac{4-1}{4+1} = \frac{3}{5} = .6, \text{ or } 60\%$$

A Summary of Mathematical Results

Let us recall our previous mathematical analysis of the AM wave. One can determine what the frequency content of an AM wave is by examining equation 3.4, given earlier in the chapter. Recall that $\omega = 2\pi f$ and that the AM wave consists of a center frequency, f_C, and two additional frequencies, the lower sideband (LSB), $f_C - f_M$ and the upper sideband (USB), $f_C + f_M$. Note that the two sidebands are mirror images of each other; their amplitudes and offset frequency from f_C are identical.

Let us also review the process of amplitude modulation. Equation 3.1 indicates that the AM wave is a sine wave with an amplitude that is varying sinusoidally. There is no such thing! A sine wave by definition has a constant frequency and amplitude.

What appears to us in the time-domain (oscilloscope) display is really the addition of several frequency components that, when added together, vectorially (by this, we mean amplitude and phase) give the appearance of a sine wave thats amplitude is varying in a sinusoidal manner. The frequency-domain (spectrum-analyzer) display of an AM wave bears this out.

The AM process produces a pair of sidebands that are symmetrically located around a center carrier frequency and are offset from the carrier by the frequency of modulation. See example 3.2 for clarity.

EXAMPLE 3.2

What is the modulation frequency and carrier frequency for the AM wave shown in Figure 3-17?

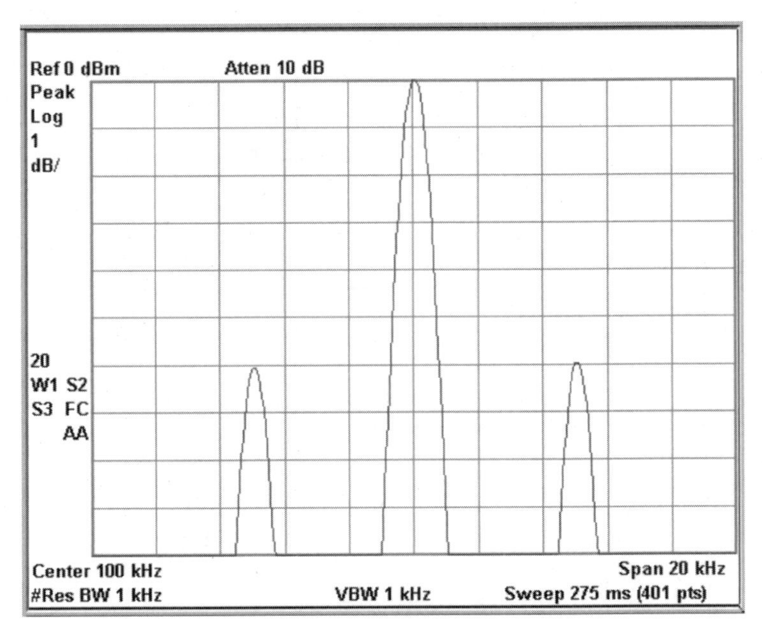

FIGURE 3-17 SA display of an AM signal

Solution From the diagram it can be seen that $f_C = 100$ kHz and $f_M = 5$ kHz because we have an LSB at 95 kHz and a USB at 105 kHz.

EXAMPLE 3.3

What will the frequencies in an AM wave be if the carrier frequency is 1 MHz, the frequency of modulation is 6 kHz, and the modulation index is less than 1?

Solution The output frequencies will be 1 MHz (the carrier), 994 kHz (LSB), and 1006 kHz (USB). It will be shown later that the 6-kHz modulating wave will most likely be rejected by the high-frequency bandpass filter circuits used in the AM process.

More Than One Modulating Tone

What if we have two or more modulating signals? In this case there will be one sideband pair per modulating tone. In the situation of modulation by voice or music, we would expect so many sideband pairs that it would be impossible to separate them from one another. For a finite number of modulating tones, one can calculate the total modulation index by using equation 3.8, where n is the total number of modulating signals:

$$m_{\text{total}} = (m_1^2 + m_2^2 + m_3^2 + \ldots + m_n^2)^{\frac{1}{2}} \qquad \textbf{3.8}$$

where the m_ns are the modulation indexes of the individual modulating tones. This total modulation index must also be restricted to equal to or less than 1 to prevent over-modulation.

Figure 3-18 on page 104 shows an example of the spectrum of an AM signal with two modulation tones of different amplitudes and frequencies.

Figure 3-19 on page 104 shows another example of two modulating tones. In this case, two equal-amplitude audio tones at 1 and 3 kHz are both modulating a 100-kHz carrier frequency. Again, the result is two pairs of sidebands with equal power.

Modualtion by Voice or Music

Figure 3-20 on page 105 shows a 500-kHz signal being modulated by a music signal. Note the numerous sidebands and how rapidly their amplitude falls off.

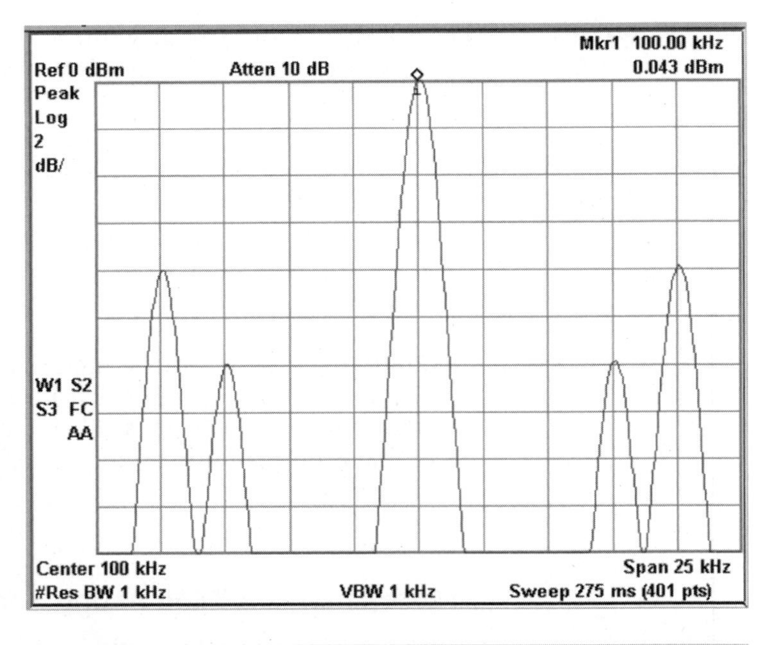

FIGURE 3-18 An AM wave with two modulating frequencies of 7.5 kHz and 10 kHz

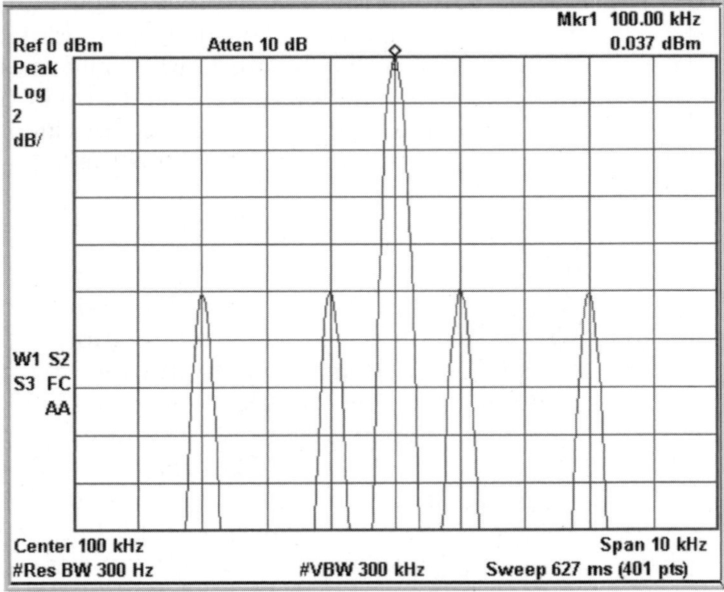

FIGURE 3-19 An AM wave with two equal-amplitude modulating tones

AM Signal Bandwidth

Although we have seen numerous frequency-domain displays of AM signals, we have not formally calculated the bandwidth of the AM signal. It is fairly straightforward to see that the range of frequencies in an AM signal

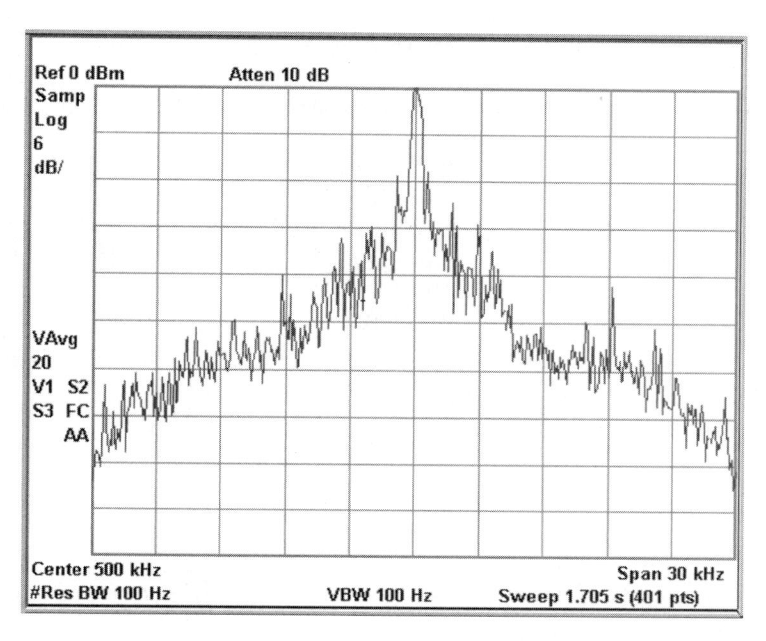

FIGURE 3-20 AM spectrum with a modulating music signal

is equal to the spacing between the lowest of the lower sidebands and the highest of the upper sidebands. In other words, the bandwidth is equal to two times the frequency of the modulating signal if there is only a single modulating tone. This is represented in equation 3.9:

$$\text{Bandwidth} = 2 \times f_{\text{mod}} \quad \textbf{3.9}$$

If there is more than one modulating tone, then the bandwidth is equal to two times the frequency of the highest modulating frequency:

$$\text{Bandwidth} = 2 \times f_{\text{highest}} \quad \textbf{3.10}$$

What if the modulating signal is voice or music? Because voice or music consists of numerous frequencies and harmonics, we show the sidebands as continuous. See Figure 3-21 on page 106 for the spectra of a typical music signal.

Note that the frequency components in the figure extend out to 15 kHz and beyond. However, their amplitude is approximately 15 to 20 dB lower than the frequency components near 0 Hz. As always, the sidebands created by a signal like this are mirror images of one another. Refer back to Figure 3-20 for the SA display of a signal modulated by music.

Bandwidth Examples

Recall Figures 3-18 and 3-19. When we previously studied these displays we did not consider the bandwidth of the signals. In Figure 3-18, we can see that the bandwidth is 20 kHz and in Figure 3-19 it is 6 kHz. In both cases, the bandwidth is given by two times the highest modulation frequency.

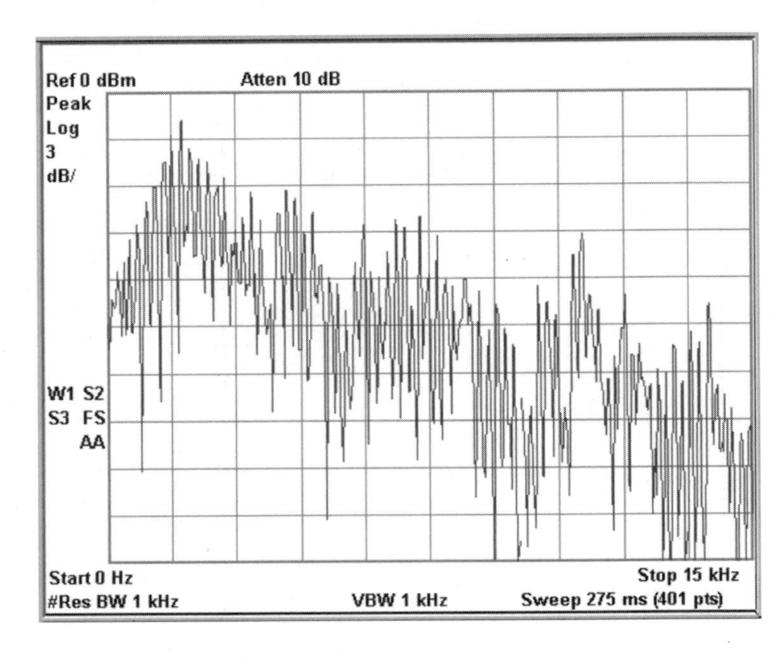

FIGURE 3-21 A spectrum-analyzer display of a typical music signal. Note the amplitude drop-off at both low and high frequencies.

Our last example will be the SA display of a signal modulated by music. For this display, it is difficult to determine the width of the signal. See Figure 3-20 again. If we include all frequency components that are less than 46 dB down from the center-carrier frequency amplitude, we would in all probability be getting a good estimate of the bandwidth. For this particular example, that estimate appears to be on the order of approximately 30 kHz. Some readers might find this result surprising, as it seems to indicate that an AM transmission includes the entire audio spectrum. However, BPFs in both the transmitting and receiving hardware that are used to select the signal also tend to attenuate sidebands that are more distant from the carrier frequency, in effect reducing the effective signal bandwidth. It should also be pointed out that the sidebands shown in Figure 3-20 do not appear to be mirror images of one another. This fact is due to the finite trace speed of the spectrum analyzer. The frequency content and amplitude of the music signal is changing faster than the spectrum analyzer can acquire and display its data points.

3.5 Power Relations in an AM Signal

Now we are ready to look at the last type of AM calculations—power relations. One can calculate the power distribution in an AM wave as follows:

$$P_{\text{total}} = P_C + P_{\text{LSB}} + P_{\text{USB}} \tag{3.11}$$

and

$$P_{total} = P_C + \frac{m^2 P_C}{4} + \frac{m^2 P_C}{4}$$
$$= P_C + \frac{m^2 P_C}{2}$$

and

$$P_{total} = P_C\left(1 + \frac{m^2}{2}\right) \qquad \textbf{3.12}$$

As one can see, the total power in an AM wave is equal to the power in the high-frequency carrier wave plus the power of the sidebands. The power of each sideband is equal to:

$$P_{LSB} = P_{USB} = \frac{m^2 P_C}{4} \qquad \textbf{3.13}$$

Therefore, the total power in an AM wave modulated to a depth of 100% is limited to

$$P_C\left(1 + \frac{m^2}{2}\right)$$

or

$$P_C\left(1 + \frac{1^2}{2}\right) = P_C(1.5) = 1.5P_C$$

or 150% of the unmodulated carrier power. As pointed out before, the amplitude of each sideband is the same, and therefore their power content is the same. Let us look at some examples.

•—EXAMPLE 3.4

See Figure 3-17. For this example, a 100-kHz carrier is being modulated at 100% ($m = 1$) by a single 5-kHz tone. Note that the bandwidth is 10 kHz ($2 \times 5 = 10$ kHz) and that the sidebands are 6 dB down from the carrier power.

•—**Solution** If we use equation 3.13 to determine the sideband power, we get:

$$P_{SB} = \frac{m^2 P_C}{4} = \frac{1^2}{4} P_C = \frac{P_C}{4} = .25P_C$$

or 25% of the power of the carrier. The display in Figure 3-17 shows the sidebands at –6 dB down from the carrier or 25% of the power of the carrier. This is the expected result!

If one uses some mathematics, it is possible to determine the value of m from an SA display. Using the relationship between carrier power and sideband power shown in Example 3.4, it can be shown that

$$m = 2\left(\frac{P_{SB}}{P_C}\right)^{\frac{1}{2}} \tag{3.14}$$

Because the SA usually displays powers as dBm, one must make the conversion from dB back to a ratio in order to use the equation. Let us look at another example.

EXAMPLE 3.5

Figure 3-22 shows a 100-kHz carrier modulated by a 6-kHz tone with $m = 0.2$, or 20% modulation. Does the SA display indicate the correct modulation index?

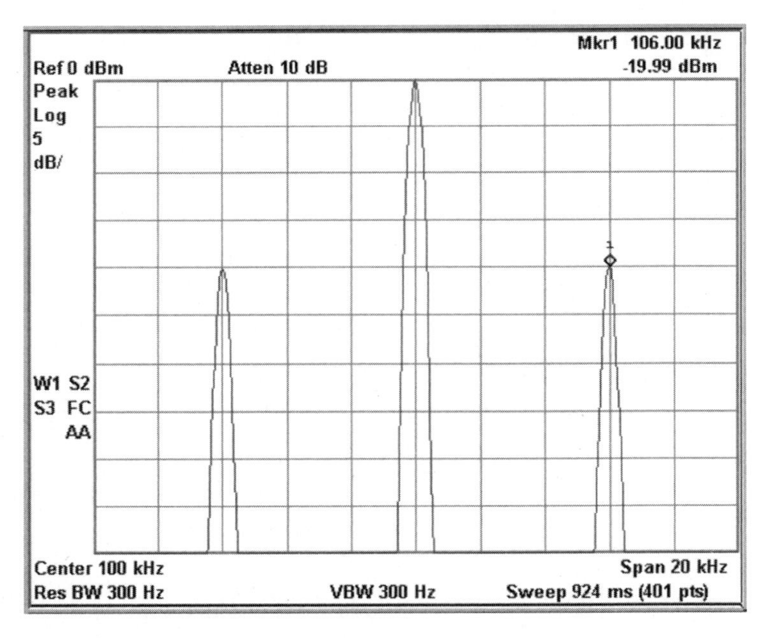

FIGURE 3-22 An AM wave modulated at $m = 20\%$

Solution As calculated, the sidebands are:

$$P_{SB} = \frac{m^2}{4}P_C = \frac{.2^2}{4}P_C = \frac{.04}{4}P_C = .01P_C$$

This yields a sideband power of 1% of the carrier power. How does this compare with the display? The sidebands are down by 20 dB (5 dB/div × 4 div = 20 dB). A value of –20 dB down is equal to one hundredth of the power. Our calculations have yet again led to success!

What about using the ratio of sideband power to carrier power, $\frac{P_{SB}}{P_C}$, to determine m? From the SA display in Figure 3-22 we see that the sidebands are 20 dB down from the carrier power. A value of –20 dB represents a ratio of .01. Plugging this value into our expression from above yields

$$m = 2\left(\frac{P_{SB}}{P_C}\right)^{\frac{1}{2}} = 2(.01)^{\frac{1}{2}} = 2(.1) = .2, \text{ or } 20\%$$

3.6 AM-Transmitter Hardware

Now we shift our attention to the process of producing AM. There are several modern methods available. The method used depends upon the desired output power and the frequency band. The two most common techniques are known as low-level and high-level modulation. For low-level modulation, off-the-shelf integrated circuit (IC) chips like a CA3080 operational transconductance amplifier (OTA) or an MC1496 balanced modulator can be used to produce output powers in the mW range at low to medium frequencies. If powers above 1 W are desired, the output of these ICs can be amplified by class-A linear amplifiers to obtain higher powers. The reader should review the different classes of amplifiers if the reference to a class-A amplifier is unfamiliar or vague.

Before starting a discussion about high-level modulation it should be pointed out that the number of high-power AM transmitters in the world is extremely low (in the tens of thousands) compared to the number of other transmitters in operation. Indeed, there are over one billion cellular telephones in operation, and counting! Most high-power AM transmitters in use are used for legacy AM or shortwave broadcasting. This area of application is relatively limited; therefore, this chapter will not spend a great deal of time on the subject of high-power AM transmitters.

Low-Level Modulation

Let us turn our attention to the block diagram of a low-level AM transmitter shown in Figure 3-23 on page 110. As can be seen, for speech or audio transmission the output of a microphone or some other source is amplified by a linear class-A amplifier that drives an RF oscillator that in turn supplies

FIGURE 3-23 A low-level AM transmitter

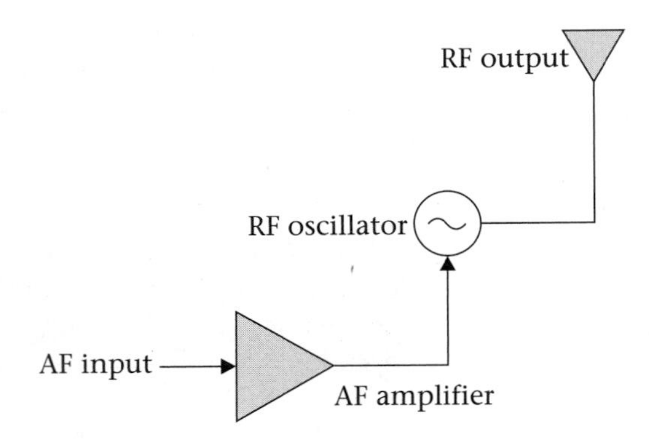

power to an antenna. The circuitry of the modulated oscillator causes its output amplitude to vary with the message signal from the microphone; hence, AM is produced. This type of transmitter is typically available on a single IC chip (MC1496).

As mentioned previously, for power outputs in the 1000-mW or +30-dBm range, one can amplify the output of the low-level transmitter just shown. See Figure 3-24 for a typical system block diagram.

FIGURE 3-24 A low-level AM transmitter with linear power amplifier

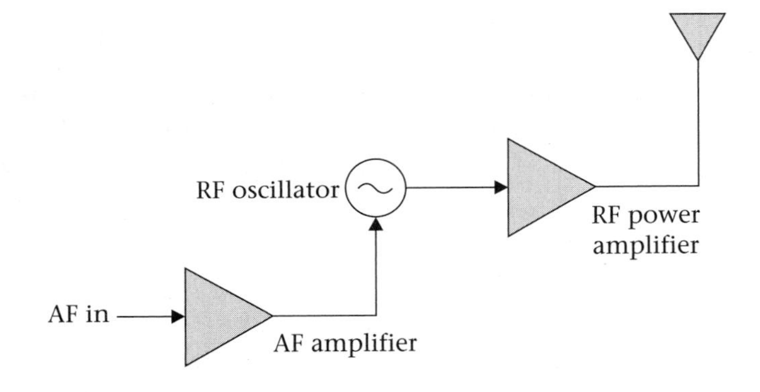

High-Level Modulation

A high-level AM transmitter block diagram is shown in Figure 3-25. This diagram shows an RF oscillator that supplies a low-level signal to a buffer amplifier, the output of the buffer amplifier (BA) is then applied to a driver amplifier (DA) or one or more intermediate power amplifiers (IPAs), and the last IPA output is applied to the final RF power amplifier (FPA), which is typically operated at class-C for high efficiency. The RF power amplifier supplies the output signal to the antenna. This section of the transmitter is called the "RF chain."

The diagram also shows an audio signal driving an audio amplifier, which in turn drives an audio power amplifier, which then drives the RF power amplifier. In this case, the audio amplifier driving or modulating the

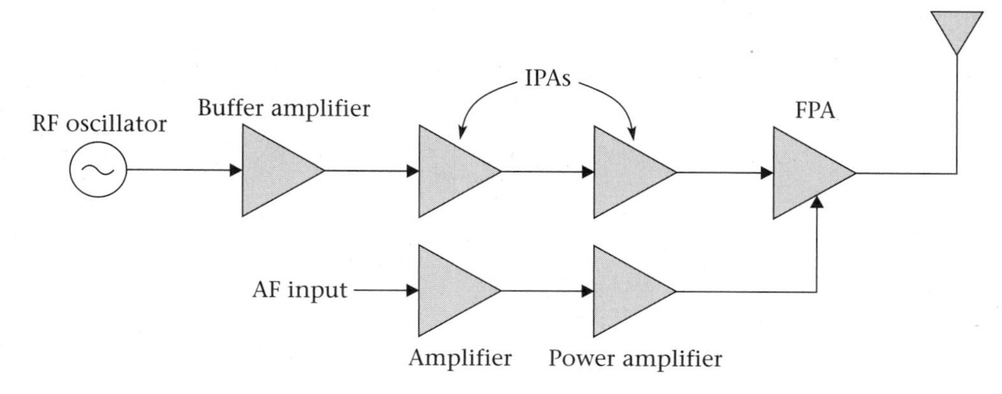

FIGURE 3-25 A high-level, high-power output AM transmitter

RF power amplifier must supply a large amount of audio power that is converted into sideband power by a mixing process in the FPA. The audio section of the transmitter is called the AF chain. In general, to obtain higher transmitter output power, amplifiers are cascaded together.

High-power AM transmitters can use either vacuum tubes or transistors to generate kilowatts (kW) of output power. Vacuum tubes (see Figure 3-26) can generate these high powers by themselves; but, the outputs of transistors have to be summed together to generate the same high-power levels. Additionally, modern high-power AM transmitters will usually employ class-D operation (an application of pulse-width modulation) to achieve even higher efficiency levels. See Figure 3-27 on page 112 for a block diagram of a typical high-power solid-state transmitter.

FIGURE 3-26 High-power vacuum tubes (*Courtesy of EIMAC Corp.*)

Although not shown in Figure 3-27, AM transmitters usually employ frequency control using a quartz crystal or a quartz-crystal referenced-frequency synthesizer based upon phase-locked loop (PLL) technology. The frequency synthesizer will allow the transmitter output frequency to be programmed to any channel (frequency) assignment in a particular service band (e.g., 1070 or 1480 kHz in the AM broadcast band).

3.7 AM-Receiver Hardware

Let us turn our attention to the receiving end of the communications channel and examine AM radio receivers. A short history of early receivers

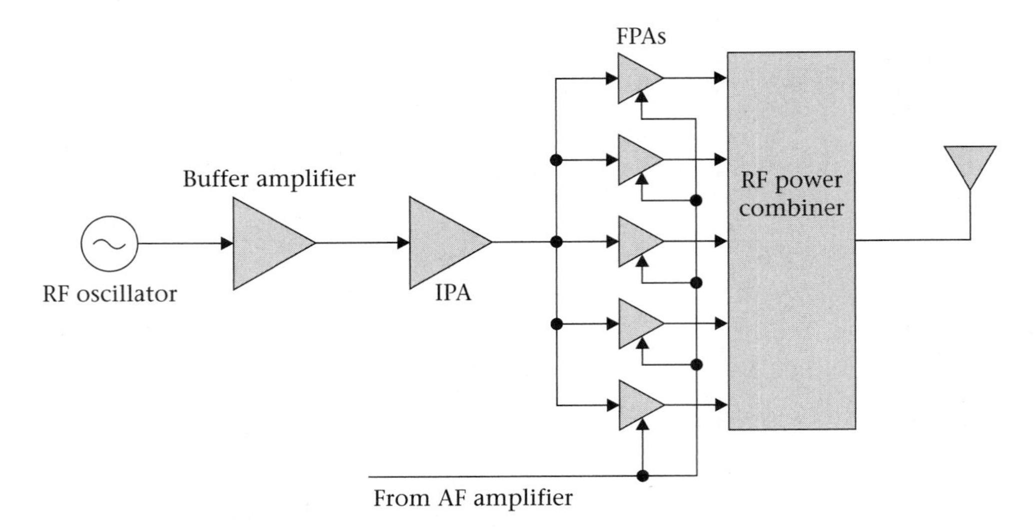

FIGURE 3-27 High-power solid-state AM transmitter

will be given first to aid in the understanding of modern receivers. The first AM receivers were developed in the pre-vacuum-tube era and used very crude detection methods. These early receivers utilized a variety of different types of detectors or what were called "coherers." Depending upon the type of transmitting apparatus in operation, the receiver would or would not use a tuning circuit. Receiver technology got a big boost with the invention of the diode detector, the vacuum-tube amplifier, and the adoption of tuneable resonant circuits. See Figure 3-28 for a block diagram of a receiver using what were at the time major technological innovations.

FIGURE 3-28 Early radio receiver in post-vacuum-tube times

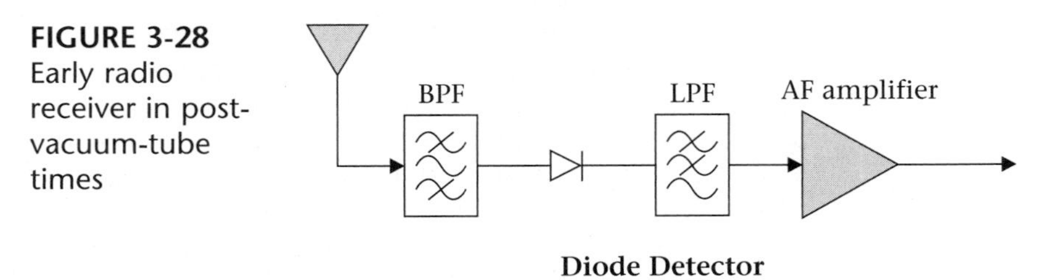

A BPF in the form of a tuned *LC* circuit would be used to select the signal, then it was detected, sent through a low-pass filter, amplified, and sent to its destination.

Early Receiver Technology

The next type of popular early radio receiver was the tuned radio-frequency receiver (see Figure 3-29). The early **tuned radio-frequency receiver** (TRF) used several cascaded, separate, or gang-tuned RF amplifiers to boost the received signal level before detection. Additionally, an audio amplifier could

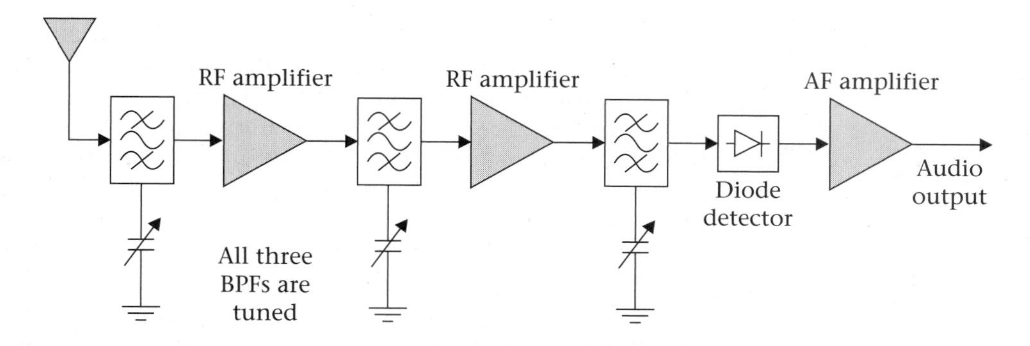

FIGURE 3-29 A tuned radio frequency (TRF) receiver

be used to drive a loud speaker or headphones. However, the TRF receiver had several flaws. The receiver was unstable due to feedback problems, and was therefore prone to go into oscillation, thus diminishing its effectiveness. The TRF receiver was also difficult to tune (most had many tuning dials) and had a receiving bandwidth that varied. The latter effect could hinder the radio's ability to effectively select between adjacent stations.

The Superheterodyne Technique

In 1913 Edwin Armstrong introduced the regenerative receiver and in 1918 patented the **superheterodyne** (superhet) receiver. The superhet receiver, not without its own subtle problems, was a vast improvement over the TRF receiver. The superheterodyne design (see Figure 3-30) is currently universally used in almost every consumer and commercial radio receiver.

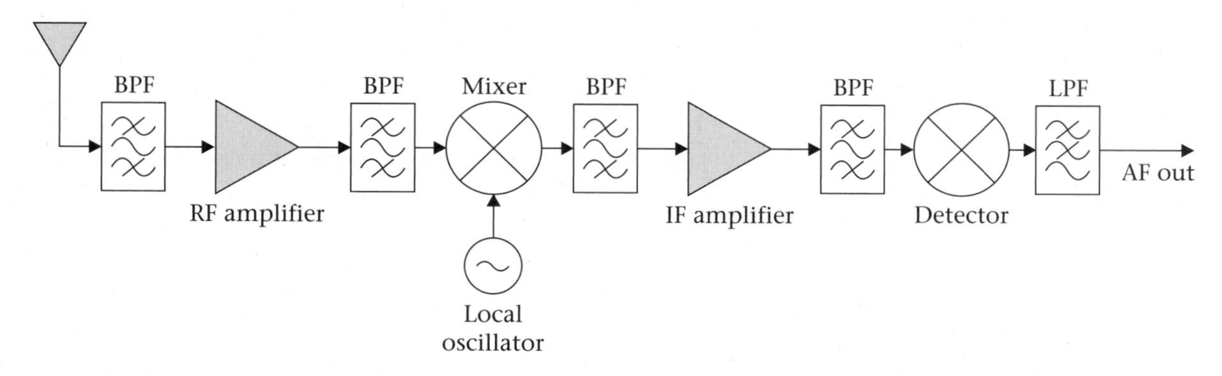

FIGURE 3-30 A block diagram of a superheterodyne receiver

The Mixer

Before we look at the operation of a superheterodyne receiver, let us discuss the operation of a mixer circuit. Mixers are integral components in superhet radios, as well as many other telecommunications systems. What does the mixer do? The **mixer** creates new frequencies or translates an existing signal to a new frequency. The mixer performs this function by effectively multiplying two signals together. We will show an example of this action first before attempting to show the mathematical model. The mixing process appears similar to modulation and signal addition. The reader might find it helpful to refer back to the section that compared those two processes.

The spectrum-analyzer display in Figure 3-31a shows two separate input signals with frequencies of 3 MHz and 5 MHz applied to a mixer. Figure 3-31b shows the frequencies that result at the output of the mixer. As can be seen from the display, there are two new frequencies: the difference frequency at 2 MHz (5 MHz – 3 MHz) and the sum frequency at 8 MHz (5 MHz + 3 MHz). The audio mixer presented earlier that added or summed signals should not be confused with the mixer that we are discussing here (see Figure 3-32).

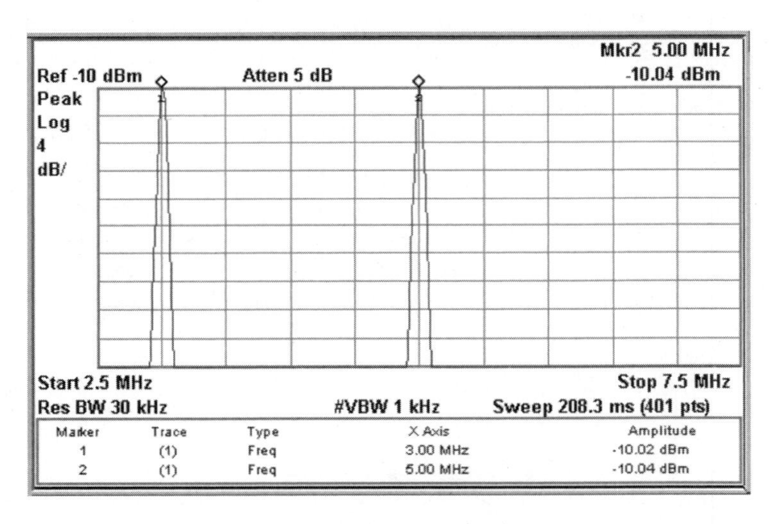

FIGURE 3-31a Two sine-wave input signals (3 and 5 MHz) applied to a mixer

Any nonlinear device can operate as a mixer; active devices such as diodes and bipolar or field-effect transistors operated in a nonlinear fashion can mix electrical signals, and certain types of crystals can be used to mix optical signals. Although not an inclusive list by any means, all are capable of performing the mixing function. As will be seen shortly, the mixing function is equivalent to the mathematical multiplication of signals.

The Mathematics of Mixing Signals

In general, a nonlinear device produces a signal at its output that can be represented by equation 3.14 (a power series). This power series models or describes a nonlinear input-output relationship.

$$v_{out} = av_{in} + bv_{in}^2 + cv_{in}^3 + dv_{in}^4 + \dots \qquad \textbf{3.15}$$

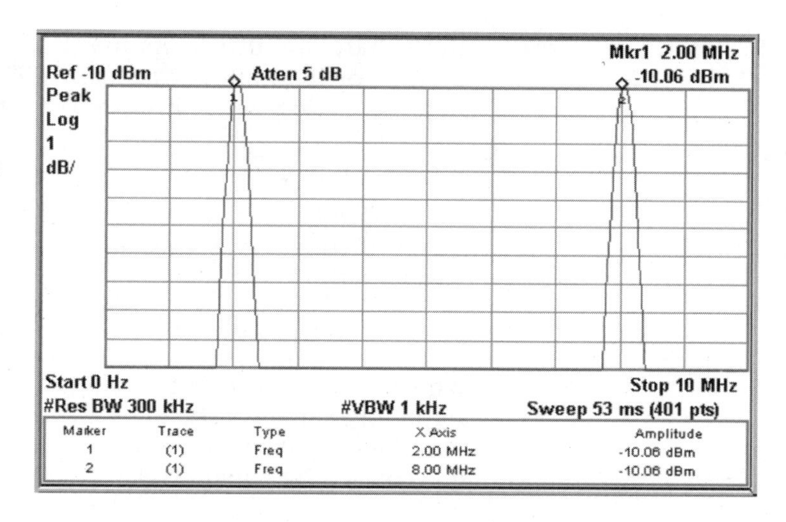

FIGURE 3-31b The output signals of the mixer at 2 MHz and 8 MHz, with input signals of 3 and 5 MHz

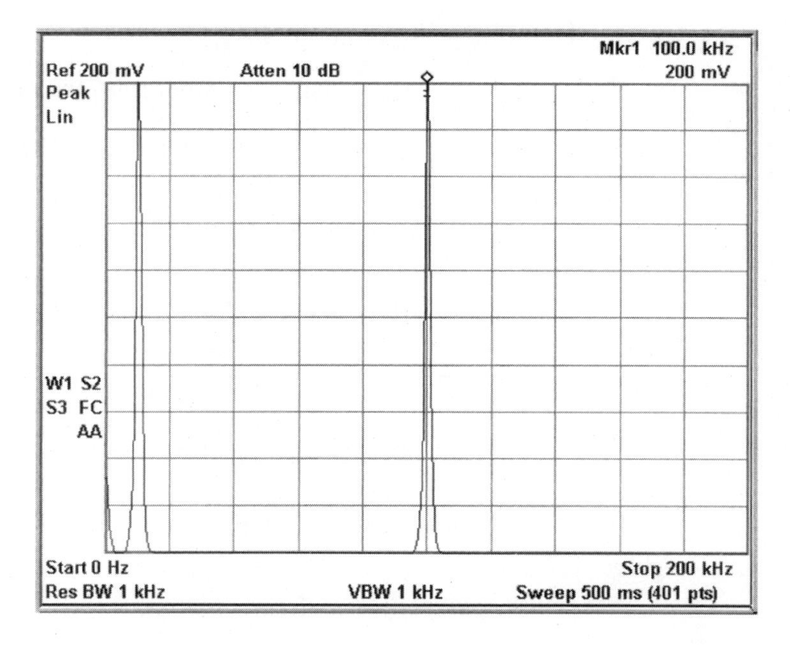

FIGURE 3-32 Adder circuit operation (the sum of two signals)

where, the coefficients, *a, b, c, d,* ... are all constants, usually with diminishing values (i.e., a > b > c > d, and so forth), determined by the characteristics of the nonlinear device.

Types of Mixers

There are basically two types of mixers, commonly known as the square-law mixer and the balanced mixer. First, we will describe the operation of the square-law mixer. As mentioned before, this type of mixer could be a diode or field-effect transistor (FET) or some other type of active device. Recall equation 3.15—if we rewrite it with only the first two terms,

$$v_{out} = av_{in} + bv_{in}^2 \tag{3.16}$$

it still fairly accurately models or describes the behavior of the square-law mixer. If the input signal to the mixer consists of two distinct frequencies, we can write v_{in} as

$$v_{in} = \sin\omega_1 t + \sin\omega_2 t \tag{3.17}$$

where $\sin\omega_1 t$ represents a signal of frequency, f_1, and $\sin\omega_2 t$ represents a signal of frequency, f_2.

The output of the square-law mixer will be given by equation 3.16 with v_{in} replaced by equation 3.17. The result of the mathematical manipulation is shown:

$$v_{out} = av_{in} + bv^2{}_{in}$$

$$v_{out} = a(\sin\omega_1 t + \sin\omega_2 t) + b(\sin\omega_1 + \sin\omega_2 t)^2$$

$$v_{out} = a\sin\omega_1 t + a\sin\omega_2 t + b\sin^2\omega_1 t + b\sin^2\omega_2 t + 2b\sin\omega_1 t\sin\omega_2 t$$

If we use the following trigonometric identities to simplify the last four terms of the equation, then

$$\sin^2 x = \frac{1}{2}(1 - \cos 2x)$$

and

$$\sin x \cdot \sin y = \frac{1}{2}[\cos(x - y) - \cos(x + y)]$$

From this we then obtain the following:

$$\begin{aligned} v_{out} = \; & a\sin\omega_1 t + a\sin\omega_2 t + \frac{b}{2}(1 - \cos 2\omega_1 t) + \frac{b}{2}(1 - \cos 2\omega_2 t) + \\ & b\left[\cos(\omega_1 - \omega_2)t + \cos(\omega_1 + \omega_2)t\right] \end{aligned} \tag{3.18}$$

Examination of Equation 3.18 reveals that at the output of the square-law mixer are the two input frequencies, f_1 and f_2, their second harmonics, $2f_1$ and $2f_2$, and the sum and difference frequency components of the two input signals, $f_1 + f_2$ and $f_1 - f_2$. At the output of the mixer the desired frequency or frequencies are selected through the use of a bandpass filter or filters.

Balanced Mixers or Balanced Modulators

Balanced mixers, or multipliers, do not reproduce their input frequencies at the output. A typical equation of operation for the balanced mixer is given by the following:

$$v_{out} = av_{in1} \cdot v_{in2} \quad \textbf{3.19}$$

where v_{in1} and v_{in2} are two separate input signals.

If $v_{in1} = \sin\omega_1 t$ and $v_{in2} = \sin\omega_2 t$, the output from the balanced mixer will be given by

$$v_{out} = a\sin\omega_1 t \times \sin\omega_2 t$$

This is the product or multiplication of two sine waves. Using the same trigonometric identities as before, we get the following output:

$$v_{out} = \frac{a}{2}\left[\cos(\omega_1 - \omega_2)t + \cos(\omega_1 + \omega_2)t\right] \quad \textbf{3.20}$$

The output frequency components are limited to the sum and difference frequencies, $f_1 + f_2$ and $f_1 - f_2$. Note that there are no input frequencies or harmonics appearing at the output of the balanced mixer. A balanced mixer or modulator is usually fabricated through the use of a "ring" of diodes and circuitry that effectively cancels out the input frequencies and their harmonics.

Figure 3-33 shows a **balanced modulator**, or balanced mixer. Again, this type of mixer only produces the product of the two input frequencies. Note the similarity of the sum and difference frequencies to the upper and lower sidebands produced in the process of amplitude modulation.

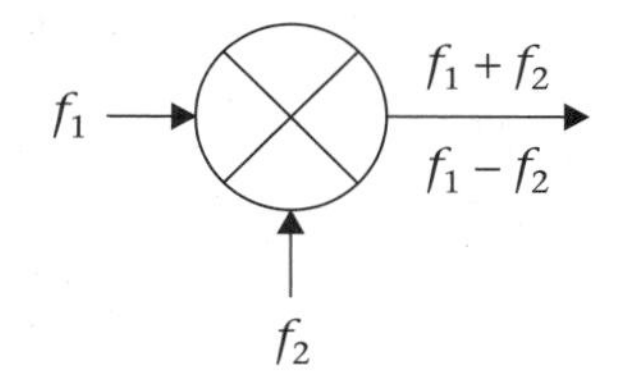

FIGURE 3-33 Block diagram of a balanced mixer or modulator

Let us look at the operation of the balanced mixer through the use of frequency displays obtained from an SA. The display shown in Figure 3-34 on page 118 shows the two input signals prior to being applied to a balanced modulator. One signal is a 25-kHz sine wave and the other is a 100-kHz sine wave. The second display, Figure 3-35 on page 118, shows the output of the balanced modulator. There are two signals evident in the display. One signal is at 75 kHz (the difference frequency) and the other signal is at 125 kHz (the sum frequency).

It is very important to note that a modulated signal, translated in frequency by a mixer, will retain the same modulation information that it started with. For an AM signal, the sidebands will have the same relationship to the center frequency both before and after the mixing process.

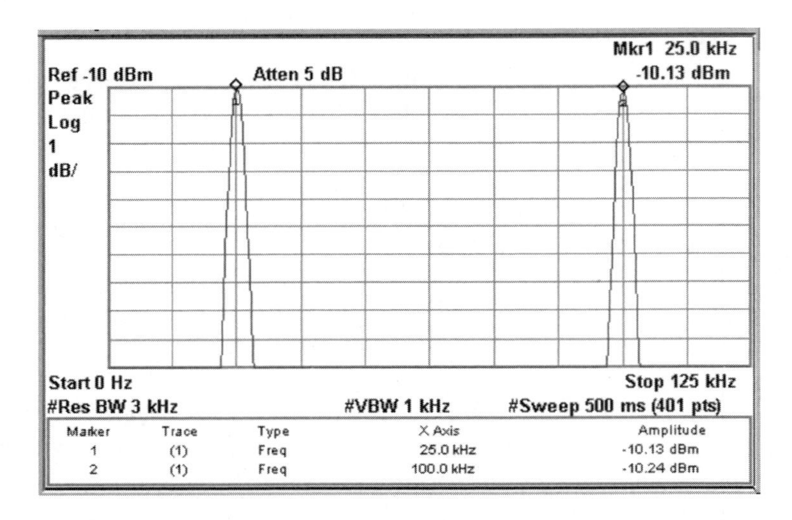

FIGURE 3-34 Input signals to a balanced modulator (25 kHz and 100 kHz)

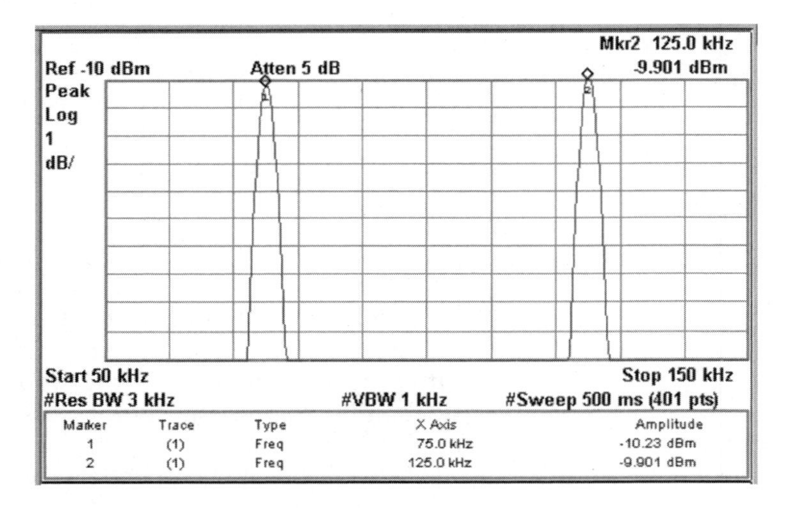

FIGURE 3-35 Output-signal spectrum from the balanced modulator (sine-wave signals at 75 kHz and 125 kHz)

Examples of Balanced Mixers and Modulators

Figure 3-36 shows an application note for the MC1496B IC. This is a typical balanced modulator in IC form. As shown in the application note, the MC1496B can be used to produce standard double-sideband full-carrier AM (DSBFC-AM) and variations of AM. At the present time this IC tends to have only limited applications. Perhaps its most useful applications would be in an introductory telecommunications laboratory course to verify the generation of AM or on a hobbyist's parts list. Modern telecommunication equipment uses application-specific ICs (ASICs) that combine the functions available from this chip with many other functions to form complete systems-on-a-chip (SOC) devices.

FIGURE 3-36 MC1496B IC balanced mixer (*Courtesy of Semiconductors Components Industries, LLC.*)

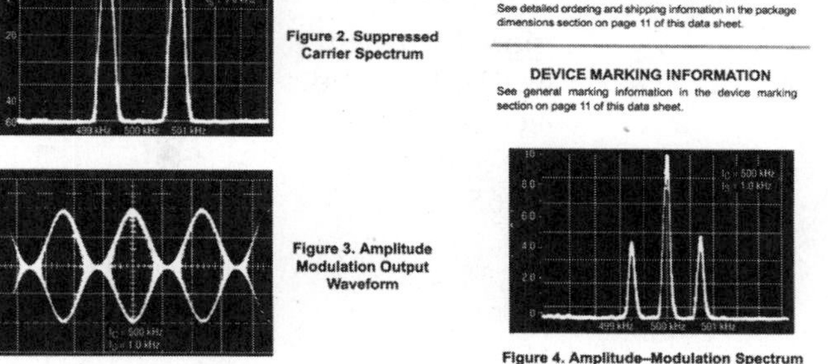

MC1496, MC1496B

Balanced Modulators/ Demodulators

These devices were designed for use where the output voltage is a product of an input voltage (signal) and a switching function (carrier). Typical applications include suppressed carrier and amplitude modulation, synchronous detection, FM detection, phase detection, and chopper applications. See ON Semiconductor Application Note AN531 for additional design information.

- Excellent Carrier Suppression –65 dB typ @ 0.5 MHz
  –50 dB typ @ 10 MHz
- Adjustable Gain and Signal Handling
- Balanced Inputs and Outputs
- High Common Mode Rejection –85 dB typical

This device contains 8 active transistors.

Figure 1. Suppressed Carrier Output Waveform

Figure 2. Suppressed Carrier Spectrum

Figure 3. Amplitude Modulation Output Waveform

ON Semiconductor™

http://onsemi.com

SO–14
D SUFFIX
CASE 751A

PDIP–14
P SUFFIX
CASE 646

PIN CONNECTIONS

Pin	Function	Pin	Function
1	Signal Input	14	V_{EE}
2	Gain Adjust	13	N/C
3	Gain Adjust	12	Output
4	Signal Input	11	N/C
5	Bias	10	Carrier Input
6	Output	9	N/C
7	N/C	8	Input Carrier

ORDERING INFORMATION
See detailed ordering and shipping information in the package dimensions section on page 11 of this data sheet.

DEVICE MARKING INFORMATION
See general marking information in the device marking section on page 11 of this data sheet.

Figure 4. Amplitude–Modulation Spectrum

September, 2001 – Rev. 6 1 Publication Order Number: MC1496/D

More on the Superheterodyne Receiver

Recall the superhet block diagram that was illustrated in Figure 3-30. After selection and amplification by the low-noise RF (**RF amplifier**) stage, the superhet will take any input signal and translate its frequency to a new, fixed **intermediate frequency** (IF). The above process lets each received signal be processed identically by the following stages. The IF stages provide a constant bandwidth with good selectivity through the utilization of modern bandpass-filter technology. The **local oscillator** (LO) signal and RF input signal mix together to produce two output signals at the mixer output. One signal is at the IF frequency (usually, LO – RF, the difference frequency) and another is at the sum frequency (LO + RF).

How does the receiver select the correct signal to apply to the IF stage? A bandpass filter (BPF) is centered at the frequency to be used by the IF stage. Let us look at two examples of this mixing process in the superhet radio. Figures 3-37a and 3-37b and Figures 3-38a and 3-38b show examples of two different input frequencies being translated to a standard IF frequency of 455 kHz. As shown in the figures, a desired input RF signal of 810 kHz is tuned by the receiver. It is mixed with the LO signal at 1265 kHz (810 kHz + 455 kHz = 1265 kHz). The result of the mixing process will yield output frequencies of 2075 kHz (the sum frequency) and 455 kHz (the difference frequency). Both of these signals contain the original modulation information (only translated in frequency). A BPF would pass the 455-kHz signal on to the next sequential stages, the IF amplifier, while rejecting the other signal and any other signals possibly produced by the mixer.

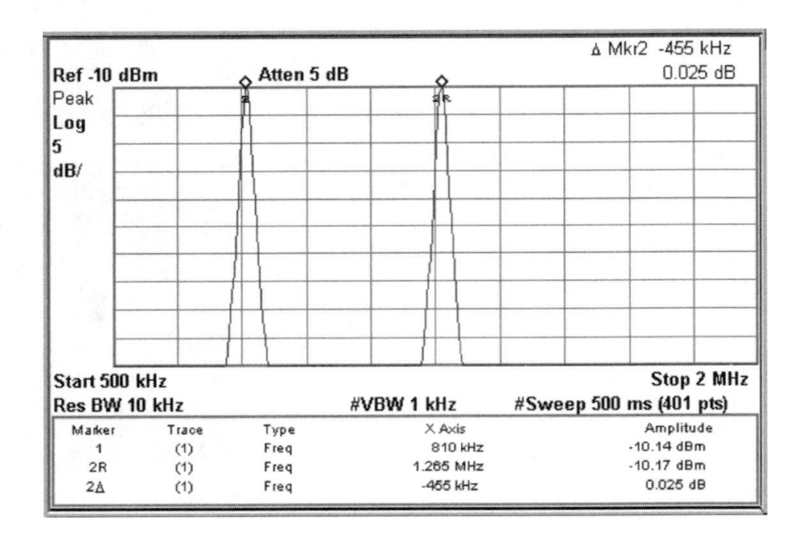

FIGURE 3-37a Example 1: frequency translation in a superhet radio. The desired signal frequency is 810 kHz; the LO produces a signal at 1265 kHz.

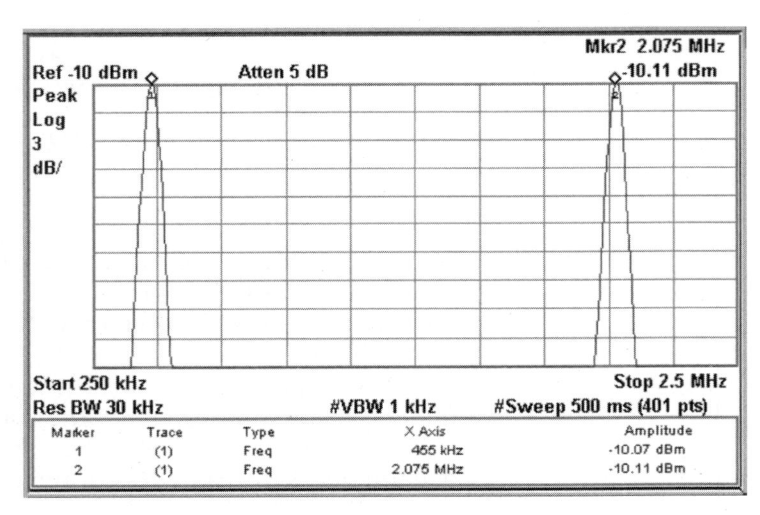

FIGURE 3-37b The mixer produces outputs at 455 kHz and 2075 kHz. A BPF will select the 455-kHz signal and pass it on to the IF amplifier stage.

For the second example, shown in Figures 3-38a and 3-38b, the desired input signal is at 1340 kHz. In this case, the LO signal is at 1795 kHz (1340 kHz + 455 kHz = 1795 kHz) and will mix with the input and yield mixer output signals at 3135 kHz and 455 kHz. Again, the latter signal will be passed by the BPF while the sum signal will be rejected by this filter.

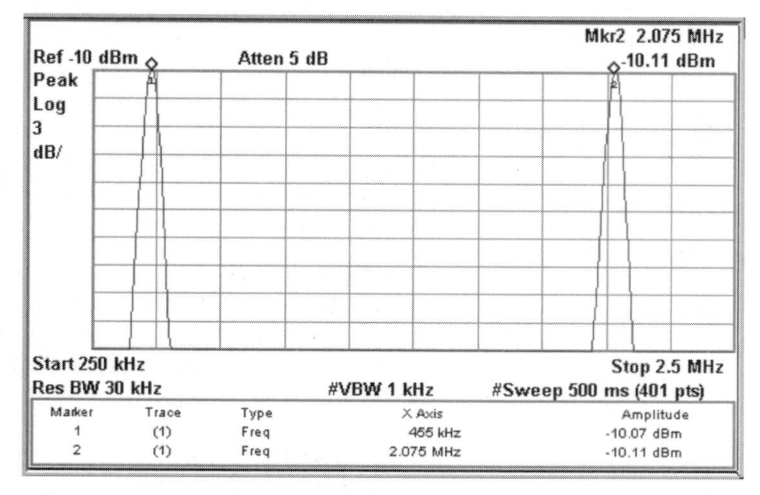

FIGURE 3-38a Example 2: frequency translation in a superhet radio. The desired signal is at 1340 kHz; the LO produces a signal at 1795 kHz.

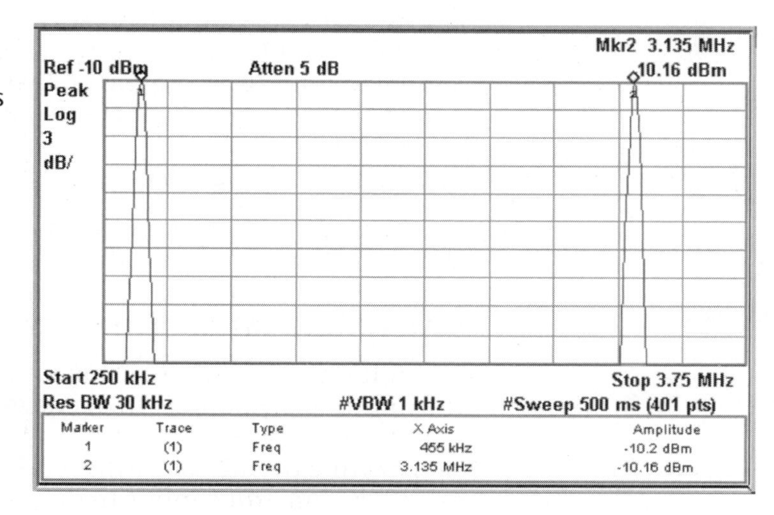

FIGURE 3-38b The mixer produces outputs at 455 kHz and 3135 kHz. The 3135-kHz signal is rejected by a BPF.

Superheterodyne-Receiver Problems

The superhet is not without its flaws. The presence of a strong signal at its input can result in the signal being received at two different input frequencies (refer back to Example 3.5) or an **image frequency** can interfere with the reception of a desired signal. The frequency of the image signal is given by the following equation, where the sign depends upon whether the local-oscillator frequency is above, high-side injection (+) or below, low-side injection (–) the desired signal (RF) frequency:

$$f_{\text{image}} = f_{\text{signal}} \pm 2f_{\text{IF}} \tag{3.21}$$

EXAMPLE 3.7

Assume an AM superhet radio with an IF of 455 kHz. From equation 3.21, we see that the image frequency is different from the desired signal by 910 kHz (2 × 455 kHz = 910 kHz). If the radio is tuned to a desired station's signal at 580 kHz, a strong signal reaching the receiver's antenna at a frequency of 1490 kHz (580 kHz + 910 kHz = 1490 kHz) might also be detected. Why is this so?

Solution The LO frequency is at 1035 kHz (580 kHz + 455 kHz). A signal at either 580 or 1490 kHz that reaches the mixer stage will mix with the LO signal to generate mixer outputs at 455 kHz. Both of these signals will be processed by the IF stage and eventually the detector stage of the superhet. Thus, a signal at 1490 kHz could be received at the correct tuning setting of 1490 kHz and also at the incorrect frequency of 580 kHz, or two places on the dial (or at two tuning points if there is not a dial in the physical sense). If the signal at 1490 kHz is not very strong, it will appear as interference to any desired signal at 580 kHz; hence, an interfering image signal. If the receiver has a high-Q BPF with which it selects input signals, the undesired and interfering **image signal** at 1490 kHz, even if it is strong, might be rejected sufficiently before the mixer stage and not interfere with the desired signal.

The topic of **image rejection** will be discussed later in the chapter with a section on multiple-conversion superheterodyne designs.

Cross-Modulation Products

In addition to the desired sum and difference frequencies, **cross-modulation products** (spurious frequencies) are produced by mixers. The production of these spurious frequencies is greatest in the presence of strong input signals. In extreme cases, this effect can cause a signal to *appear* to be received at numerous input frequencies and interfere with the reception of signals actually being received on those frequencies. The following equation indicates the frequencies that a receiver might respond to depending upon the local oscillator frequency and the intermediate frequency:

$$f_{\text{spurious}} = \left(\frac{m}{n}\right) f_{\text{lo}} \pm \frac{f_{\text{IF}}}{n} \tag{3.22}$$

where m and n are integers.

The SA display in Figure 3-39 shows a portion of the frequencies present in a mixer output when strong input signals are applied to it. Note all the new frequencies produced.

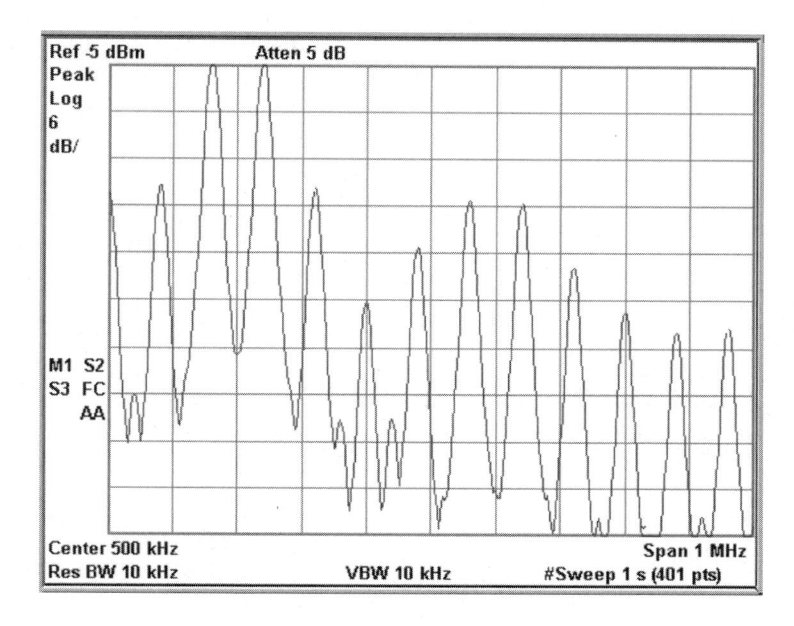

FIGURE 3-39 Mixer cross-modulation products resulting from two strong inputs to a nonlinear component (mixer)

A Functional Description of the Superheterodyne Stages

Let us examine the operation of the various subsystems in the superhet receiver in more detail. The RF amplifier is used to provide a tuned low-noise amplifier (LNA) to increase the sensitivity of the receiver and also to provide for some preselection of input signals (recall the image-frequency problem). Without an RF amplifier, a receiver will have difficulty picking up low power signals; strong signals at undesired frequencies will be able to reach the mixer and possibly produce unwanted cross-modulation products. An RF amplifier is almost always needed at high frequencies to reduce receiver noise and to prevent reradiation of the local oscillator signal away from the receiver via the receiver's own antenna. This type of unwanted radiation is prohibited by Section 15 of the FCC's Rules and Regulations.

The Local Oscillator

As already outlined, the mixer's function is to translate the input signal frequency to the fixed IF. The LO provides the signal power to operate the mixer, forcing it into its nonlinear range of operation, and to produce the correct frequency translation. Today LOs are typically very accurate (crystal controlled) phase-locked loop (PLL) *frequency synthesizers* controlled by microprocessors to produce the correct frequency automatically. Recall that usually $f_{LO} = f_{RF} + f_{IF}$. This is known as high-side injection, as the LO frequency is greater than the RF frequency. The intermediate frequency amplifier provides gain and, most importantly, selectivity to the received signal. Filter technology has matured to the point that filter shape factors can approach 1. The IF stage can therefore selectively amplify the desired signal while effectively excluding adjacent channel signals.

The Detector

The **detector** subsystem demodulates or decodes the received signal. Demodulation is basically the reverse of the modulation process that took place at the transmitter. For AM, an easy way to understand what happens in the detector stage is to consider that the detector is performing the mixing function. Figure 3-40 illustrates AM detector operation.

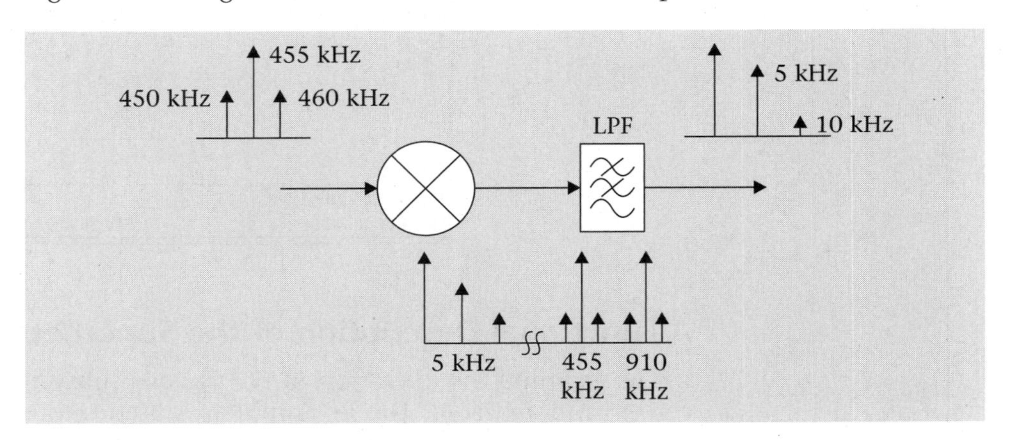

FIGURE 3-40 AM detector operation

As the figure shows, the input signal to the detector/mixer is the same as the original transmitted signal; however, it is now translated to the receiver's intermediate frequency (typically 455 kHz). When this signal is applied to the detector/mixer, the various frequency components undergo the mixing process. This process yields sum and difference frequency components at the output of the detector. The difference in frequency between the center carrier frequency and the sidebands is equal to the frequency of the original modulating signal (5 kHz for this example). Also, the amplitude of the output at the modulating frequency will be proportional to the original modulating signal amplitude. A low-pass filter located after the detector/mixer will allow the recovered modulating signal to pass on to the receiver's destination, but attenuate all other detector output-frequency components. As shown in Figure 3-40, there will be a small frequency component at the second harmonic of the original modulating signal. A detector implemented with a simple nonlinear active device like a diode will introduce a large amount of distortion into the detected output.

Automatic gain control (AGC) compensates for large variations in input-signal power by feeding back control signals to the early stages of the superhet. These control signals are able to modify overall system gain in an attempt to keep the output level constant regardless of input-signal strength. The output stages of a superhet process the signal for delivery to its destination.

Receiver Characteristics

Wireless receivers have specifications that indicate their operational characteristics. Receiver sensitivity is usually given as the number of femtowatts (1 fW = 10^{-15} W) of signal power needed to provide a certain level of signal-to-noise ratio (SNR) at the receiver input terminals. Noise figure (NF) and noise temperature (NT) are used to describe the quality of the receiver. Bit error rate (BER) or intersymbol interference (ISI) can be used to indicate the performance of digital receiver systems. Other receiver characteristics are **selectivity** or bandwidth specifications (the ability to select only one signal while rejecting other nearby signals), receiver distortion levels, an acceptable range of received signals (**dynamic range**), and spurious response (image-frequency rejection).

What effect does noise have on the correct demodulation of AM signals? As pointed out in chapter 2, noise will add to a signal. Because AM is an analog process, any additive noise will distort the demodulated signal. It has been determined in a qualitative fashion that an SNR of 60 dB yields perfect signal reconstruction, meaning that the added noise is imperceptible to the typical listener of an AM transmission. As the SNR is lowered, perceptible static is introduced into the demodulated signal, and at some point the signal becomes unintelligible.

IC Superheterodyne Receivers

Present-day technology is such that several manufacturers produce linear ICs that perform the entire AM superheterodyne function on a single chip, with the exception of the RF amplifier, filtering, and AF power-amplification functions. As system-on-a-chip (SOC) technology advances and microelectromechanical systems (MEMS) technology improves allowing RF filtering functions, one will eventually be able to envision the entire receiver process residing on a single IC chip.

Superheterodyne Variations

An improvement to the standard superhet is the utilization of double, or in some cases triple, conversion. A **double-conversion** superhet is shown in Figure 3-41 on page 126. It typically uses a high first IF frequency and a low second IF frequency.

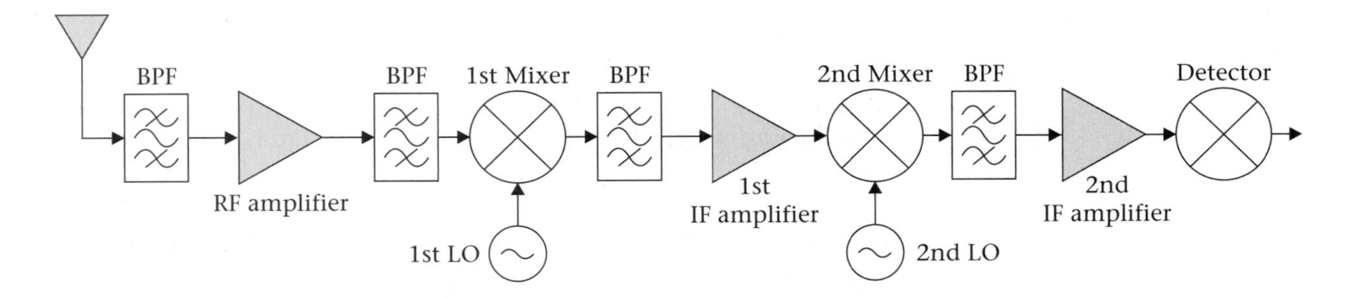

FIGURE 3-41 A double-conversion superhet receiver

The double-conversion superhet improves upon many of the single-conversion superhet's deficiencies. Note that the design allows for greatly improved image rejection due to a relatively high first IF frequency. It also allows for better selectivity due to a very low second IF frequency. See equation 3.23 for the image rejection (IR) of a single tuned circuit.

$$IR = \frac{\text{Voltage Gain at } f_{\text{signal}}}{\text{Voltage Gain at } f_{\text{image}}} = \sqrt{1 + Q^2 x^2} \qquad \textbf{3.23}$$

where, Q is the quality factor, or Q, of the tuned circuit (BPF) and x is given by

$$x = \frac{f_{\text{image}}}{f_{\text{signal}}} - \frac{f_{\text{signal}}}{f_{\text{image}}} \qquad \textbf{3.24}$$

It can be shown that the value of x, and therefore the image rejection (IR), will increase as the value of f_{image} increases. Recall that the value of f_{image} is directly proportional to the intermediate frequency ($f_{\text{image}} = f_{\text{signal}} \pm 2\, f_{\text{IF}}$). Therefore, for a double-conversion superhet a high value of system image rejection is provided by the relatively high first IF frequency.

What about the low second IF frequency? Why is a very low IF good for selectivity? The answer is that it is easier to achieve a shape factor close to 1 at lower frequencies. To recap, the double-conversion superhet uses a high frequency for the first IF frequency, which provides a high value of image rejection. The second mixer uses a fixed local oscillator to bring the signal down to a very low second IF frequency. As indicated before, this will allow the receiver to have excellent selectivity.

Other receiver variations are usually due to special receiver requirements, such as the reception of extremely broadband RF signals or specialized modulation techniques. One example of such a variation would be the receiver used by the standard Doppler-shift police-RADAR speed detectors. These receivers have no tuneable local oscillator; they are designed to have the internal-transmitter signal source and the Doppler-shifted received signal mix together to produce the low-frequency output that represents the target's speed (this receiving technique gives rise to the term *homodyne* receiver). Another example would be the use of *up conversion* or a high-frequency IF to provide enough IF bandwidth to process the received signal.

The Software Radio

With advances in microelectronics, high-speed analog-to-digital conversion (ADC) has become practical at RF frequencies and digital-signal-processing (DSP) techniques have become faster and more sophisticated. These advances have impacted the traditional superhet receiver in that the long-established detector can now be replaced by a high-speed ADC converter. This ADC converts the IF signal directly into digital. Once converted to a digital signal, the radio can then process the digital information using high-speed DSP technology and application-specific ICs (ASICs). Thus, we can now add *spurious-free dynamic range* (SFDR) to our list of receiver characteristics. This specification is used to reflect the linearity of the ADC process. See Figure 3-42 for a block diagram of a digital radio.

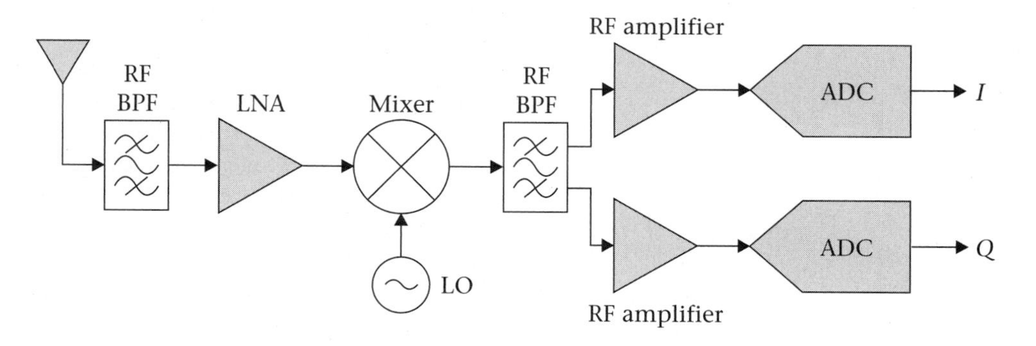

FIGURE 3-42 Digital IF/baseband software radio

The processing of the information in Figure 3-42 is under software control, hence the term **software radio**. A software radio can be reconfigured through programming, and is thus capable of numerous modes of operation. By this statement we mean to imply that different modulation schemes, multiplexing formats, and access technologies can be accommodated by a single reconfigurable piece of equipment. Some of these topics will be covered in future chapters. Note the *I* and *Q* outputs from the receiver. These two outputs are used for more sophisticated digital-modulation schemes that will be introduced in chapter 6.

The holy grail of the software-radio world is a direct RF conversion scheme in which the desired RF signal is converted directly into the digital world without the need for down conversion. See Figure 3-43 on page 128 for an example.

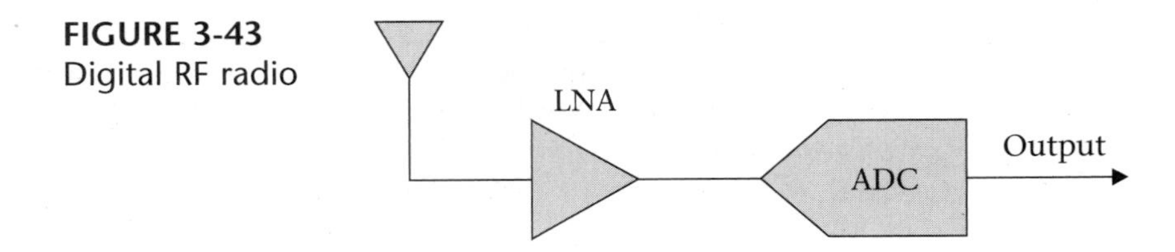

FIGURE 3-43
Digital RF radio

3.8 Alternative AM Schemes

There are variations to standard amplitude modulation. Standard AM, also known as **double-sideband full-carrier** (DSBFC) **AM** (FCC designation A3) or **large-carrier AM** (LCAM) is extremely inefficient. We can eliminate the frequency component that is always present and contains the most power (the carrier) and not lose any information about the modulating signal. One might ask how is this possible. Isn't the carrier signal necessary to carry the information? The answer is *no*! When the early pioneers of radio communications first implemented AM schemes, they used processes that inherently retained the center carrier signal in the final transmitted signal. It was demonstrated fairly early on that either **double-sideband suppressed-carrier** (DSBSC) **AM** or **single-sideband suppressed-carrier** (SSBSC) **AM**, or some variation of these themes, would be much more efficient. This is not to say that these forms of AM are easy to implement. Indeed, the circuitry needed early on for these variations on standard AM was more complex than what the technology of the day would allow. Today's technology is a different story. Recall that the MC1496B IC introduced earlier in the chapter is capable of producing an acceptable form of DSBSC AM.

We have seen block diagrams of standard low-level and high-level AM schemes in Figures 3-23 and 3-24. Now let us examine the generation of DSBSC AM using a balanced modulator. Producing DSBSC AM is the first step toward producing SSBSC AM. Note that SSBSC is preferred to DSBSC because it only occupies one-half of the bandwidth needed for DSBSC, an important savings of a precious resource, either RF spectrum or bandwidth! An SSBSC system allows two signals to be transmitted in the same bandwidth as a standard AM signal.

At this time we will show the frequency and time displays of a DSBSC AM signal that is produced by the balanced modulator in Figure 3-44a. As can be seen on the SA display in Figure 3-44b, this process yields a suppressed signal at the carrier frequency, f_C, and only a single pair of sidebands (the LSB and USB) located symmetrically around, f_C. The resulting DSBSC signal is shown in the time domain in Figure 3-45.

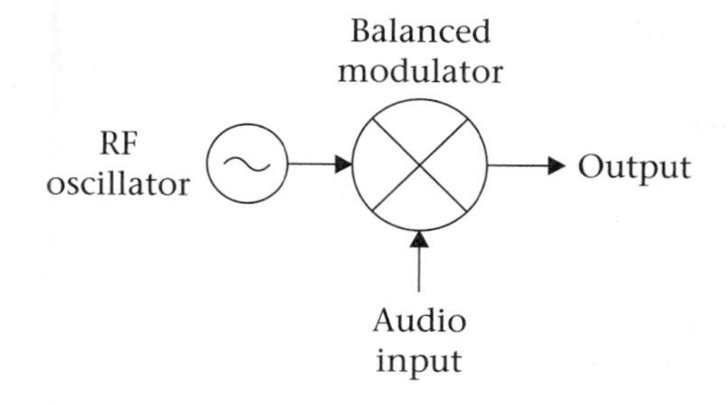

FIGURE 3-44a A balanced modulator used to produce DSBSC AM

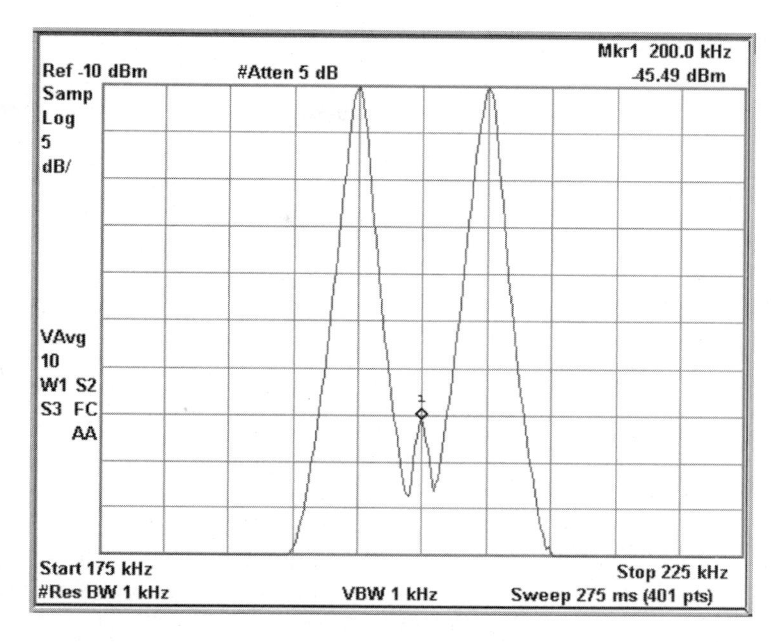

FIGURE 3-44b The frequency-domain display of a DSBSC signal

FIGURE 3-45 A DSBSC AM signal as displayed on an oscilloscope

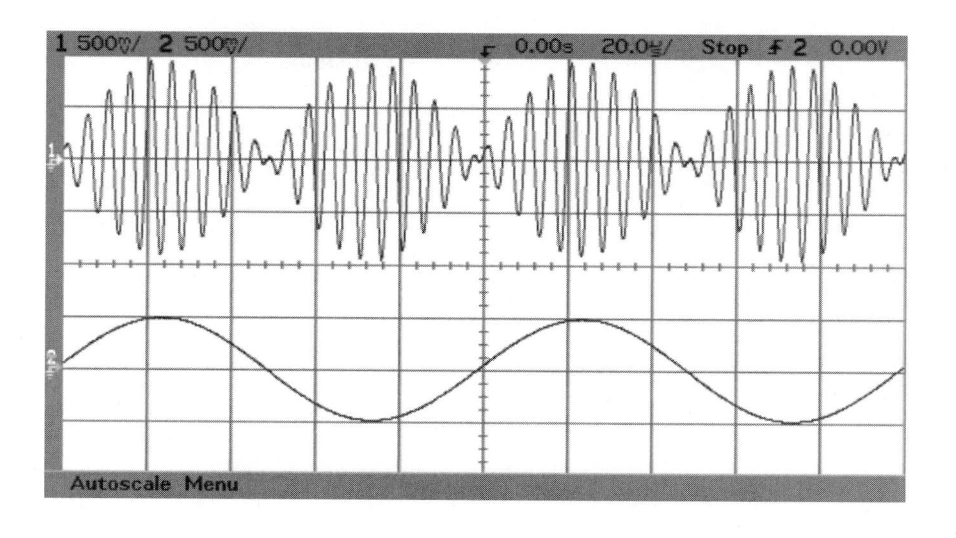

Note that the DSBSC signal envelope increases with the absolute magnitude of the modulating signal and that it is a maximum whenever the modulating signal is at a maximum, and zero whenever the modulating tone passes through zero. If there is no modulating signal, the output of

the balanced modulator would ideally go to zero. Note also that only the sidebands of an AM signal exhibit any variation in their characteristics. If the modulating frequency increases, the sidebands spread apart, taking on a new frequency; if the modulating frequency decreases, the sidebands move closer together, again taking on new frequencies relative to the center frequency. If the modulating signal becomes higher in amplitude, the sidebands increase in power level; if the modulating signal becomes lower in amplitude, the power level of the sidebands becomes smaller. The carrier-frequency component for LCAM or DSBFC, on the other hand, has consistent frequency and power characteristics, and therefore actually contains no information about the modulating signal.

Generation of SSB Using a Filter

Presently, the most common method of producing SSBSC is to convert an easily generated DSBSC signal to an SSBSC signal by filtering out one of the sidebands. If the transmitter removes the lower sideband and retains the upper sideband, this is called upper-sideband (USB) SSB. The opposite scenario would yield lower-sideband (LSB) SSB. The basic transmitter is shown in Figure 3-46 and the filtering process is depicted in Figure 3-47.

FIGURE 3-46 A basic SSBSC AM transmitter

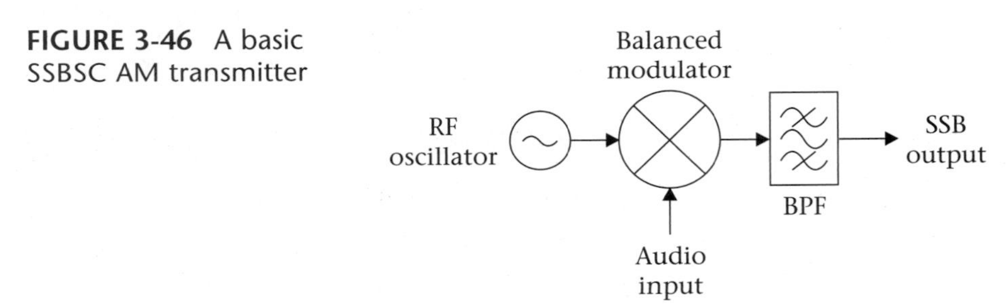

FIGURE 3-47 Generation of upper- or lower-sideband SSB using a BPF to eliminate the other sideband

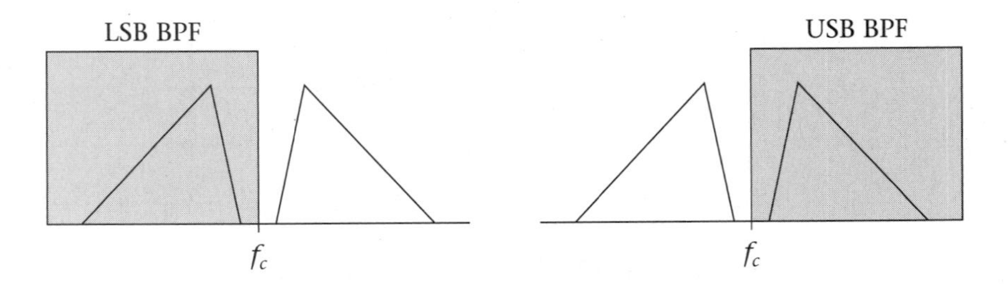

The SSB Receiver

At the detector of an SSB receiver, we must somehow reinsert the missing frequency component at f_C. The detector of the SSB receiver is usually a balanced modulator (also called a **product detector**), which is driven by a beat-frequency oscillator (BFO). If the BFO is at the correct frequency, which implies some type of frequency coordination between the transmitter and receiver, these two signals mix together and yield the original modulating signal or signals (see Figure 3-48).

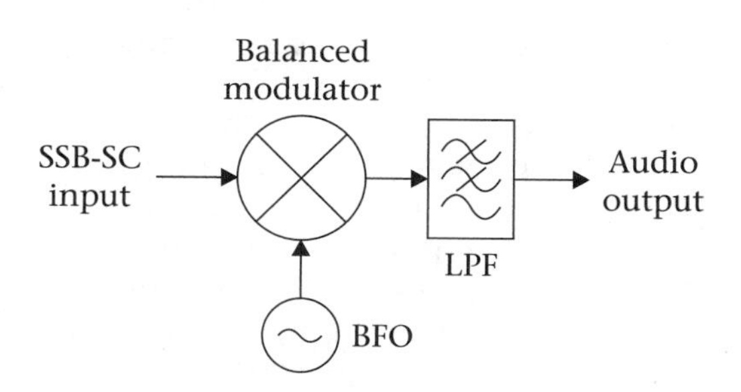

FIGURE 3-48 An SSBSC receiver using a BFO and a product detector to recover the modulating signal

SSB Advantages

SSB has one-half the bandwidth of DSBFC. This translates into twice the traffic on the same channel. Furthermore, there is no wasted power at f_C. AT&T made extensive use of SSB for their "L-carrier" long-distance systems before the advent of fiber-optic cables. There are other forms of SSB in use, as mentioned before. A form of **pilot-subcarrier** SSB transmits a reduced-power carrier to allow carrier regeneration at the receiver. This greatly enhances the SSB receiver operation. **Vestigial sideband** (VSB) is used by legacy TV broadcasting to save bandwidth. When the first TV standards were being created and frequency bands allocated, it was determined that the use of VSB would allow the transmission of three standard NTSC TV channels in the same bandwidth needed to broadcast two DSBFC channels. Hence, VSB modulation was adopted as the standard. Again, the RF spectrum is a limited and valuable resource.

Figure 3-49 shows an example of VSB TV Broadcasting. Only a vestige of the lower sideband is retained. This has the effect of lowering the needed bandwidth of the TV signal from 9 MHz to 6 MHz.

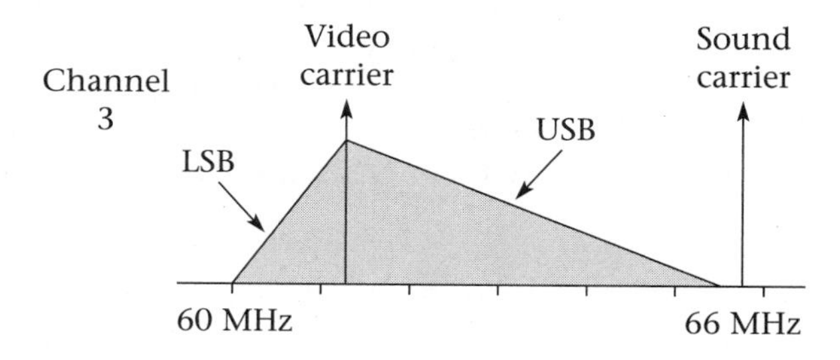

FIGURE 3-49 VSB AM for TV-signal transmission (channel 3 signal is shown)

Summary

The process of producing AM is a signal-mixing or multiplying scheme where a high-frequency signal is usually mixed with a low-frequency signal. This process therefore produces several new frequencies—the upper and lower sidebands that correspond to the sum and difference frequencies in the mixing process. Once this concept is understood, the AM signal bandwidth and the power relations between the signal components can be worked out mathematically. The value of the modulation index, m, is defined as the ratio of modulating-signal amplitude to high-frequency carrier amplitude, and is fundamental to the AM-power relations and the absence or presence of signal distortion. Signal bandwidth of an AM wave is equal to twice the modulating signal bandwidth when the index of modulation is limited to 100%.

There are two basic methods of producing LCAM—low- and high-level modulation. In each case, an RF signal is effectively mixed with a lower-frequency audio signal. As mentioned before, the resulting AM signal consists of three frequency components.

To receive amplitude modulation, a superheterodyne receiver is typically used. The superhet receiver employs various stages to amplify and then translate the received input signal to a fixed lower frequency to be further amplified, filtered, and then detected. Filters, RF amplifiers, mixers, intermediate frequency amplifiers, and detector stages are used in the superhet receiver. A superhet radio will have various specifications that characterize its operation. The more common of these specifications, such as noise figure, sensitivity, image rejection, selectivity, dynamic range, and distortion give an overall picture of the system operation.

For demanding applications, the superhet can employ double conversion or up conversion to enhance its operation. For more sophisticated modulation and access schemes, the classic superhet radio is undergoing a transformation. This transformation is bringing digital signal processing closer and closer to the received RF signal. Presently, ADC converters can be used to digitize the IF signal, and future plans are to eliminate signal down conversion entirely. Direct digitization or digital RF radio is not here yet, but it is not too far off in the future either.

Modern AM variations include single sideband (SSB) and vestigal sideband (VSB). SSB systems only transmit one sideband while suppressing the carrier and the other sideband. VSB transmits one sideband, the carrier, and a small portion of the remaining sideband. These modulation schemes allow for increased efficiency in both power and bandwidth usage.

Questions and Problems

Section 3.1

1. Do an Internet search to determine where and when the first AM broadcast stations went on the air in the United States.

Sections 3.2 and 3.3

2. Describe how an AM wave transports information.

Section 3.4

3. For the oscilloscope displays shown in Figures 3-50a and 3-50b, determine the approximate value of the index of modulation, m.
4. For the spectrum-analyzer displays shown in Figures 3-51a and 3-51b on page 134, determine the modulating frequencies, the overall signal bandwidth, and the total depth of modulation.

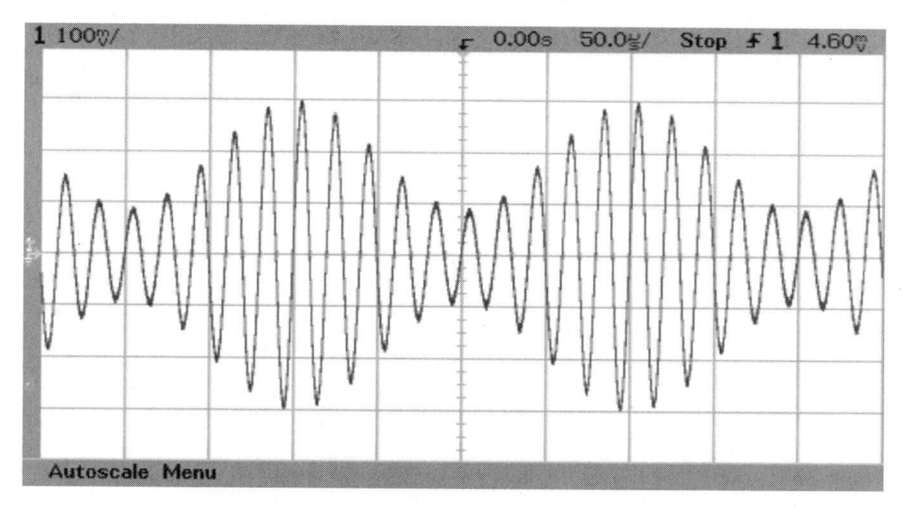

FIGURE 3-50a
Question 3

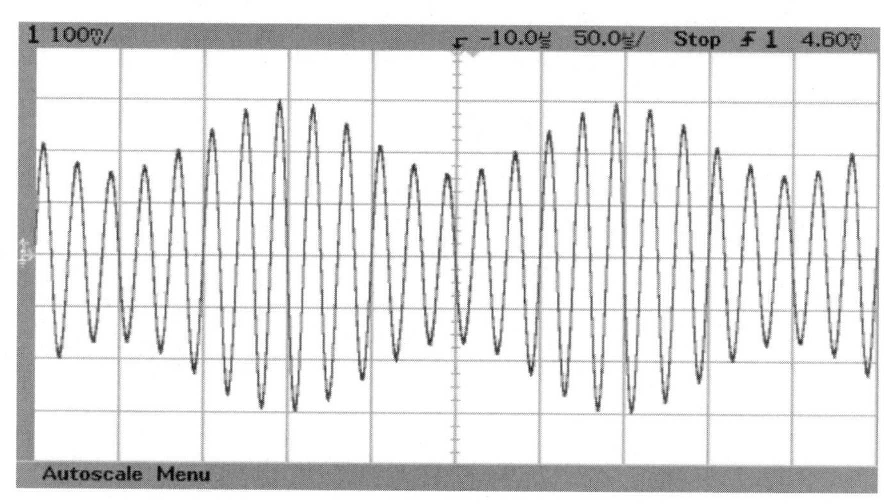

FIGURE 3-50b
Question 3

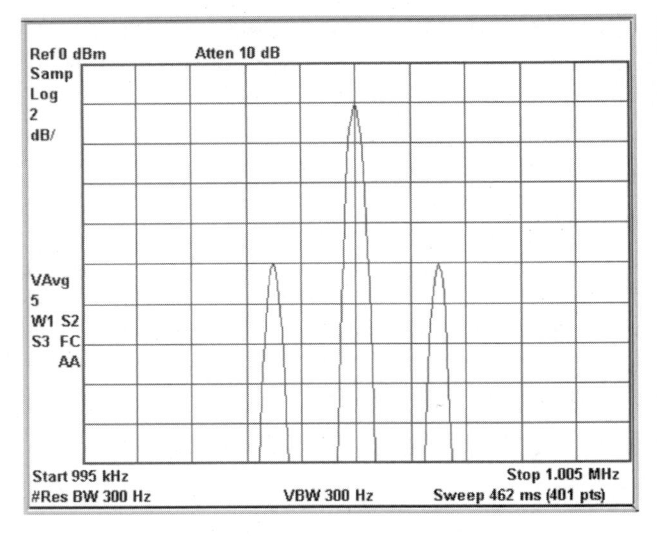

FIGURE 3-51a Question 4

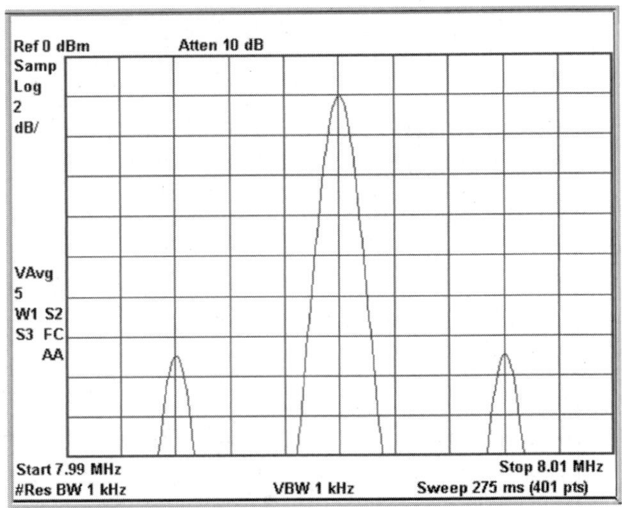

FIGURE 3-51b Question 4

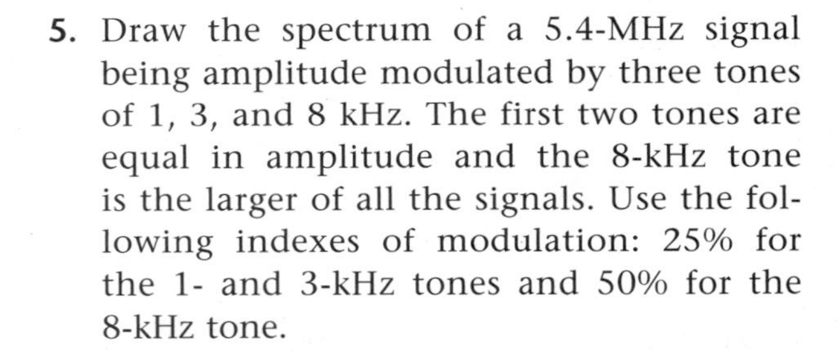

5. Draw the spectrum of a 5.4-MHz signal being amplitude modulated by three tones of 1, 3, and 8 kHz. The first two tones are equal in amplitude and the 8-kHz tone is the larger of all the signals. Use the following indexes of modulation: 25% for the 1- and 3-kHz tones and 50% for the 8-kHz tone.

Section 3.5

6. Determine the spectra of a 1-MHz AM signal with 25 W of carrier power and a modulating signal with $m = 0.7$ at a frequency of 2.5 kHz. Your answer should include the values of the sideband powers.
7. For the spectrum-analyzer display shown in Figure 3-52, determine the frequency of the modulating signals and the percentages of modulation.
8. Determine the percentage change in AM-transmitter RF current for 80% modulation versus 0% modulation.

Section 3.6

9. Do an Internet search to learn how a vacuum tube amplifies and describe your findings.
10. Explain the concept of frequency synthesis.

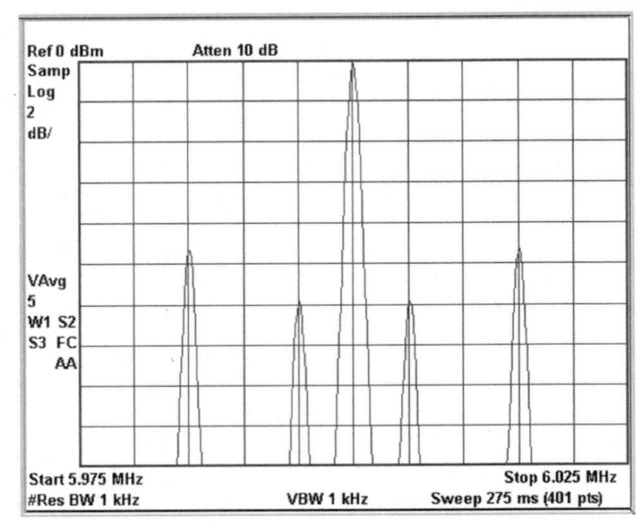

FIGURE 3-52 Question 7

Section 3.7

11. Explain the function of a mixer. Use two frequencies to illustrate your point.
12. Why does a balanced modulator give a cleaner output signal?
13. Come up with your own "Example 3" of the operation of tuning a superhet radio. Assume the standard 455-kHz IF frequency. (See pages 120 and 121.)
14. What is the purpose/function of the local oscillator?
15. What are cross-modulation products? When do they commonly occur?
16. Go on the Internet to learn what the mixer-specification conversion loss means. Give an example for an RF input signal level of –28 dBm.
17. Describe what happens to a signal (the processes it goes through) as it travels through a superheterodyne receiver.

Section 3.8

18. Describe the advantages of SSB AM. Can you detect any disadvantages? If so, what are they?
19. Explain why over-the-air NTSC broadcasting uses VSB AM.
20. Many people call the detector of an AM radio the second mixer. Why?

Chapter Equations

$$e(t) = E_p \sin(\omega t + \phi) \quad \textbf{3.1}$$

$$\text{Amplitude} = E_C + e_{\text{mod}} = E_C + E_M \sin \omega_M t \quad \textbf{3.2}$$

$$e(t) = (E_C + E_M \sin \omega_M t) \times \sin \omega_C t \quad \textbf{3.3}$$

$$e_{\text{AM}}(t) = E_C \sin(\omega_C t) + \frac{mE_C}{2} \cos(\omega_C - \omega_M)t - \frac{mE_C}{2} \cos(\omega_C + \omega_M)t \quad \textbf{3.4}$$

$$m = \frac{E_M}{E_C} \quad \textbf{3.5}$$

$$m = \frac{E_M}{E_C} \times 100\% \quad \textbf{3.6}$$

$$m = \frac{E_{\text{max}} - E_{\text{min}}}{E_{\text{max}} + E_{\text{min}}} \quad \textbf{3.7}$$

$$m_{\text{total}} = (m_1^2 + m_2^2 + m_3^2 + \ldots + m_n^2)^{\frac{1}{2}} \quad 3.8$$

$$\text{Bandwidth} = 2 \times f_{\text{mod}} \quad 3.9$$

$$\text{Bandwidth} = 2 \times f_{\text{highest}} \quad 3.10$$

$$P_{\text{total}} = P_C + P_{\text{LSB}} + P_{\text{USB}} \quad 3.11$$

$$P_{\text{total}} = P_C\left(1 + \frac{m^2}{2}\right) \quad 3.12$$

$$P_{\text{LSB}} = P_{\text{USB}} = \frac{m^2 P_C}{4} \quad 3.13$$

$$m = 2\left(\frac{P_{\text{SB}}}{P_C}\right)^{\frac{1}{2}} \quad 3.14$$

$$v_{\text{out}} = av_{\text{in}} + bv_{\text{in}}^2 + cv_{\text{in}}^3 + dv_{\text{in}}^4 + \ldots \quad 3.15$$

$$v_{\text{out}} = av_{\text{in}} + bv_{\text{in}}^2 \quad 3.16$$

$$v_{\text{in}} = \sin\omega_1 t + \sin\omega_2 t \quad 3.17$$

$$v_{\text{out}} = a\sin\omega_1 t + a\sin\omega_2 t + \frac{b}{2}(1 - \cos 2\omega_1 t) + \frac{b}{2}(1 - \cos 2\omega_2 t) + b\left[\cos(\omega_1 - \omega_2)t + \cos(\omega_1 + \omega_2)t\right] \quad 3.18$$

$$v_{\text{out}} = av_{\text{in1}} \cdot v_{\text{in2}} \quad 3.19$$

$$v_{\text{out}} = \frac{a}{2}\left[\cos(\omega_1 - \omega_2)t + \cos(\omega_1 + \omega_2)t\right] \quad 3.20$$

$$f_{\text{image}} = f_{\text{signal}} \pm 2f_{\text{IF}} \quad 3.21$$

$$f_{\text{spurious}} = \left(\frac{m}{n}\right)f_{\text{lo}} \pm \frac{f_{\text{IF}}}{n} \quad 3.22$$

$$IR = \frac{\text{Voltage Gain at } f_{\text{signal}}}{\text{Voltage Gain at } f_{\text{image}}} = \sqrt{1 + Q^2 x^2} \tag{3.23}$$

$$x = \frac{f_{\text{image}}}{f_{\text{signal}}} - \frac{f_{\text{signal}}}{f_{\text{image}}} \tag{3.24}$$

Frequency Modulation

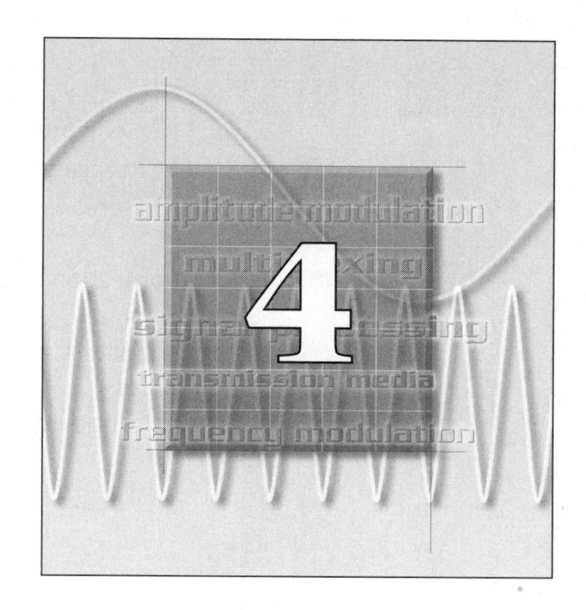

Objectives Upon completion of this chapter, the student should be able to:

- Discuss the differences between AM and angle modulation.
- Explain the advantages and disadvantages of the different analog modulation techniques.
- Discuss the relationship between FM and PM.
- Calculate modulation index, signal bandwidth, sideband frequencies and power levels, and be able to sketch the frequency spectra of an FM signal.
- Explain the effect of noise upon an FM signal and relate its effect to deviation and bandwidth.
- Discuss the typical FM transmitter and explain the difference between direct and indirect generation of FM.
- Discuss the differences between AM and FM superheterodyne receivers, including capture effect, noise threshold, and pre-emphasis and de-emphasis.
- Discuss FM demodulation techniques, including the use of a phase-locked loop (PLL).
- Describe FM stereo broadcasting and discuss typical FM-receiver specifications.

Outline

Key Terms

angle modulation
Bessel functions
capture effect
de-emphasis
deviation sensitivity
direct FM
frequency deviation
index of modulation
indirect FM
modulation sensitivity
narrow-band FM
pre-emphasis
quieting sensitivity
rest frequency
threshold effect
wideband FM

Introduction

This chapter will introduce the student to the other forms of analog modulation—frequency modulation (FM) and phase modulation (PM)—both of which are commonly known as angle modulation. Both FM and PM are used extensively in communications systems. FM is used in radio broadcasting, for the transmission of the sound signal in standard (NTSC) TV, for private land-mobile radio systems, for direct-satellite broadcasting, and for cordless and cellular telephone systems, just to name a few common applications. PM by itself and in combination with AM is used extensively in modern data-communications systems. Angle modulation has a very important advantage over AM in its ability to provide increased immunity to noise. Angle-modulation systems typically require a larger bandwidth than AM systems, a necessary trade-off for its improved resistance to noise.

Topics covered in this chapter will include:

- theory of FM operation
- frequency deviation
- modulation index
- theory of PM operation
- a comparison of FM and PM
- Bessel functions
- the spectrum of an angle-modulated signal
- signal bandwidth
- sideband power relations
- noise effects on FM

- FM-stereo multiplex operation
- frequency-modulation transmission and reception hardware

A system familiar to most readers of this chapter—standard FM broadcasting, a mature technology—will serve as the vehicle used to present this material. Finally, FM-stereo multiplex operation will be introduced as an example of an early form of combinational modulation used to send more than one signal over the same channel (multiplexing).

4.1 An Introduction to the Development of FM

This chapter will introduce the reader to the second analog form of modulation. This type of modulation scheme is known as angle modulation. **Angle modulation** can be further subdivided into two distinct types: frequency modulation (FM) and phase modulation (PM).

The history and evolution of angle modulation basically revolves around one man, Major Edwin Armstrong. Armstrong, a radio pioneer who invented first the regenerative and then the superheterodyne receiver in the 1910s, worked on the principles of frequency and phase modulation starting in the 1920s. It was not until the 1930s, however, that he finally completed work on a practical technique for wideband frequency-modulation broadcasting. For further information, visit a Web site devoted to Armstrong's work at http://users.erols.com/oldradio/.

As a historical footnote, it should be pointed out that at the turn of the last century, the very early Paulson arc transmitter actually used the simplest form of FM, frequency-shift keying (FSK), to transmit a wireless telegraph signal. With this type of wireless transmitter, a continuous electrical arc would have its fundamental output frequency altered by closing a telegraph key. When the key was closed, it would short out several turns of a tuning inductor, thus changing the transmitter output frequency. For this reason it was a form of FSK.

Despite Armstrong's efforts, the implementation of FM broadcasting was fought by RCA and NBC through 1945, only becoming popular in the United States during the late 1960s and early 1970s when technological advances reduced the cost of equipment and improved the quality of service. Many public-safety departments were early adopters of FM for their fleet communications. AMPS cellular-telephone service, an FM-based system, was introduced in the United States in 1983. Today FM is used for the legacy FM broadcast band, standard TV-broadcasting sound transmission, direct-satellite TV service, cordless telephones, and just about every type of business band and mobile-radio service. FM is capable of much more noise immunity than AM, and is now the most popular form of analog modulation.

4.2 Frequency-Modulation Theory

We will start our discussion of angle modulation by first examining frequency modulation. The classic definition of *FM* is that the instantaneous output frequency of a transmitter is varied in accordance with the modulating signal. Recall that we can write an equation for a sine wave as follows:

$$e(t) = E_P \sin(\omega t + \phi) \qquad \textbf{4.1}$$

While amplitude modulation is achieved by varying E_P, frequency modulation is realized by varying ω in accordance with the modulating signal or message. Notice that one can also vary ϕ to obtain another form of angle modulation known as phase modulation (PM). Later we will examine the relationship between FM and PM. See Figure 4-1 for a time display of a typical FM signal.

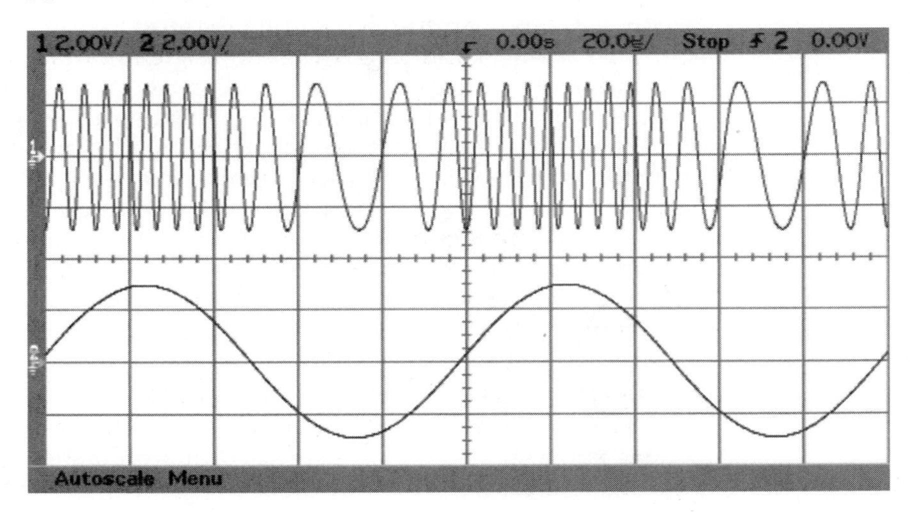

FIGURE 4-1 A typical FM signal shown with the modulating signal

Definitions

An important concept in the understanding of FM is that of **frequency deviation**. The amount of frequency deviation a signal experiences is a measure of the change in transmitter output frequency from the **rest frequency** of the transmitter. The rest frequency of a transmitter is defined as the output frequency with no modulating signal applied. For a transmitter with linear modulation characteristics, the frequency deviation of the carrier is directly proportional to the amplitude of the applied modulating signal. Thus an FM transmitter is said to have a **modulation sensitivity**, represented by a constant, k_f, of so many kHz/V,

$$k_f = \text{frequency deviation/V} = k_f \text{ kHz/V}$$

For a single modulating tone of $e_M(t) = e_M \sin(\omega_M t)$, the amount of frequency deviation is given by

$$\delta(t) = k_f \times e_M(t)$$

where $\delta(t)$ is the instantaneous frequency deviation and $e_M(t)$ represents the modulating signal. The peak deviation is given by

$$\delta = k_f \times E_M \quad \textbf{4.2}$$

where both δ and E_M are peak values.

EXAMPLE 4.1

A certain FM transmitter has a modulation sensitivity, k_f, of 10 kHz/V. If a 5-kHz sine wave of 2 $V_{p\text{-}p}$ is applied to this transmitter, determine the frequency deviation that occurs.

Solution The applied modulating sine wave varies between ±1 V at a rate of 5000 times per second; therefore, the output of the transmitter will appear to deviate ±10 kHz from its rest frequency at a rate of 5000 times per second. The term *appear* is used here to indicate that one would most likely envision a time display of the signal to visualize this operation. Mathematically, this is represented as

$$\delta = 10 \text{ kHz/V} \times E_M = \pm 10 \text{ kHz}$$

It is instructive to pause and think about the FM process at this time. As the example above indicates, for frequency modulation, the amount of frequency deviation is proportional to the amplitude of the applied modulating signal, similarly, the number of times per second that the transmitter signal deviates is equal to the frequency of the applied modulating signal. Note that the amplitude of the FM wave remains constant during the modulation process.

FM versus PM

Let us return to equation 4.1, repeated here:

$$e(t) = E_P \sin(\omega t + \phi)$$

As previously mentioned, it is possible to vary both ω (FM) and ϕ (PM) in the above equation. Each term is part of the argument of the sine wave. So, what is the difference between varying one versus the other? A simple answer is that there is no difference, as either one will change the sine wave's frequency. However, closer inspection and the employment of more mathematical rigor reveal that there are some subtle differences between the two forms of angle modulation. However, due to the complex mathematics involved these differences will not be discussed here. Practically speaking, it is possible to obtain FM from PM, as depicted in Figure 4-2; but most present-day FM systems do not generate FM by this method. This process of generating FM is known as **indirect FM**.

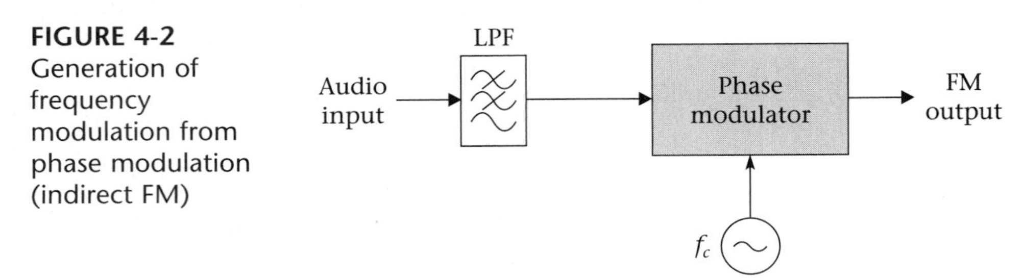

FIGURE 4-2 Generation of frequency modulation from phase modulation (indirect FM)

If we compare the two forms of modulation in the time domain, shown in Figures 4-3a and 4-3b on page 145, we would observe that the FM wave and PM wave appear quite similar; however, their timing appears to be out of sync. Indeed, the FM wave has its maximum frequency deviation during the peaks of the input signal, while the PM wave has its maximum frequency deviation during zero crossings of the input signal. Without showing the phase relationship of the input wave to the modulated wave, it would be impossible to tell the difference between the two forms of angle modulation if one simlpy looked at the resulting modulated waveform.

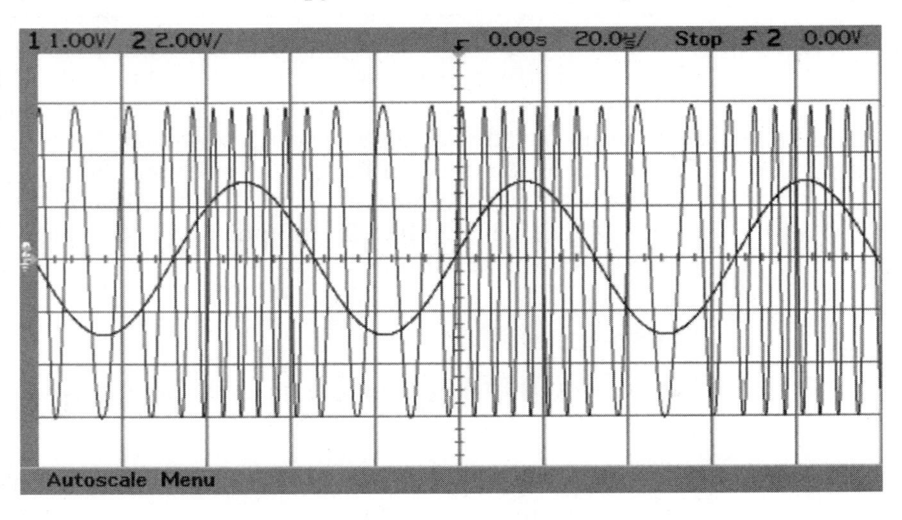

FIGURE 4-3a A frequency-modulated waveform and the modulating wave

The Subtle Difference between FM and PM

One way to tell the difference between FM and PM is to observe the following: If the instantaneous *frequency* of the signal is directly proportional to

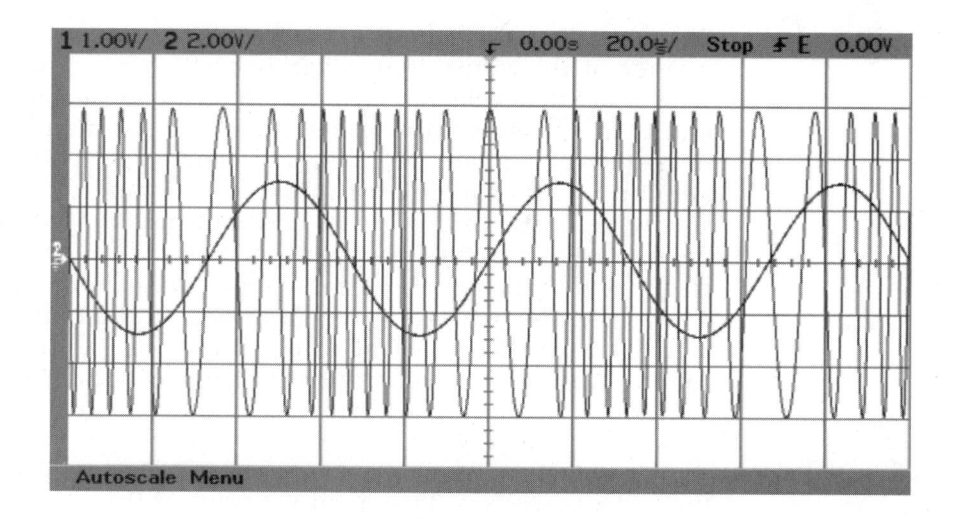

FIGURE 4-3b A phase-modulated wave and the modulating wave

the amplitude of the input signal, it is FM. On the other hand, if the instantaneous *phase* of the signal is proportional to the amplitude of the input signal, it is PM. This last statement, although correct, is unclear because the term *instantaneous phase* is undefined at present. Another way of expressing this last statement is to say that for PM, the transmitter output frequency is at the rest frequency when the input signal is at either its most positive or most negative voltage. The subtle difference between FM and PM is not really very important, for the vast majority of wireless angle modulation transmitters use FM. That being the case, our comments will focus almost entirely on FM from this point on. Several different forms of PM will be discussed in much more detail in chapter 6 in relation to digital modulation.

4.3 Mathematical Analysis of FM

As was done with AM, a mathematical analysis of a high-frequency sine wave, modulated by a single tone or frequency, will be used to yield information about the frequency components in an FM wave, FM power relations, and the bandwidth of an FM signal.

From the definition of frequency deviation, an equation can be written for the signal frequency of an FM wave as a function of time:

$$f_{\text{signal}} = f_C + k_f e_M(t) = f_C + k_f E_M \sin\omega_M t \qquad \textbf{4.3}$$

and substitution of $\delta = k_f \times E_M$ yields:

$$f_{\text{signal}} = f_C + \delta \sin\omega_M t \qquad \textbf{4.4}$$

But what does this equation indicate? It seems to be saying that the frequency of the transmitter is varying with time. This brings up the same type of problem that was observed when we looked at a time display of AM and then performed a mathematical analysis in an attempt to determine its frequency content. With AM, the signal appeared to be a sine wave thats amplitude was changing with time. At the time, it was pointed out that a sine wave, by definition, has a constant peak amplitude, and thus cannot have a peak amplitude that varies with time. What about the sine wave's frequency? It also must be a constant and cannot be varying with time. As was the case with AM, where it turned out that our modulated wave was actually the vector sum of three sine waves, a similar situation is true for FM. An FM wave will consist of three or more frequency components vectorially added together to give the appearance of a sine wave thats frequency is varying with time when displayed in the time domain.

A somewhat complex mathematical analysis will yield an equation for the instantaneous voltage of an FM wave of the form shown here:

$$e_{\text{FM}}(t) = E_C \sin(\omega_C t + m_f \sin\omega_M t) \tag{4.5}$$

where E_C is the rest-frequency peak amplitude, ω_C and ω_M represent the rest and modulating frequencies, and m_f is the index of modulation.

This equation represents a single low-frequency sine wave, f_M, frequency modulating another high-frequency sine wave, f_C. Note that this equation indicates that the argument of the sine wave is itself a sine wave.

The Index of Modulation

The **index of modulation**, m_f, is given by the following relationship:

$$m_f = \frac{\delta}{f_M} \tag{4.6}$$

A few more comments about the index of modulation, m_f, are appropriate. As can be seen from the equation, m_f is equal to the peak deviation caused when the signal is modulated by the frequency of the modulating signal; therefore, m_f is a function of both the modulating signal amplitude and frequency. Furthermore, m_f can take on any value from 0 to infinity. Its range is not limited as it is for AM.

Percentage of Modulation

At this point the question might be asked, Is there anything in the FM process similar to the percentage of modulation of an AM signal? The answer is *yes*. However, unlike AM, it has nothing to do with the index of modulation. The practical implementation of FM communication systems in a limited bandwidth-channel environment, such as cellular radio, requires a limitation upon the maximum frequency deviation to prevent adjacent channel interference. For example, the FCC's Rules and Regulations limit FM broadcast-band transmitters to a maximum frequency deviation of ±75 kHz. The maximum allowable deviation will be assigned the value of 100% modulation. Therefore, in equation form, the percentage of modulation is given by:

$$\%\ \text{Modulation} = \frac{\delta}{\delta_{\max}} \times 100\% \tag{4.7}$$

EXAMPLE 4.2

An FM broadcast-band transmitter has a peak deviation of ±60 kHz for a particular input signal. Determine the percentage of modulation.

Solution From the equation,

$$\%\ \text{Modulation} = \frac{\pm 60\ \text{kHz}}{\pm 75\ \text{kHz}} \times 100\%$$

$$\%\ \text{Modulation} = .8 \times 100\% = 80\%$$

If one were to visit a local FM radio station, this would be the reading that is monitored in the studio by the on-the-air employee.

Bessel Functions and their Relationship to FM

We now return to our mathematical analysis of the FM wave. Equation 4.5 cannot be solved with algebra or trigonometric identities. However, certain Bessel-function identities are available that will yield solutions to equation 4.5 and allow us to determine the frequency components of an FM wave. As a note of explanation, **Bessel functions** appear as solutions in numerous physical problems, quite often involving cylindrical or spherical geometries.

Without going through the rather detailed mathematics to solve equation 4.5, the resulting equation is shown here. It itemizes the various signal components in an FM wave and their amplitudes.

$$e_{FM}(t) = E_C \left\{ \begin{array}{l} J_0(m_f)\sin\omega_C t - J_1(m_f)\left[\sin(\omega_C - \omega_M)t - \sin(\omega_C + \omega_M)t\right] + \\ J_2(m_f)\left[\sin(\omega_C - 2\omega_M)t + \sin(\omega_C + 2\omega_M)t\right] - \\ J_3(m_f)\left[\sin(\omega_C - 3\omega_M)t - \sin(\omega_C + 3\omega_M)t\right] + \\ J_4(m_f)\left[\sin(\omega_C - 4\omega_M)t + \sin(\omega_C + 4\omega_M)t\right] - \ldots \end{array} \right\} \tag{4.8}$$

What this equation indicates is that there are an infinite number of sideband pairs for an FM wave. Each sideband pair is symmetrically located about the transmitter's rest frequency, f_C, and separated from the rest frequency by integral multiples of the modulating frequency, $n \times f_M$, where $n = 1, 2, 3, \ldots$. The magnitude of the rest frequency and sideband pairs is dependent upon the index of modulation, m_f, and given by the Bessel-function coefficients, $J_n(m_f)$, where the subscript n of J_n is the order of the sideband pair and m_f is the modulation index. Note that $J_n(m_f)$ is all one term and not the product of two numbers.

Several examples might provide some insight to the meaning of $J_n(m_f)$:

$J_0(1.0)$ represents the rest-frequency amplitude of an FM wave with an index of modulation equal to 1.0.

$J_1(2.5)$ is the amplitude of the first pair of sidebands for an FM wave with $m_f = 2.5$.

$J_7(m_f)$ is the amplitude of the seventh pair of sidebands with an unknown index of modulation, m_f.

Because most technology students have never dealt with Bessel functions before, equation 4.8 looks extremely complex and difficult to work with. However, after several examples, dealing with the equation should become a less daunting task.

Before considering the examples, let us see how we determine the value of the term $J_n(m_f)$. From very complex mathematics, the values of the $J_n(m_f)$ terms can be calculated from an "infinite series." Therefore, the results of the numerical computation of the values of $J_0(m_f)$, $J_1(m_f)$, $J_2(m_f)$, and so forth are usually plotted on a graph like the one shown in Figure 4-4.

There are several points that can be made about the Bessel functions plotted on the graph. For small values of m_f, the only Bessel functions with any significant amplitude are $J_0(m_f)$ and $J_1(m_f)$ (the rest frequency and the first sideband pair), while the amplitude of the higher-order ($n > 1$) sideband pairs is very small. As m_f increases, the amplitude of the rest freuency decreases and the amplitude of the higher-order sidebands increases, which would seem to indicate an increasing signal bandwidth. Further inspection of the graph indicates that as m_f keeps increasing, the sideband pairs are essentially zero amplitude until about $m_f = n$, at which point they increase in amplitude to a maximum and then decrease again. In all cases, as m_f keeps increasing, each Bessel function appears to act like an exponentially decaying sine wave. Therefore, the amplitudes of the higher-order sideband pairs eventually approach zero.

An extremely interesting point is also observed about the Bessel-function amplitudes from the graph. In all cases, including the rest frequency $J_0(m_f)$, the amplitude of the Bessel function goes to zero for numerous values of m_f, meaning that the rest-frequency component of the FM wave can disappear. This fact also holds for each pair of sidebands.

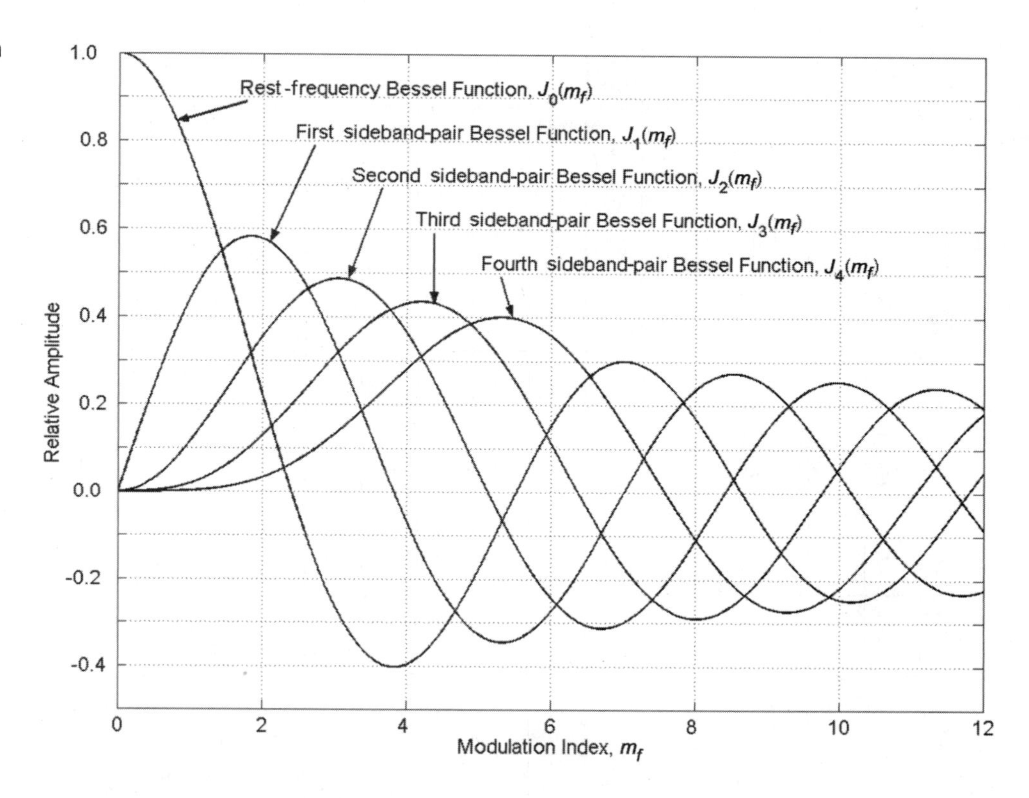

FIGURE 4-4 A graph of the Bessel coefficients

Bessel-Function Table

The information from the graph is usually put into table form for integer or fractional values of m_f. A table of Bessel-function values is shown in Table 4-1 on page 150. Note that amplitude values with minus signs represent phase shifts of 180 degrees and that amplitude values less than 0.01 have been left out of the table because they represent component frequencies with insignificant power content.

Examples of Different Values of Index of Modulation

At this time, several examples of frequency displays of FM waves with different values of m_f will be shown.

Figure 4-5 on page 151 shows the frequency-domain display of the resultant FM signal resulting from a modulating signal of 10 kHz, an index of modulation of 0.25, and rest frequency of 500 kHz. For this case, there is only one pair of sidebands with appreciable power. This type of FM signal meets the strict definition of **narrow-band FM** (NBFM), where $m_f \leq 0.5$. The bandwidth used by an NBFM signal is approximately equal to that of an AM signal. FM transmitters commonly used by business band and other mobile FM radio services for voice transmission will typically have

x	Bessel-function order, n																
m_f	J_0	J_1	J_2	J_3	J_4	J_5	J_6	J_7	J_8	J_9	J_{10}	J_{11}	J_{12}	J_{13}	J_{14}	J_{15}	J_{16}
0.00	1.00	—	—	—	—	—	—	—	—	—	—	—	—	—	—	—	—
0.25	0.98	0.12	—	—	—	—	—	—	—	—	—	—	—	—	—	—	—
0.5	0.94	0.24	0.03	—	—	—	—	—	—	—	—	—	—	—	—	—	—
1.0	0.77	0.44	0.11	0.02	—	—	—	—	—	—	—	—	—	—	—	—	—
1.5	0.51	0.56	0.23	0.06	0.01	—	—	—	—	—	—	—	—	—	—	—	—
2.0	0.22	0.58	0.35	0.13	0.03	—	—	—	—	—	—	—	—	—	—	—	—
2.41	0	0.52	0.43	0.20	0.06	0.02	—	—	—	—	—	—	—	—	—	—	—
2.5	–.05	0.50	0.45	0.22	0.07	0.02	0.01	—	—	—	—	—	—	—	—	—	—
3.0	–.26	0.34	0.49	0.31	0.13	0.04	0.01	—	—	—	—	—	—	—	—	—	—
4.0	–.40	–.07	0.36	0.43	0.28	0.13	0.05	0.02	—	—	—	—	—	—	—	—	—
5.0	–.18	–.33	0.05	0.36	0.39	0.26	0.13	0.05	0.02	—	—	—	—	—	—	—	—
5.53	0	–.34	–.13	0.25	0.40	0.32	0.19	0.09	0.03	0.01	—	—	—	—	—	—	—
6.0	0.15	–.28	–.24	0.11	0.36	0.36	0.25	0.13	0.06	0.02	—	—	—	—	—	—	—
7.0	0.30	0.00	–.30	–.17	0.16	0.35	0.34	0.23	0.13	0.06	0.02	—	—	—	—	—	—
8.0	0.17	0.23	–.11	–.29	–.10	0.19	0.34	0.32	0.22	0.13	0.06	0.03	—	—	—	—	—
8.65	0	0.27	0.06	–.24	–.23	0.03	0.26	0.34	0.28	0.18	0.10	0.05	0.02	—	—	—	—
9.0	–.09	0.25	0.14	–.18	–.27	–.06	0.20	0.33	0.31	0.21	0.12	0.06	0.03	0.01	—	—	—
10.0	–.25	0.04	0.25	0.06	–.22	–.23	–.01	0.22	0.32	0.29	0.21	0.12	0.06	0.03	0.01	—	—
12.0	0.05	–.22	–.08	0.20	0.18	–.07	–.24	–.17	0.05	0.23	0.30	0.27	0.20	0.12	0.07	0.03	0.01

TABLE 4-1 A table of Bessel Functions of the first kind

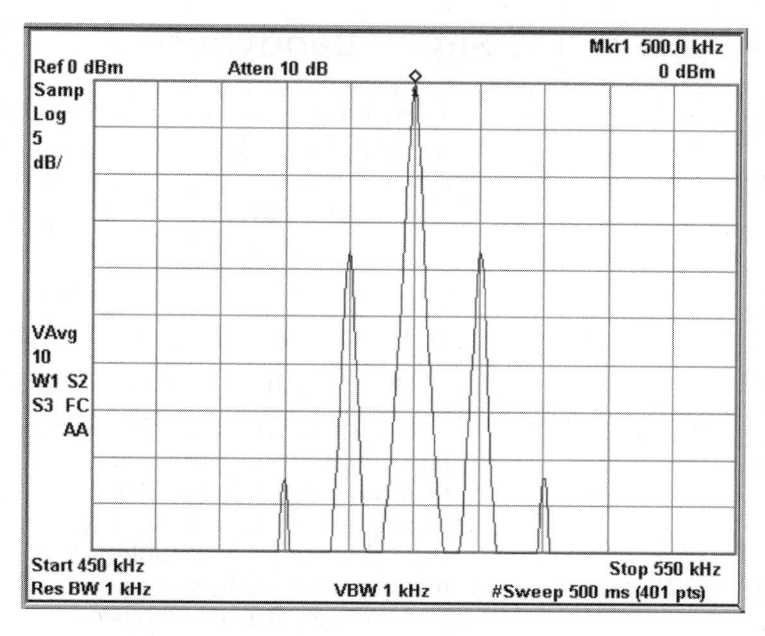

FIGURE 4-5 An SA display for $m_f = 0.25$ and $f_M = 10$ kHz

maximum frequency deviations of less than 10 kHz and $m_f \geq 0.5$. These applications do not meet the strict definition of NBFM, though they are generally referred to as NBFM systems.

For the example of an FM signal shown in Figure 4-6, with $m_f = 1.0$, there are several pairs of sidebands with appreciable power.

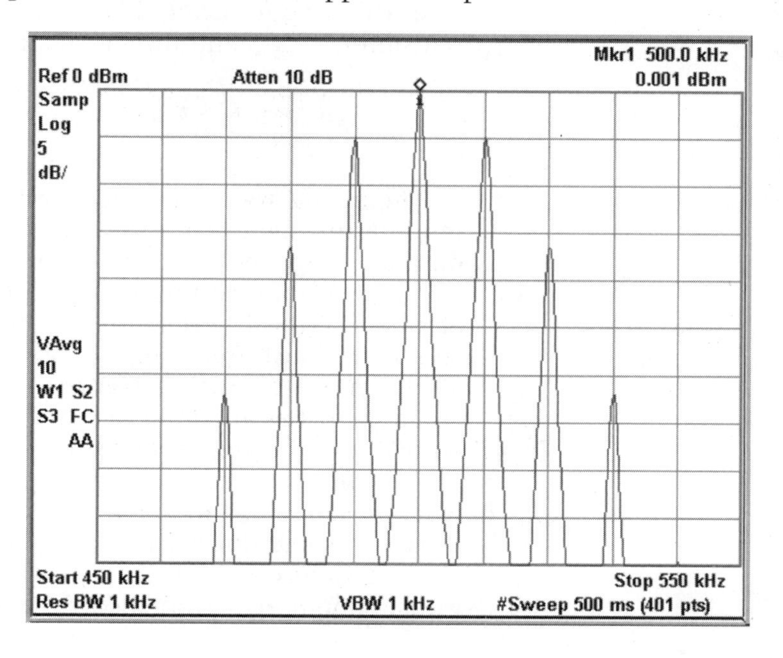

FIGURE 4-6 An SA display of $m_f = 1$ and $f_M = 10$ kHz

4.4 FM Signal Bandwidth

At this time, we should consider how to measure or determine the bandwidth of FM signals. From the SA display shown in Figure 4-6, we saw that there are three pairs of sidebands, each spaced 10 kHz apart. There are many more sideband pairs for this signal, but their amplitudes (and therefore their power content) are negligible. The total bandwidth is given by:

$$\text{Bandwidth} = f_M \times \# \text{ of sideband pairs} \times 2 \qquad \textbf{4.9}$$

or, for this example,

$$\text{Bandwidth} = 10 \text{ kHz} \times 3 \times 2 = 60 \text{ kHz}.$$

This is three times the bandwidth of an AM signal having the same modulating tone. The bandwidth of an FM signal is usually determined by the number of significant sidebands. For the example just cited, over 99% of the signal power is contained in the three pairs of sidebands. If one uses the values shown in Table 4-1, any sideband pair with a relative amplitude less than .05 could be ignored, and the other frequency components would still contain at least 99.5% of the total power in the FM signal.

We might be interested in how our SA displays correlate with the mathematical analysis presented earlier. The plot of the Bessel funnctions shown as Figure 4-4 and the table of Bessel-function values (Table 4-1) are both normalized to 1.0. This would be the relative amplitude of the rest frequency with no modulation present. For a transmitter with no modulation, the output power at the rest frequency would be equal to the power transmitter, P_{trans}, and the signal amplitude (rms value) could be found from

$$E_{\text{signal}} = \left(\frac{P_{\text{trans}}}{R}\right)^{1/2} \qquad \textbf{4.10}$$

Figure 4-7 was generated by the following sequence of steps: First, the modulating signal amplitude was reduced to zero and the rest-frequency output voltage set to a convenient value (100 mV), which also becomes the reference level for the spectrum analyzer. Then the signal was modulated at $m_f = 1.0$ ($\delta = 10$ kHz and $f_M = 10$ kHz). The sideband and rest-frequency levels were measured using the spectrum-analyzer marker function.

From the display in Figure 4-7 it can be seen that the rest-frequency component has an amplitude of 77 mV, the first pair of sidebands is about 44 mV, and the amplitude of the second pair of sidebands is approximately 11 mV (from marker measurements). Table 4-2 shows the values for $J_n(m_f)$ when $n = 0, 1, 2,$ and 3 and $m_f = 1.0$. Values are shown from Table 4-1 and those measured directly from the SA display. How do these values compare?

The values from the table and those measured with the SA are within measurement error and should help to confirm the theoretical analysis of

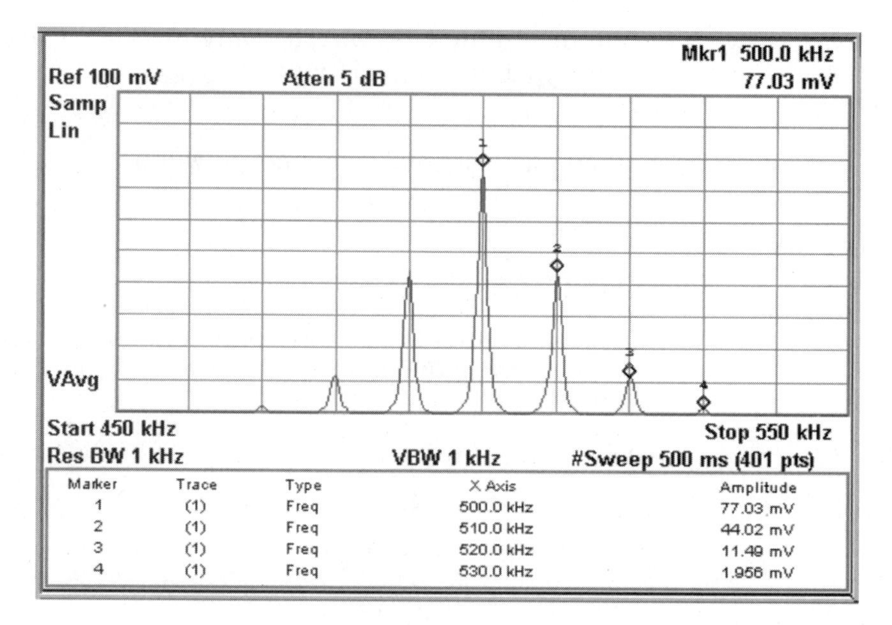

FIGURE 4-7 An SA display of $m_f = 1$ with a linear vertical scale. The reference level of the SA (top line) is set to 100 mV.

TABLE 4-2 Comparison of $J_n(m_f)$ values fromTable 4-1 and direct SA measurement

Amplitude	$J_0(m_f)$	$J_1(m_f)$	$J_2(m_f)$	$J_3(m_f)$
From Table	.77	.44	.11	.02
From SA	.7703	.4402	.1149	.01956

FM presented earlier. The next section will show how to calculate the power in a particular sideband or at the rest frequency of an FM wave.

4.5 FM Power Relations

Recall that for an FM wave the amplitude of the signal, and hence the power, remains constant. This means that the power in the individual frequency components of the wave must add up to the transmitter output power. Furthermore, if the modulation index changes, the total power must redistribute itself over the resulting frequency components.

If there is no modulation, then $m_f = 0$ and $J_0 = 1.0$. Mathematically, this can be shown by the following:

$$P_{\text{rest freq}} = J_0^2 \times P_{\text{trans}}$$

or

$$P_{\text{rest freq}} = P_{\text{trans}}$$

for $m_f = 0.0$.

To determine the power for any individual frequency component, we can use the following relation:

$$P_n = J_n^2(m_f) \times P_{\text{trans}} \tag{4.11}$$

Furthermore, the total signal power will be given by:

$$P_{\text{total}} = (J_0^2 + 2J_1^2 + 2J_2^2 + 2J_3^2 + \ldots) \times P_{\text{trans}} \tag{4.12}$$

EXAMPLE 4.3

An FM transmitter has a power output of 10 W. If the index of modulation is 1.0, determine the power in the various frequency components of the signal.

Solution From the row for $m_f = 1.0$ in Table 4-1, we have the following:

$$J_0 = .77, \quad J_1 = .44, \quad J_2 = .11, \quad \text{and } J_3 = .02$$

Using equation 4.11,

$$P_0 = J_0^2(P_{\text{trans}}) = (.77)^2 \times 10 = 5.929 \text{ W}$$

at the rest frequency,

Similarly,

$$P_1 = 1.936 \text{ W}, \quad P_2 = 0.121 \text{ W}, \quad \text{and } P_3 = 0.004 \text{ W}$$

The total power in the FM wave is the sum of all the powers of the different frequency components, therefore,

$$\begin{aligned} P_{\text{total}} &= P_0 + 2P_1 + 2P_2 + 2P_3 = 5.929 + 2(1.936) + 2(.121) + 2(.004) \\ &= 10.051 \text{ W} \end{aligned}$$

Rounding error accounts for the extra 51 mW.

Let us compare our calculations to an SA display of $m_f = 1.0$. Figure 4-8 is an appropriate SA display to use. After normalizing the rest-frequency power to the reference level, the index of modulation was set to $m_f = 1.0$ and the power of each frequency component was measured relative to the reference level using the SA marker function. This time the vertical scale is logarithmic.

One might want to compare the values measured in Figure 4-8 with those that can be calculated using equation 4.11. In each case, the signal level of the frequency components have been compared to the reference level in terms of their value of dB down from the reference. See Table 4-3.

Again, the calculated and measured values are well within measurement error and rounding error from the table values. This comparison should

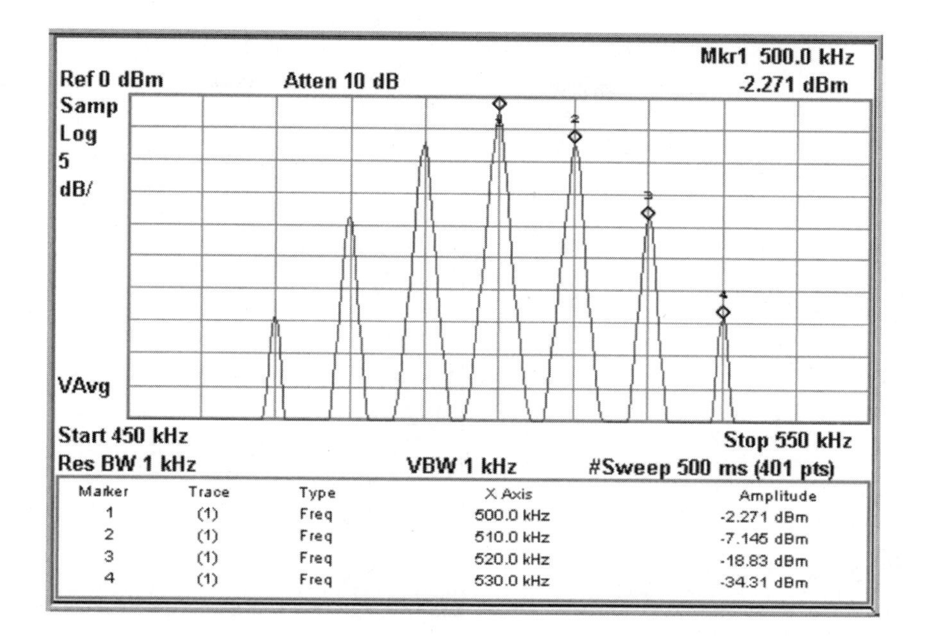

FIGURE 4-8 An SA display of an FM signal with $m_f = 1$

TABLE 4-3 A comparison of calculated and measured power levels below the reference

dB down from the reference	$J_0(m_f = 1)$	$J_1(m_f = 1)$	$J_2(m_f = 1)$	$J_3(m_f = 1)$
Calculated dB down	−2.27 dB	−7.13 dB	−19.17 dB	−33.98 dB
Measured dB down	−2.271 dB	−7.145 dB	−18.83 dB	−34.31 dB

help to confirm that the mathematical theory previously presented about FM is indeed an accurate portrayal of what is actually happening in the process know as FM!

Wideband FM

Let us look at an FM wave with a large value of m_f. See Figure 4-9 on page 156. For this signal, with $m_f = 20$, there are too many pairs of sidebands for the SA to resolve. This is a wideband signal and would fall into the category of what is known as **wideband FM** (WBFM). The modulating signal is 1.5 kHz and the bandwidth of the FM signal is approximately 60–70 kHz, *if* all frequencies are included that are less than approximately 20 dB down (–20 dB) from the apparent rest-frequency amplitude. What occurs if the modulating signal is a voice or music signal? Figure 4-10 on page 156 shows a typical SA display of an FM signal produced by a modulating music signal. There is a continuous range of frequencies in a music signal

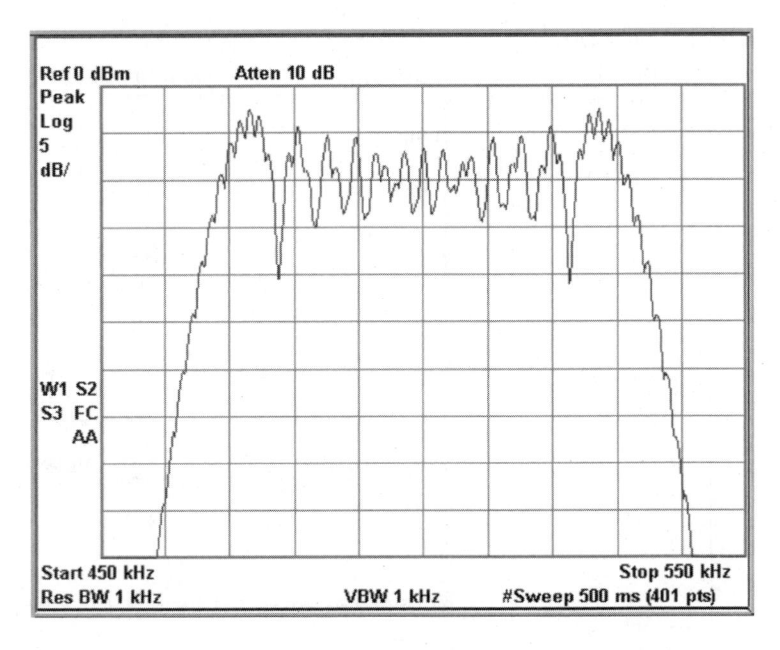

FIGURE 4-9 An SA display of an FM signal with $m_f = 20$, $\delta = 30$ kHz, and $f_M = 1.5$ kHz

and the resultant FM signal contains, for all practical purposes, a continuous range of sidebands. Figure 4-10 has a bandwidth of approximately 150 kHz. Again, to determine the signal bandwidth we have included all frequency components with an amplitude less than 20 dB down from the

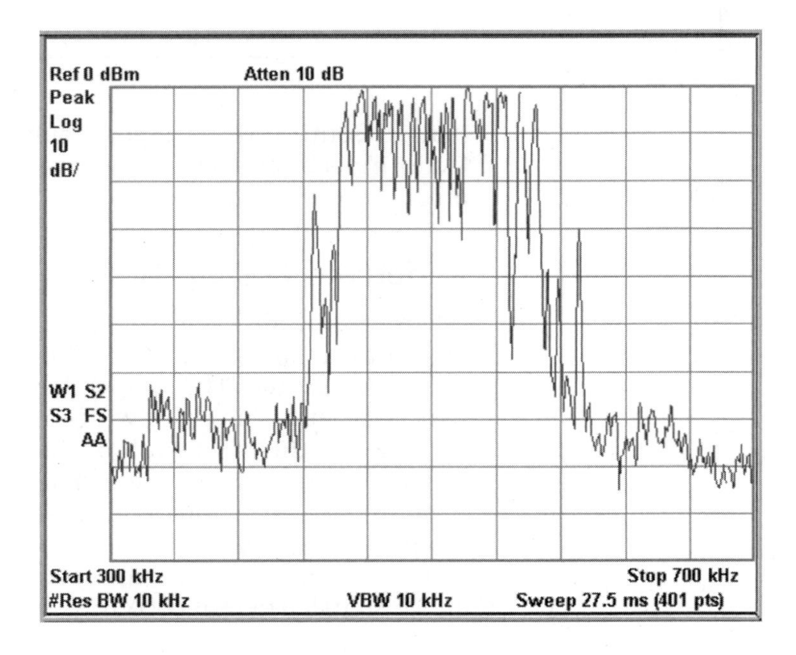

FIGURE 4-10 An FM signal spectrum. The modulating signal is music.

apparent rest-frequency amplitude. One might be tempted to ask, Why is the SA display not symmetrical? The answer lies in the finite time it takes for the analyzer trace to sweep across the screen. During the time required for the trace sweep (27.5 msec, in this case) the frequency content and amplitude of the music has changed significantly, yielding changing sideband pairs during the sweep time when the SA is acquiring its data points.

FM Rest Frequency and Sideband Nulls

As mentioned before, an interesting aspect of FM is that for certain values of modulation index the rest-frequency component of the FM wave can disappear! See Table 4-4, Column 2 for Bessel-function "zeros" of the rest frequency. A term given to the m_f zero values is *eigenvalues*.

TABLE 4-4 Zeros of the Bessel Functions

Number of Zero	$J_0(m_f)$	$J_1(m_f)$	$J_2(m_f)$	$J_3(m_f)$
0	2.41	3.83	5.14	6.38
1	5.53	7.00	8.42	9.76
2	8.65	10.17	11.62	13.02
3	11.79	13.32	14.80	16.22
4	14.93	16.47	17.96	19.41

This fact also extends to each pair of sidebands, as can be seen by looking at the Bessel-function graph that was shown in Figure 4-3. The zeros of the first three sideband pairs are tabulated in Table 4-4 in the columns to the right of the rest-frequency zeros given by $J_0(m_f)$. Of what importance is this phenomena? This information can be used to calibrate or determine the **deviation sensitivity**, k_f, of an FM transmitter using standard test equipment. Recall the relationship between the index of modulation and frequency deviation:

$$m_f = \frac{\delta}{f_M} \text{ or } \delta = m_f \times f_M$$

Using a spectrum analyzer, one can set the value of m_f to one of its eigenvalues by adjusting for a null in the amplitude of a frequency component. With an accurate frequency generator or frequency counter, one can set or measure the value of f_M and calculate the deviation sensitivity, k_f, of the transmitter in kHz/V. An example of a rest-frequency null or zero is shown in Figure 4-11 on page 158.

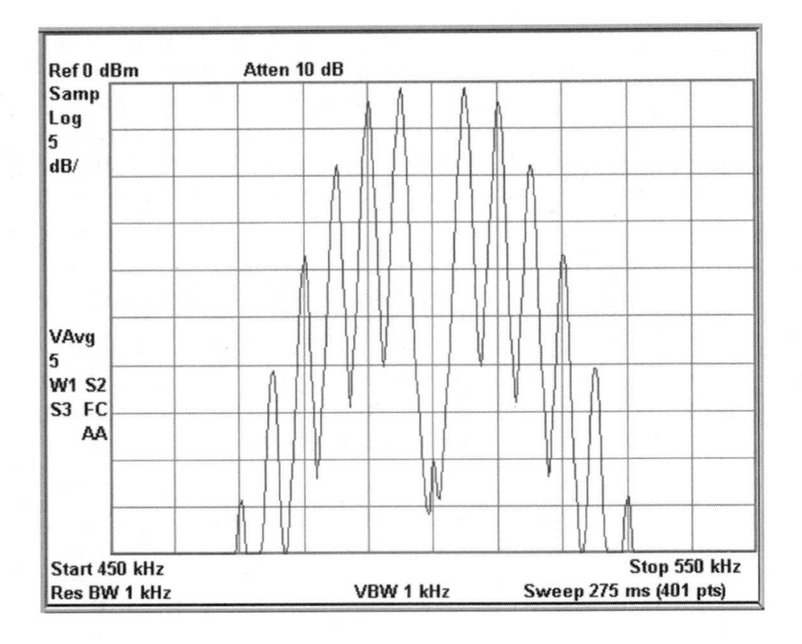

FIGURE 4-11 An FM wave with $m_f \cong 2.405$. Note the null at the center rest frequency.

As can be seen in the SA display, the frequency component at $J_0(m_f)$, the rest frequency has been "nulled" out. For this display, the modulation frequency was set to 10 kHz and the frequency deviation was set to 24.06 kHz, hence $m_f \cong 2.406$ and $J_0 \cong 0$. For any FM system, if one starts with $m_f = 0$ and gradually increases the index of modulation by raising the input-signal level, the rest-frequency component will go through a series of nulls as m_f becomes larger. The same effect will happen to the first sideband pair, then the second pair, and so on. See Figure 4-12 for an example of a sideband null.

FM Services

Before looking at the hardware used to produce and receive FM, let us examine several FM services. The first example will be the legacy FM broadcast band. Figure 4-13 is an SA display of the FM broadcast band from 88 to 108 MHz as received off air. This band uses 200-kHz channel assignments starting at 88.1 MHz and ending at 107.9 MHz. Guard bands exist between adjacent channels, and the FCC's rules regarding assignment of station frequencies will preclude any co-channel or adjacent-channel interference.

The next example is the FM business band in the 150- to 174-MHz range. Figure 4-14 on page 160 shows an SA display of the off-air signals in

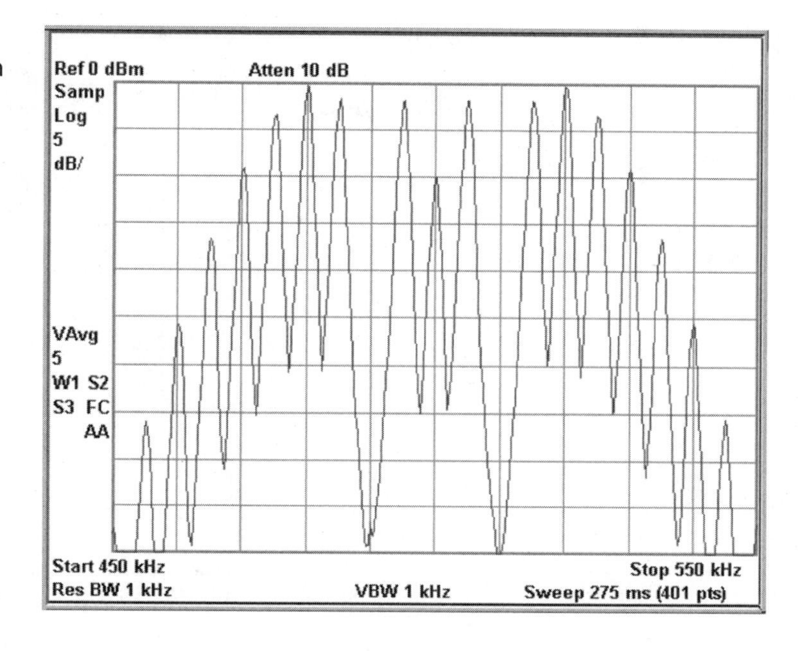

FIGURE 4-12 An FM spectrum showing a null in the second sideband pair, $m_f = 5.12$

this band. These channels are only 30-kHz wide and the transmitter deviation is limited. As users activate their microphones, these off-air signals are displayed as blips of RF energy on the SA display.

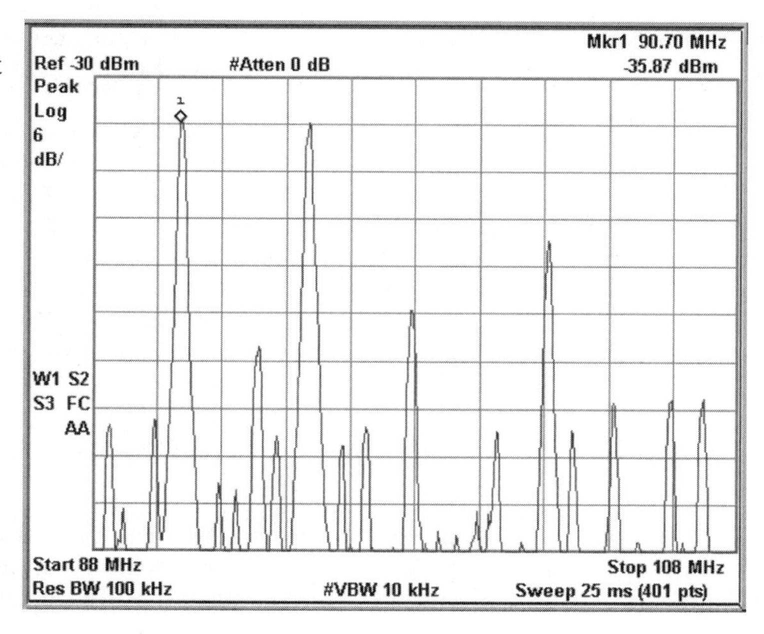

FIGURE 4-13 The FM broadcast band, off-air spectrum

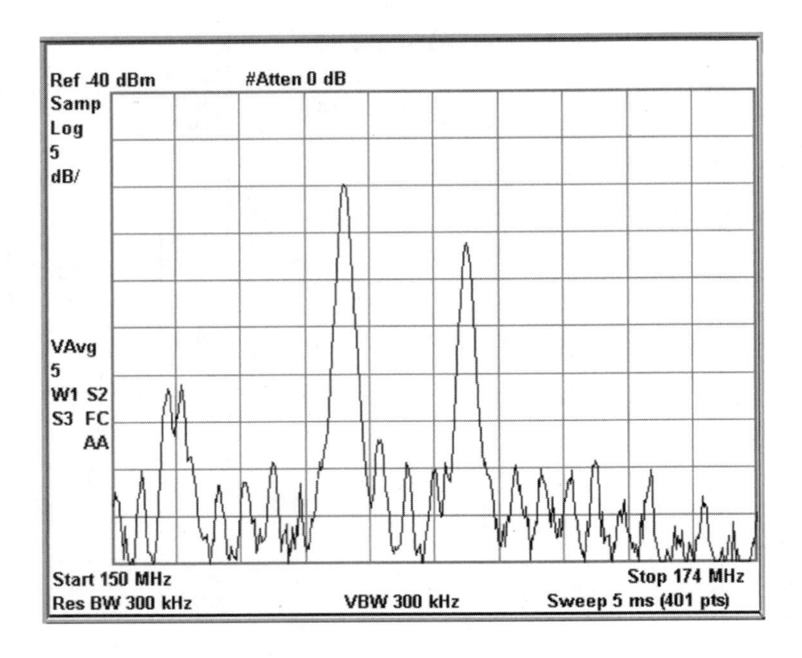

FIGURE 4-14 The FM business band, 150–174 MHz

4.6 The Effect of Noise on FM

Recall AM and the effect of noise on it. Random electrical variations added to the AM signal altered the original modulation of the signal. For FM, noise still adds to the signal, but because the information resides in frequency changes instead of amplitude changes, the noise tends to have less of an effect.

Expanding upon this idea a bit, one notes that the random electrical variations encountered by the FM signal will indeed cause distortion by "jittering" the frequency of the FM signal. However, the change in frequency modulation caused by the jittering usually turns out to be less than the change in the amplitude modulation caused by the same relative amplitude noise variations on an AM signal. Also unlike AM, the effect of the frequency jittering becomes progressively worse as the modulating frequency increases. In other words, the effect of noise increases with modulation frequency.

Pre-Emphasis and De-Emphasis

To compensate for this last effect, FM communication systems have incorporated a noise-combating system of pre-emphasis and de-emphasis. How is this done? **Pre-emphasis** gives added amplitude to the higher modulating frequencies prior to modulation under a well-defined pre-emphasis

(HPF) curve. This added amplitude will serve to make the higher frequencies more immune to noise by increasing their index of modulation. **De-emphasis** is just the opposite operation (using an LPF) and it is done at the receiver. The net effect of the two filtering processes is to cancel one another out. However, the benefit of increased noise immunity is retained. See Figure 4-15 for the filter characteristics for FM broadcasting (the filter time constant = 75 μsec).

FIGURE 4-15
Pre-emphasis and de-emphasis filter characteristics

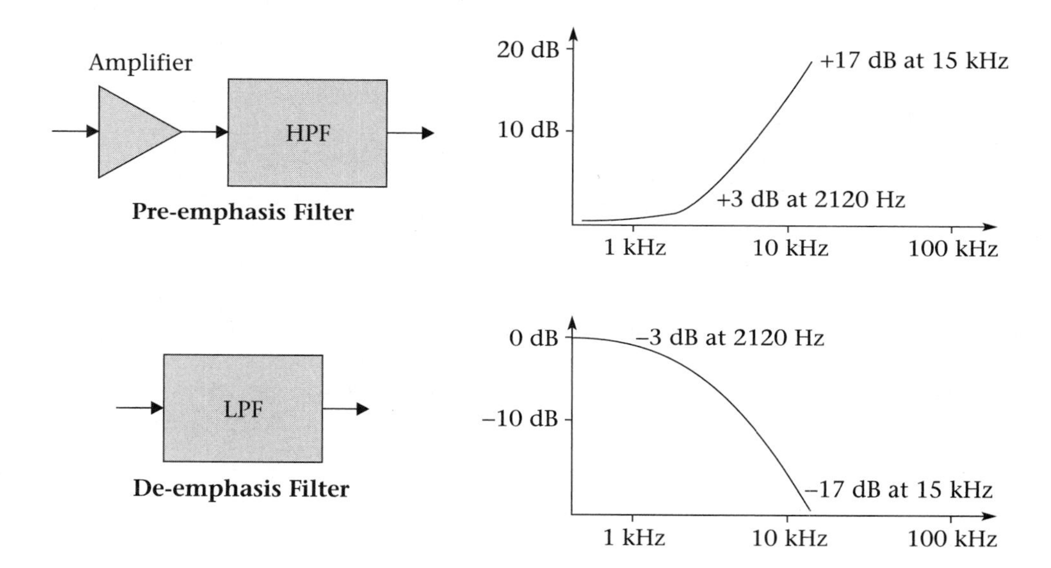

Threshold and Capture Effect

Another interesting effect of noise on FM operation is what is called the **threshold effect**. As long as there is a sufficient signal-to-noise ratio (SNR) at the input to the FM receiver, the FM system has substantially better noise performance than an AM system. However, there is a point below which FM-system performance is no longer better than AM. As a matter of fact, beyond this point performance can be even worse than AM. This effect can be traced to the use of limiters in the FM receiver. The purpose of the limiter is to remove any AM noise on the signal because it contains no information. If the signal strength is high enough, the limiter performs its function. If the input signal strength is not sufficient, the limiter does not perform its function, and the noise performance of the receiver is similar to an AM receiver.

The **capture effect** refers to the case of two co-channel or adjacent-channel FM signals being received at the same time by an FM receiver. When this occurs, the FM receiver will treat the weaker signal as interference and the stronger signal is said to have captured the receiver. The amplitude limiter in the receiver will tend to suppress the weaker signal in the same manner as it suppressed AM noise. If the two FM signals are of almost identical strength, the FM receiver will switch back and forth between the two signals. If the interfering signal is stronger than the desired signal, the FM receiver will lock onto the interfering signal. This effect can be experienced as one drives down an interstate between two metropolitan areas and receives signals from two stations that are nearby in the frequency spectrum but geographically separate.

4.7 FM Generation

The most common method used to generate FM is called **direct FM**. This method uses an active device that can be set up to implement a voltage-to-frequency (V/F) function. One such device is a varactor diode. Recall that the varactor diode is equivalent to a capacitor when reverse biased. Furthermore, the equivalent capacitance of the varactor diode varies as the reverse-bias voltage applied to it is increased or decreased. Used in conjunction with a tuned circuit, the varactor diode can convert an input-signal voltage into a varying oscillator output frequency. This type of system is often called a frequency-modulated oscillator, or FMO. See Figure 4-16.

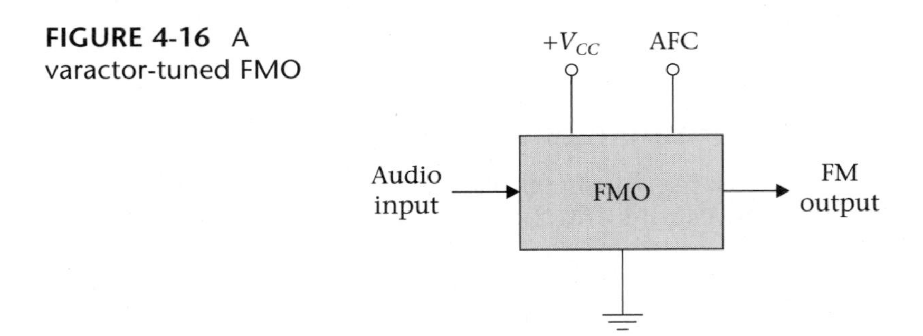

FIGURE 4-16 A varactor-tuned FMO

Generation of FM with ICs

There are currently off-the-shelf IC chips that can be used to produce low-power (up to about +20 dB or 100 mW) FM signals over a wide frequency range (typically up to 1 GHz). If power in the range of 1 W is desired, the output power of these ICs can be amplified by class-A linear amplifiers.

As mentioned in chapter 3 when discussing AM IC modulators, the semiconductor industry has evolved to offer application-specific ICs (ASICs) that are effectively systems on a chip (SOC). Manufacturers of low-power devices like cellular telephones have a host of ICs to choose from that provide complex modulation schemes with onboard amplifiers and frequency synthesizers. If higher final antenna power is desired, amplifier stages can be added. See a typical manufacturer's Web page at either http:\\www.motorola.com or http:\\www.rfmd.com for data sheets of their RF products.

Typical High-Power FM Transmitters

Figure 4-17 shows a block diagram of the most common type of FM transmitter designed for output powers in excess of 5 to 10 W. An audio signal is amplified and applied to a pre-emphasis network (HPF). It is then applied directly to a frequency-modulated oscillator (FMO), which provides a small output signal in the mW range. A buffer amplifier and successive driver amplifiers raise the output power to the required level. Radio-frequency power transistors with forced-air cooling can easily supply output powers of over 100 W and air-cooled vacuum tubes can supply kilowatts of output power at frequencies into the UHF range. Refer back to Figure 3-26 for an example. Low-power FM transmitters (approximately 20 to 50 W maximum) are sometimes called FM exciters if they are used to drive high-power amplifiers. In either case, the FM transmitter will use driver or intermediate power amplifiers (IPAs) and a final power amplifier (FPA) to achieve their final output power. These high-power amplifier stages can be run as class-C amplifiers to obtain higher efficiency.

FIGURE 4-17 A typical high-power FM transmitter

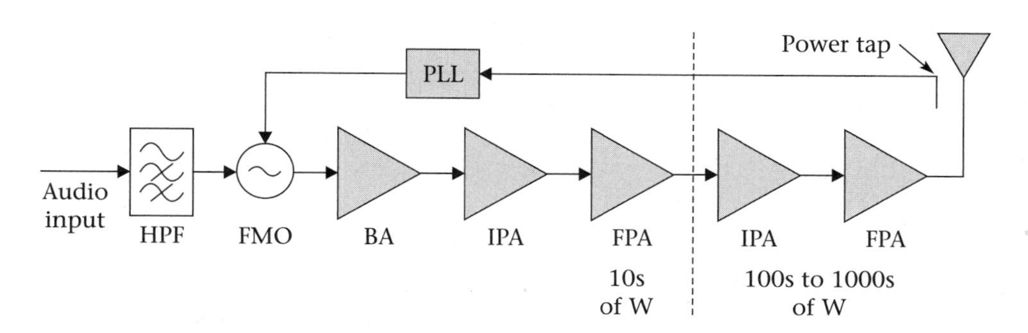

To set the FM-transmitter rest frequency, the FM transmitter will normally incorporate a programmable-PLL frequency synthesizer that can cover the entire band of operation for the particular type of service (i.e. the FM broadcast band is from 88.1 to 107.9 MHz by 200 kHz steps). The average transmitter rest frequency will be compared to the PLL output frequency and an error voltage will be applied to the FMO to correct any frequency drift from the assigned transmitter output frequency.

4.8 Frequency Modulation Receivers

Let us now take a look at a typical FM broadcast-band receiver, as pictured in Figure 4-18. The FM receiver is very similar to the AM superhet receiver, with only slight modifications. The FM receiver will usually always have an RF amplifier (usually a low-noise MOSFET transistor) and a separate mixer and local oscillator. These three subsystems, if constructed with discrete components, will usually be shielded to prevent reradiation of high-frequency energy from the receiver and to prevent oscillations resulting from unwanted feedback. The intermediate frequency (IF) will usually be the standard value of 10.7 MHz with a bandwidth of approximately 200 kHz to allow for a WBFM signal to be amplified and passed on to the FM-detector subsystem. Notice that because of the IF frequency chosen, image frequencies are 21.4 MHz from the desired signals, which puts them outside the 20-MHz WBFM broadcast band. As a result of this, a large part of the superhet-receiver image problem due to other FM stations has been eliminated. The detector will be quite different from the type used in the AM receiver and must reverse the process that occurred at the transmitter. Therefore, a frequency-to-voltage (F/V) converter is needed to demodulate the received signal.

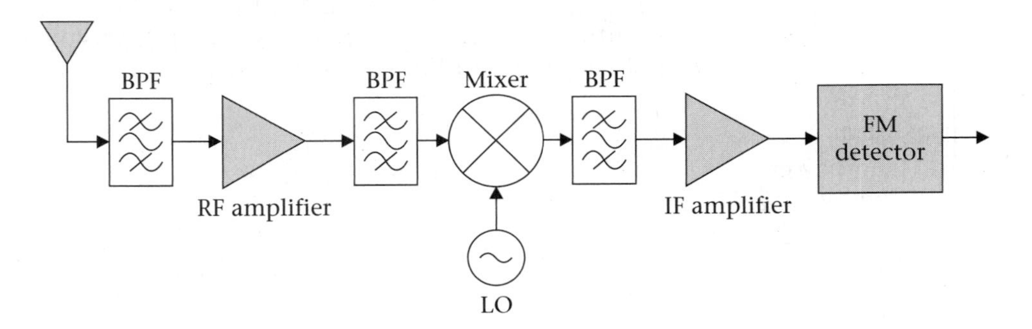

FIGURE 4-18 A typical FM superhet receiver

Review of the Superhet's Characteristics

Before examining the FM detector in more detail, we should remind ourselves of some of the characteristics of superhet operation. After selection of the desired signal by the BPFs used for tuning, the signal is amplified by the RF amplifier. All received signals are then translated in frequency to one set, intermediate frequency (10.7 MHz) by the mixer in conjunction with the local oscillator. The signal at 10.7 MHz is amplified by the IF stages and then sent to the detector stage. The detector type depends upon the kind of analog modulation being received.

The superhet receiver has specifications of sensitivity, selectivity, distortion, dynamic range, and spurious responses. Superhet receivers often have enhancements such as double conversion, notch filters, squelch, and signal-strength indicators, depending upon the application. For FM receivers, sensitivity is usually given in terms of "useable sensitivity." This is the sensitivity required for a certain value of SINAD. For portable FM equipment, the sensitivity is specified for a 12-dB SINAD. This is sufficient for voice communications, but for FM broadcast receivers, the sensitivity is specified for a 30-dB SINAD. The other specification peculiar to FM receivers is that of quieting sensitivity. This is a measure of the effectiveness of the FM receiver's ability to reduce noise in the presence of a signal.

The FM Detector

A typical modern FM receiver uses a phase-locked loop (PLL) subsystem as the detector. The PLL is insensitive to amplitude variations and can perform the F/V function; it can therefore be used as an FM detector. The FM stereo receiver must also include further demultiplexing circuitry, which we will discuss later. A PLL detector is shown in Figure 4-19.

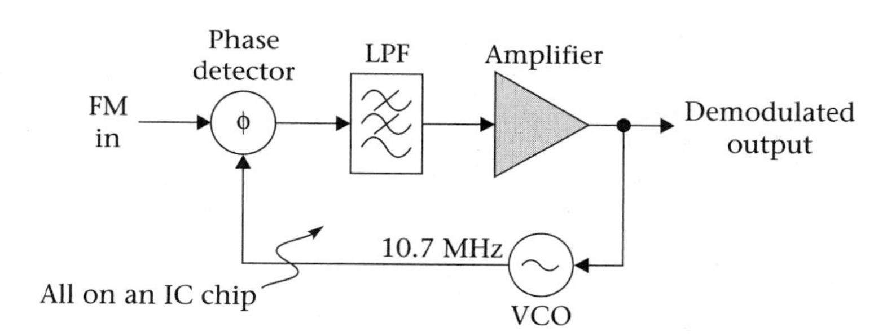

FIGURE 4-19 A typical FM detector using a phase-locked loop

The PLL detector works in the following manner: The voltage-controlled oscillator (VCO) free runs at the 10.7-MHz intermediate frequency. The incoming signal, if unmodulated, locks up with the VCO signal, causing there to be no signal (ac) output from the LPF stage.

Now let us assume that the incoming signal has been modulated by a single audio tone. The phase detector will output an error voltage to the VCO in an attempt to drive the VCO into lock-up with the incoming signal. Because the incoming signal frequency is deviating both above and below the 10.7-MHz rest frequency at a certain number of cycles per second, the VCO will do the same, following the input-signal frequency variations. The error voltage from the LPF, which drives the VCO, will be identical to the original modulating signal, and hence is taken as the demodulated output.

4.9 FM Stereo

FM stereo broadcasting was introduced during the early 1960s. The scheme that was adopted was chosen to be compatible with the monaural FM radios that were in existence at the time. Essentially, the system performs the multiplexing of two signals and further combines them into a complex baseband signal that modulates the FM carrier. Figure 4-20 shows a block diagram of the typical legacy analog-stereo generator used to drive an FM transmitter. A left and right source of information are first pre-emphasized and then fed to adder circuits. The output of one adder is the sum of the two signals, or the L + R signal (the monaural signal), and the output of the other adder is the difference of the two signals, or L – R. The L – R signal is applied to a balanced modulator along with a 38-kHz signal. The output of the balanced modulator is a DSBSC AM signal centered at 38 kHz. A portion of the 38-kHz signal is divided in frequency to become 19 kHz, and all three signals are applied to a summer/adder circuit at the output of the generator.

The resulting stereo-generator output-signal spectrum is shown in Figure 4-21. The L + R signal, which contains baseband frequencies from near 0 Hz to 15 kHz, occupies that portion of the frequency spectrum. There are guard bands around the 19-kHz tone, which is called a pilot subcarrier signal, that will be used at the receiver to aid in the demodulation of the received signal. The L – R signal, which has been DSBSC amplitude modulated by a 38-kHz tone, occupies the frequency range from 23 to 53 kHz. A state-of-the-art stereo generator will use digital techniques to synthesize the 19- and 38-kHz tones and the composite final baseband signal. The stereo-demodulator block diagram is shown in Figure 4-22.

FIGURE 4-20 A typical stereo-generator block diagram

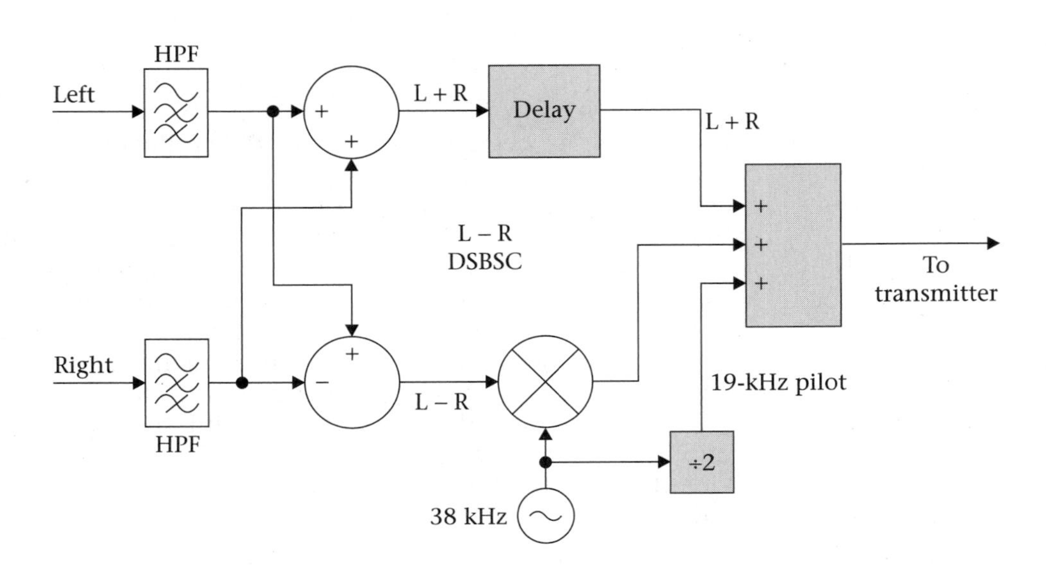

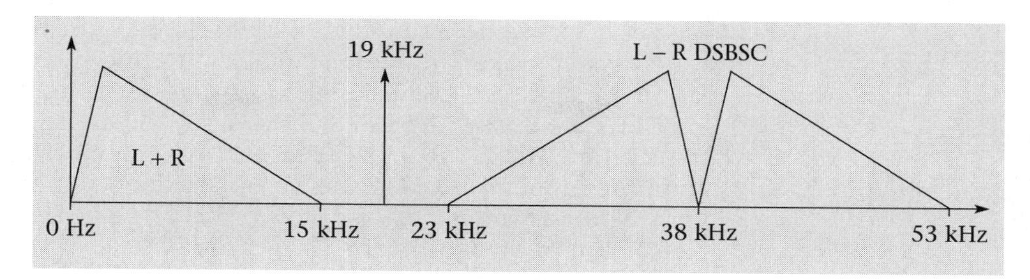

FIGURE 4-21 The output spectrum of a stereo generator

The composite signal, as shown in Figure 4-21, is recovered at the FM detector after transmission and reception and then fed to the stereo decoder shown in Figure 4-22. Usually, this is a single IC chip or part of the detector IC chip.

The composite signal is separated by two BPFs and one LPF into three separate signals. The L + R signal, which occupies the 0- to 15-kHz range does not undergo any additional signal processing, with the exception of the addition of a small time delay. The 19-kHz pilot tone is recovered by a narrow BPF centered at 19 kHz. This signal undergoes a frequency doubling to 38 kHz, and is then applied to a balanced modulator. The L – R DSBSC AM signal is recovered by another BPF centered at 38 kHz, and it is also applied to the previously mentioned balanced modulator. The balanced-modulator output consists of the original left-minus-right (L – R) signal. At this point the L + R and L – R signals are applied to adder circuits (sometimes referred to as a matrix) that yield the separate left and right signals. The signals are then fed to identical audio amplifiers.

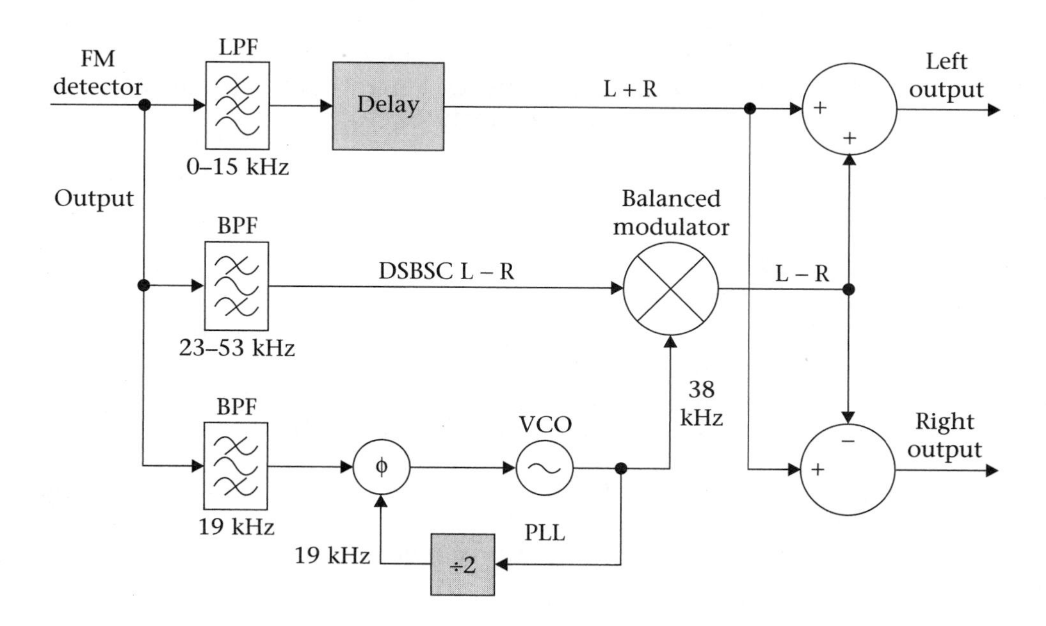

FIGURE 4-22 A typical stereo-demodulator block diagram

Let us pause here to reflect upon what has just been explained. The FM stereo system takes an audio signal (L – R) and performs amplitude modulation on it. A composite signal consisting of the unmodulated L + R signal, a 19-kHz tone, and the DSBSC AM L – R signal are now applied to an FM transmitter. The L – R signal has been modulated twice! First, it was DSBSC amplitude modulated, and then it was frequency modulated. Is this a problem? The answer is *no*. To recover the original signal, one must simply reverse the prior steps in the correct order.

At the receiver the FM detector reverses the FM process, yielding exactly the same signal that was produced by the stereo generator at the transmitter. The DSBSC AM L – R signal and a 38-kHz tone are then applied to a balanced modulator and the two signals are effectively mixed together, yielding the difference frequencies, which are the original left-right signal. As explained before, adder circuits then convert the L + R and L – R signals into just the left and right signals. The demodulation process has first reversed the FM process and then the DSBSC AM process to recover the original information. The question might be asked, Is there a limit to the number of times a signal can be modulated and to the type of modulation used? The simple answer is *no*. However, each time a signal undergoes the modulation-demodulation process, noise and nonlinear effects are introduced into the signal. So practically speaking, there is a finite limit to the number of times a signal can be modulated before distortion becomes discernible.

Several manufacturers produce ICs that are complete systems on a chip (SOC). Shown in Figure 4-23 is a simplified block diagram of an IC chip that can be used anywhere on the planet to receive any form of AM or FM broadcasting service presently being used. It also has the ability to receive specialized weather broadcasts and medium-wave transmissions. It contains a dual-conversion AM receiver, an FM receiver, a PLL frequency synthesizer, and many other features.

Summary

The process of producing FM involves the instantaneous variation of the output frequency of a transmitter in accordance with the modulating signal. This process produces an infinite number of sidebands, which are related to the modulating frequency. The number of significant sidebands, and the power distribution over the sidebands, is dependent upon the modulating frequency and the amount of frequency deviation. For specific values of the index of modulation, one can use Bessel-function tables to determine the sideband amplitude distribution and the signal bandwidth. In general, unless the index of modulation is restricted, FM signals

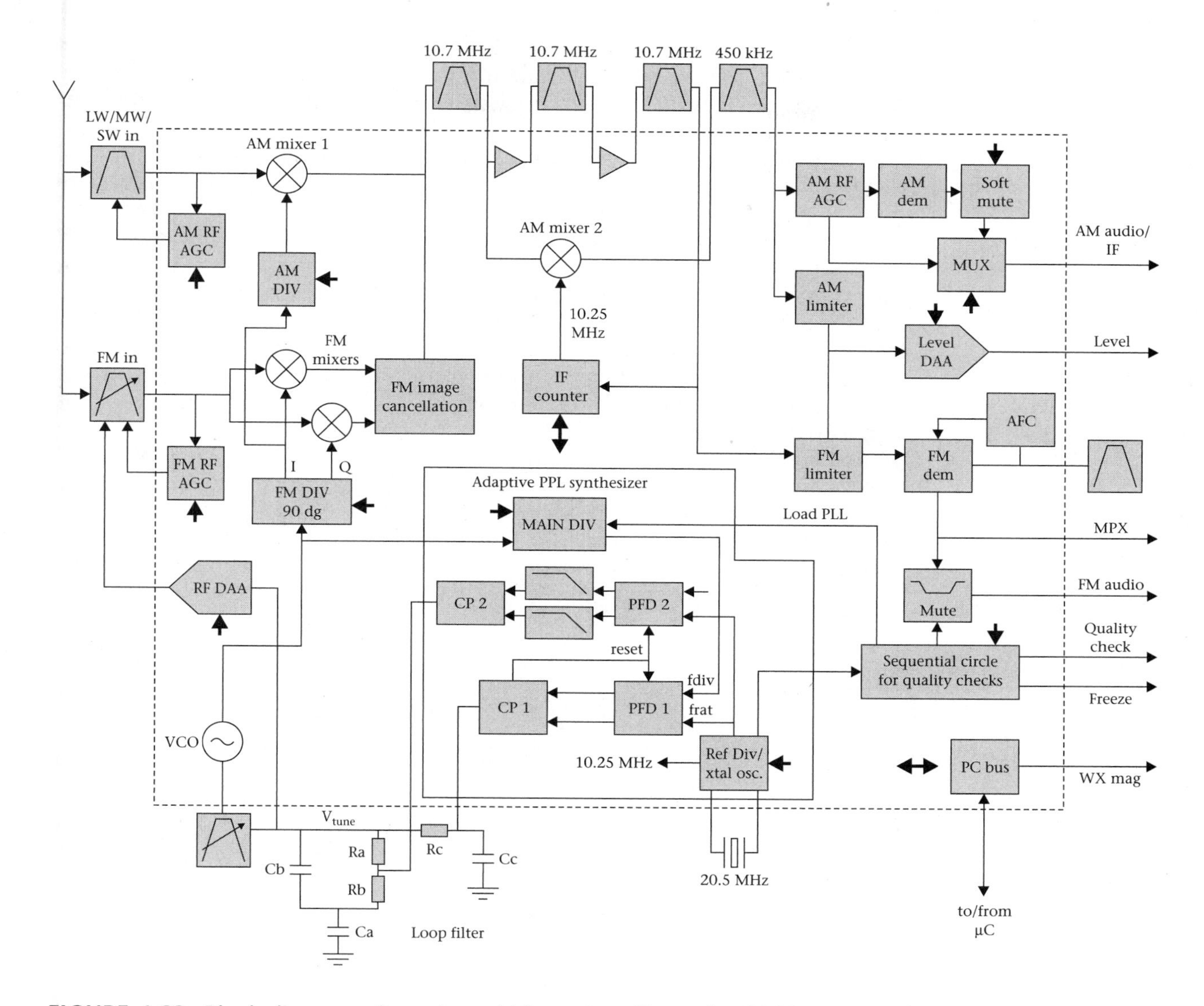

FIGURE 4-23 Block diagram of a universal IC receiver (*From the ISSCC Digest,* Volume 41 © 1998 IEEE)

can be many times wider than the baseband signals that performed the modulation in the first place.

The usual method of producing an FM signal is to use a frequency-modulated oscillator (FMO) that converts an input voltage into a frequency deviation. At the heart of these systems (buried on an IC) is the varactor diode. This device is used to perform the voltage-to-frequency conversion needed for the production of FM. If a high-power FM signal is needed, one simply amplifies the low-power FM signal (repeatedly if necessary) to obtain the desired power. FM transmitters require some form of frequency control for stability. This is usually achieved through the use of a PLL synthesizer and a feedback circuit to correct for any frequency drift. Today, ASICs are used in the design of transmitter/receiver hardware utilizing FM as the modulation technique. If large amounts of transmitting power are needed, power amplifiers are used.

To receive FM, a superhet receiver is used. This receiver will have an RF or low-noise amplifier (LNA) stage, separate mixer and local-oscillator stages, an IF at 10.7 MHz, and an FM detector. The detector is usually a PLL that does a frequency-to-voltage conversion of the received signal. The local oscillator will usually employ a PLL frequency synthesizer to develop the local-oscillator signal and supply front-panel display information that indicates the receiver's current tuning status. Additional features, such as signal-strength indicators, are added on an as needed basis.

FM stereo is a system used with legacy FM broadcasting and TV sound transmission that allows two separate audio signals to be sent between the transmitter and receiver. This multiplexing system allows for the transmission of stereophonic information and enhances the information by adding a spatial quality to the audio signal.

Questions and Problems

Section 4.1

1. Do an Internet search on Major Edwin Armstrong. Write a short paragraph about his legal battles with RCA over the new modulation technology, FM.

Sections 4.2 and 4.3

2. Describe the difference between amplitude modulation and frequency modulation.
3. What does the term *modulation sensitivity* refer to?
4. How does FM embed the modulation signal's information into the FM carrier wave?
5. What is the meaning of the FM index of modulation, m_f, and how does it vary from the AM index of modulation?
6. Define the FM *percentage of modulation*.
7. What are Bessel functions and what is their relationship to FM?
8. What is the relationship of Bessel functions to the frequency spectrum of an FM wave?
9. Draw a sketch of the frequency spectra of an FM wave with $m_f = 2.5$.
10. Explain the difference between NBFM and WBFM.

Sections 4.4 and 4.5

11. Draw the spectrum of an FM wave with a modulating signal of 5 kHz and an index of modulation of 2. Determine the bandwidth of this signal.
12. What is different about the rest frequency of an FM wave and the carrier frequency of an AM wave?

13. What eventually happens to the sideband amplitude of an FM signal as the sidebands become further removed from the rest frequency?

14. At what index of modulation does the second pair of sidebands in an FM wave disappear?

15. Prove that all the individual sideband powers add up to the total power for an FM wave with $m_f = 2$.

Sections 4.6 to 4.8

16. Describe why FM is more resistant to noise than AM.

17. Describe the process of pre-emphasis and de-emphasis for FM modulation. What does it accomplish?

18. Describe several differences between FM and AM superhet receivers. Why are these differences needed?

19. Why is the IF frequency of an FM receiver so high?

20. Explain the purpose of the stereo pilot carrier at 19 kHz.

Chapter Equations

$$e(t) = E_P \sin(\omega t + \phi) \qquad 4.1$$

$$\delta = k_f \times E_M \qquad 4.2$$

$$f_{\text{signal}} = f_C + k_f e_M(t) = f_C + k_f E_M \sin \omega_M t \qquad 4.3$$

$$f_{\text{signal}} = f_C + \delta \sin \omega_M t \qquad 4.4$$

$$e_{\text{FM}}(t) = E_C \sin(\omega_C t + m_f \sin \omega_M t) \qquad 4.5$$

$$m_f = \frac{\delta}{f_M} \qquad 4.6$$

$$\% \text{ Modulation} = \frac{\delta}{\delta_{\text{max}}} \times 100\% \qquad 4.7$$

$$e_{FM}(t) = E_C \left\{ \begin{array}{l} J_0(m_f)\sin\omega_C t - J_1(m_f)\left[\sin(\omega_C - \omega_M)t - \sin(\omega_C + \omega_M)t\right] + \\ J_2(m_f)\left[\sin(\omega_C - 2\omega_M)t + \sin(\omega_C + 2\omega_M)t\right] - \\ J_3(m_f)\left[\sin(\omega_C - 3\omega_M)t - \sin(\omega_C + 3\omega_M)t\right] + \\ J_4(m_f)\left[\sin(\omega_C - 4\omega_M)t + \sin(\omega_C + 4\omega_M)t\right] - \ldots \end{array} \right\} \quad \textbf{4.8}$$

$$\text{Bandwidth} = f_M \times \#\text{ of sideband pairs} \times 2 \quad \textbf{4.9}$$

$$E_{\text{signal}} = \left(\frac{P_{\text{trans}}}{R}\right)^{\frac{1}{2}} \quad \textbf{4.10}$$

$$P_n = J_n^2(m_f) \times P_{\text{trans}} \quad \textbf{4.11}$$

$$P_{\text{total}} = (J_0^2 + 2J_1^2 + 2J_2^2 + 2J_3^2 + \ldots) \times P_{\text{trans}} \quad \textbf{4.12}$$

Pulse Modulation

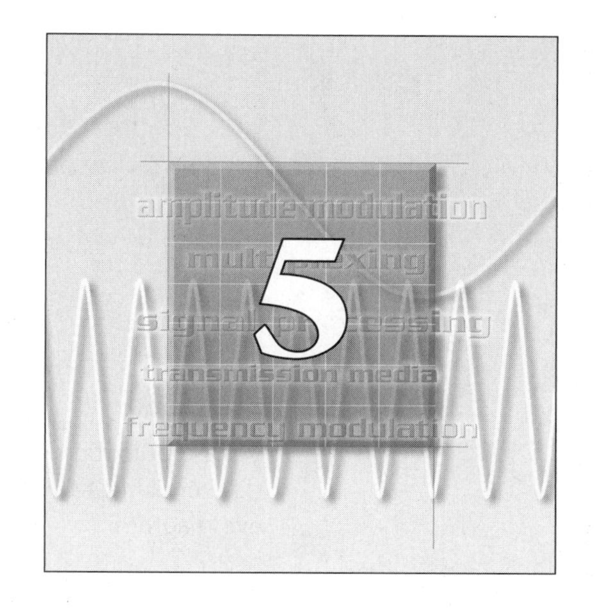

Objectives Upon completion of this chapter, the student should be able to:

- Compare analog and digital communication techniques and discuss the relative merits of each.
- Calculate channel information capacity and discuss the Nyquist sampling rate.
- Describe the common types of analog pulse-modulation schemes.
- Describe pulse-code modulation (PCM), including the concepts of quantizing levels, resolution, bit rate, and adaptive PCM.
- Describe the coding and decoding of PCM.
- Discuss the merits and disadvantages of common-line codes.
- Discuss the effect of noise on PCM.
- Describe delta modulation (DM) and adaptive DM.
- Discuss the frequency spectra of a pulse-modulated signal.
- Discuss the use of TDM for digital signals.
- Discuss the use of PCM in the present telephone system.
- Discuss digital audio- and video-compression techniques.

Outline

Key Terms

adaptive DPCM
A-law
aliasing
bipolar
bit time
companding
delta modulation
differential PCM
eye diagram
intersymbol interference
line codes
MPEG
non-return-to-zero
periodic
quantizing levels
regenerator
repeater
resolution
sample-and-hold
sampling theorem
signal-to-quantizing-noise ratio
sinc pulse
spectral fold-over
μ-law
unipolar

Introduction

Increasingly, signals used in modern telecommunications systems are digital in nature. With increased computer-to-computer traffic and the digitizing of analog signals to obtain both improvements in noise immunity and bandwidth reduction through compression, we are seeing a shift to an all-digital telecommunications-network infrastructure that delivers voice, data, and multimedia content in a seamless fashion. This chapter will look mainly at the digital transmission of analog signals; chapter 6 will examine the digital transmission of both digitized analog signals and data. The advantages of digital transmission will be reviewed, the required sampling rates discussed, the concept of a repeater introduced, and channel capacity analyzed. Topics covered in this chapter include analog forms of pulse modulation, pulse-code modulation, signal companding, delta modulation, the coding and decoding of PCM signals, line-encoding techniques, intersymbol interference, eye patterns, and pulse-modulation frequency spectra. Additionally, the student will briefly be introduced to the T1 carrier system and given a short overview of digital-compression techniques.

5.1 Introduction to Pulse-Modulation Theory

Recall the classic analog-modulation forms of AM, FM, and PM. A short review of AM and FM will be given before we look at pulse-modulation techniques. A time-domain display of amplitude modulation is shown in Figure 5-1 and a frequency-domain display is shown in Figure 5-2.

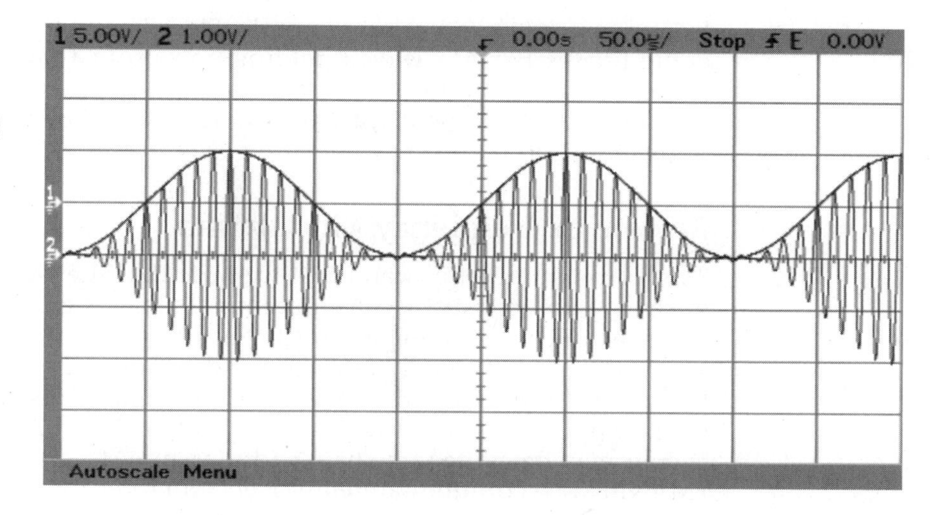

FIGURE 5-1 A time-domain display of a modulating signal and the resulting AM waveform

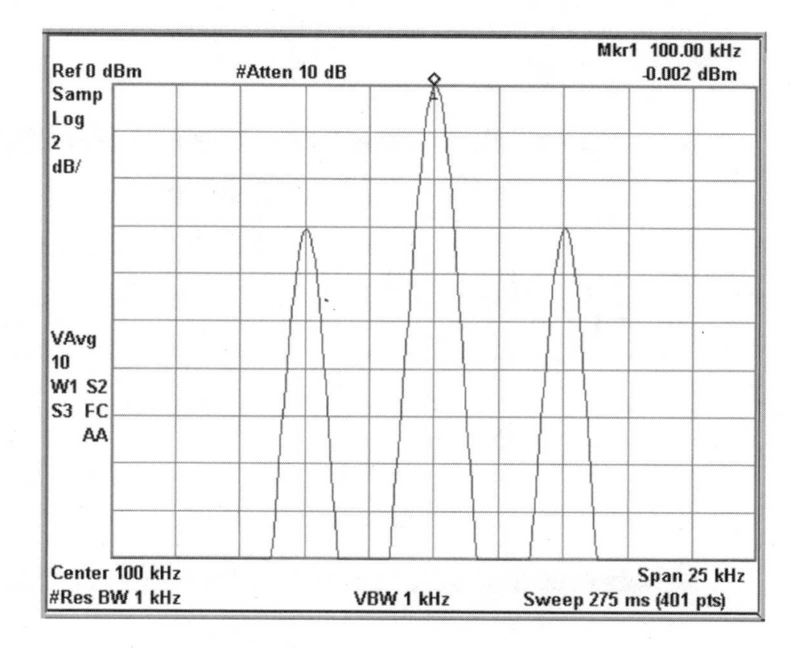

FIGURE 5-2 A frequency-domain display of an AM signal

Review of Amplitude Modulation

Note that the AM process produces a signal consisting of the carrier frequency and two symmetrical sidebands. The total bandwidth of this AM signal is two times the frequency of the modulating signal and the information about the modulating signal's amplitude and frequency is continuously embedded in the sideband characteristics (i.e., frequency separation from the carrier and relative amplitude). The receiver in an AM system demodulates the AM wave, yielding an output that is a combination of the received signal and the noise added to the signal from the transmission process. Additionally, there is noise and distortion introduced by the transmitter/receiver pair.

Review of Frequency Modulation

As shown in Figures 5-3 and 5-4, we note that the FM (or PM) process produces a signal consisting of numerous sidebands and a continuously varying spectral component at the transmitter rest frequency, f_C. The total bandwidth of the signal varies and is dependent upon the index of modulation, m_f. In any case, the FM-signal bandwidth is usually many times that of the applied modulating frequency if wideband FM is employed. However, it may have the same bandwidth as an AM signal if strict narrow-band FM (NBFM) techniques are employed. In either case, the noise introduced to an FM system during the transmission and reception of the signal has less effect upon the FM signal than an equivalent AM signal if the signal power is above a certain threshold value.

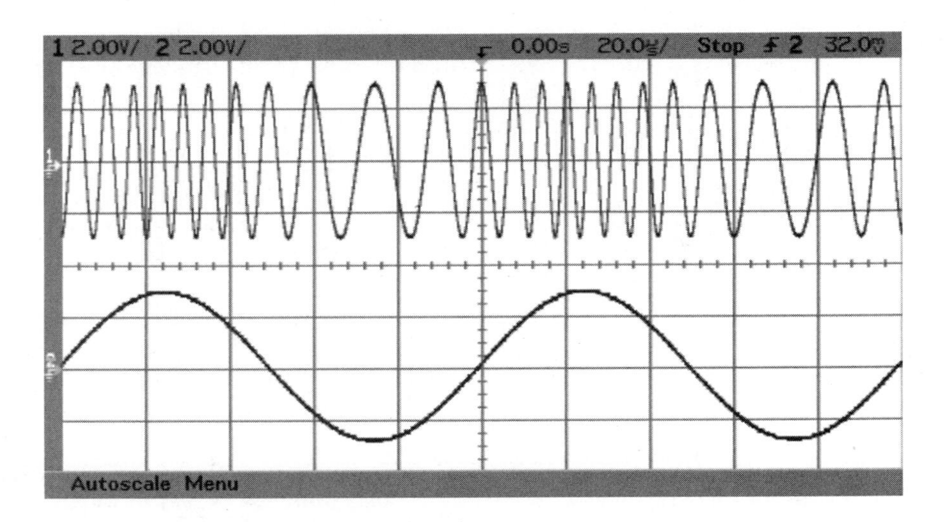

FIGURE 5-3 The time-domain display of an FM signal

5.2 Analog Pulse Modulation

Analog pulse modulation is implemented by varying some characteristic of a regularly occurring pulse signal with the modulating signal. This means that we have to sample the modulating signal and impress the acquired

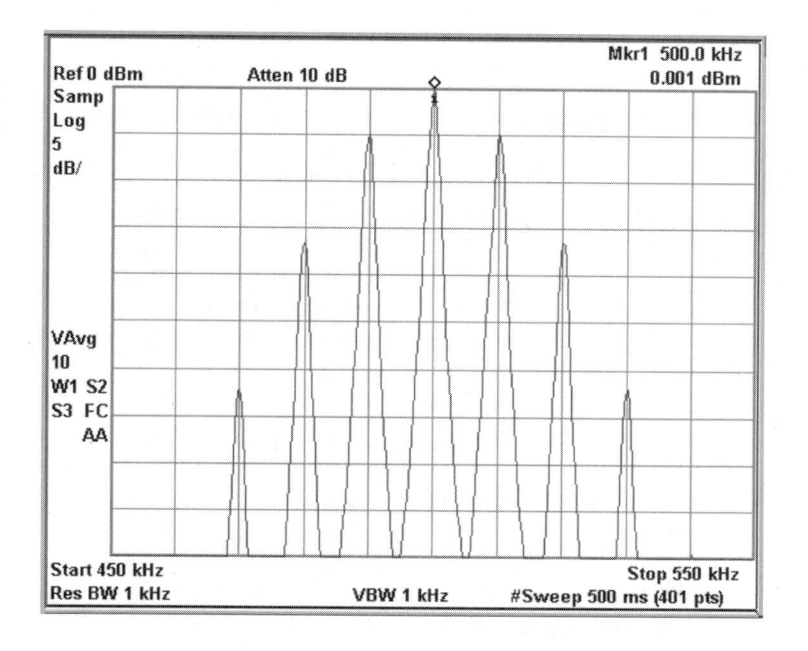

FIGURE 5-4 The frequency-domain display of a typical FM signal

sample values upon the pulses. Typically, a low duty cycle pulse train is used to carry the information through the communications channel. See Figure 5-5 for an example of a pulse train with a 5% duty cycle.

By a regularly occurring pulse signal, we mean a **periodic** pulse signal such that the pulses occur at regular time intervals. The pulse parameter to be varied could be the pulse's amplitude, width or duration, and so forth.

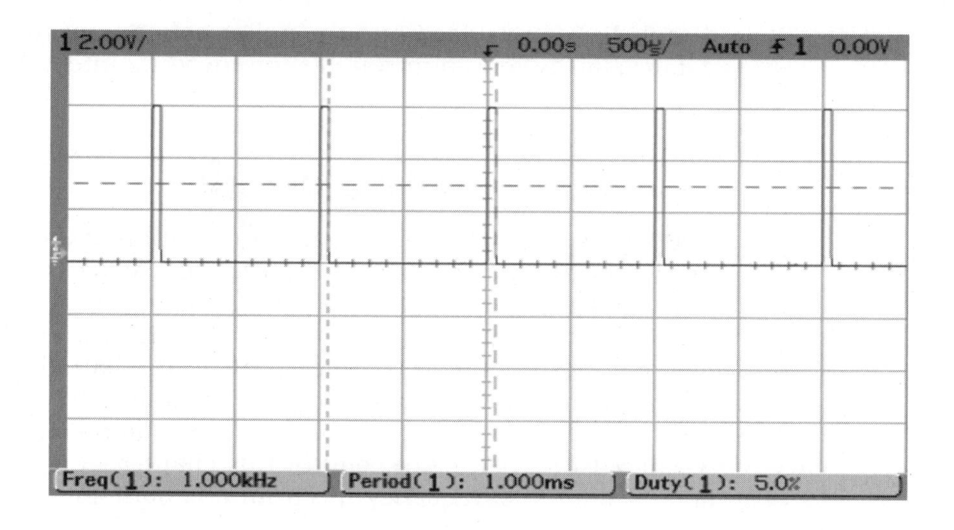

FIGURE 5-5 Typical low duty cycle (5%) pulse train

The Sampling Theorem

There are some fundamental restrictions upon this modulation process. The **sampling theorem** by Nyquist tells us that we must sample the signal at least two times faster than the highest frequency component in the signal. Furthermore, each of these samples must be transmitted in order to reconstruct the signal at the receiver without any distortion.

Let us give an example of this concept: Today, pulse-code modulation encoded telephone calls are routinely transmitted by *T-carriers* over the PSTN facilities. A person's voice with a normal frequency range of 300 to 3400 Hz is sampled at a rate of 8,000 times per second (slightly above what is theoretically required) and the value of each sample is embedded into a sequence of pulses that are transmitted every 125 μsec (1 sec/8000 = 125 μsec) in a short burst. The received pulses are decoded and applied to a digital-to-analog converter. The decoded digital signal contains the spectral components necessary to reconstruct the person's voice at the receiving end.

Advantages of Pulse Modulation

Increased noise immunity and the ability to perform time-division multiplexing (TDM) are the two most important advantages of using certain types of pulse modulation. Pulse modulation allows for the transmission of numerous separate signals or messages over the same channel. Pulse-modulation techniques also lend themselves to signal regeneration and the ability to implement extensive error detection and correction coding techniques. Furthermore, pulse modulation can be used to transmit both analog and digital information. Obviously, analog information must be sampled and embedded into the pulses before transmission, whereas digital data is already in a form suitable for encoding into a pulse stream.

Disadvantages of Pulse Modulation

The biggest disadvantage to pulse modulation is the increased bandwidth needed to transmit the signal. For the pulse signals, Fourier analysis shows that a pulse has a large amount of harmonic energy, and therefore a bandwidth much larger than the pulse's repetition rate. If the pulse is sent through a band-limited channel, some of the harmonic energy will be lost or attenuated and the pulse will be "rounded off" or distorted. In severe cases the receiver will be unable to determine the value of the signal (i.e., was it a 1 or a 0?). At the end of this chapter we will return to this topic and examine the frequency spectrum of a pulse-modulated signal.

Sampling of a Signal

Modern microelectronics has given us several solutions for sampling a signal. **Sample-and-hold** (S/H) circuits, IC timer chips, and analog-to-digital (ADC) converters that perform the encoding task for us are readily available.

Let us look at the first forms of pulse modulation. The earliest forms to be implemented are sometimes known as "analog" pulse modulation because they allow for a continuous variation in some parameter of the pulse signal. Pulse amplitude modulation (PAM) uses an S/H circuit for its implementation. This circuit is triggered at the sample rate, captures the value of the sample at that instant, and maintains the sample value until the next sampling trigger pulse. As shown by Figure 5-6, the resulting output signal is a pulse train with the amplitudes of the pulses proportional to the signal level when it was sampled. Notice that the output pulse signal in Figure 5-6 is not from a true S/H circuit, but rather from an analog switch that is being turned on for a short time at each sample time.

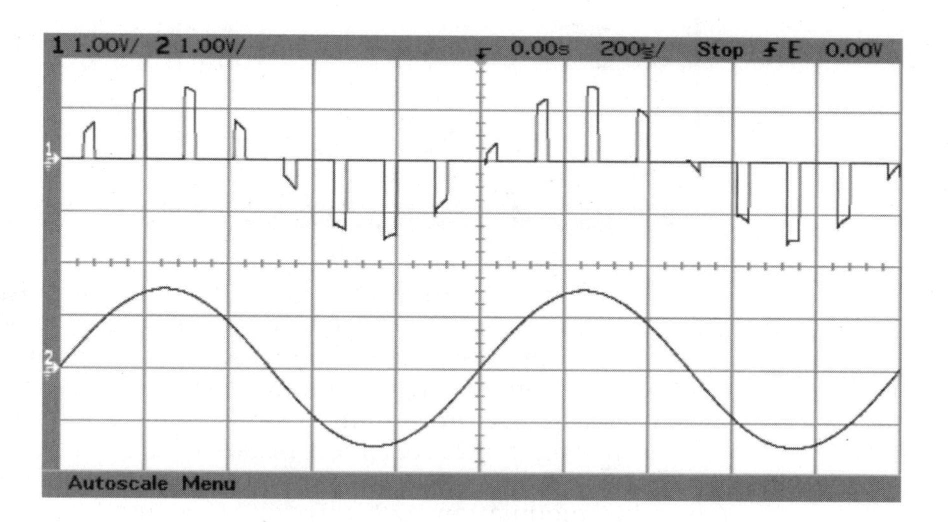

FIGURE 5-6 Typical PAM signal for a sampled sine wave

Pulse Amplitude Modulation Uses

Pulse amplitude modulation can be utilized in a time-division multiplexing scheme. If very short duration (low duty-cycle) pulses are used, there will be a relatively large amount of time before the next pulse occurs. During this unused time, the samples from another signal, or *several* other signals, could be interleaved with the original signal and sent over the same transmission path with a gain in effective channel capacity and/or efficiency. The disadvantage of PAM is that it suffers the same noise problems as regular AM. The receiver must still output a signal that is proportional to what it receives; this type of signal will be corrupted by noise added during the transmission process and by the receiver itself.

Pulse-Width Modulation

Our next form of analog pulse modulation is pulse-width, or pulse-duration, modulation (PWM, or PDM). Figure 5-7 shows a time display of a pulse-width modulated signal.

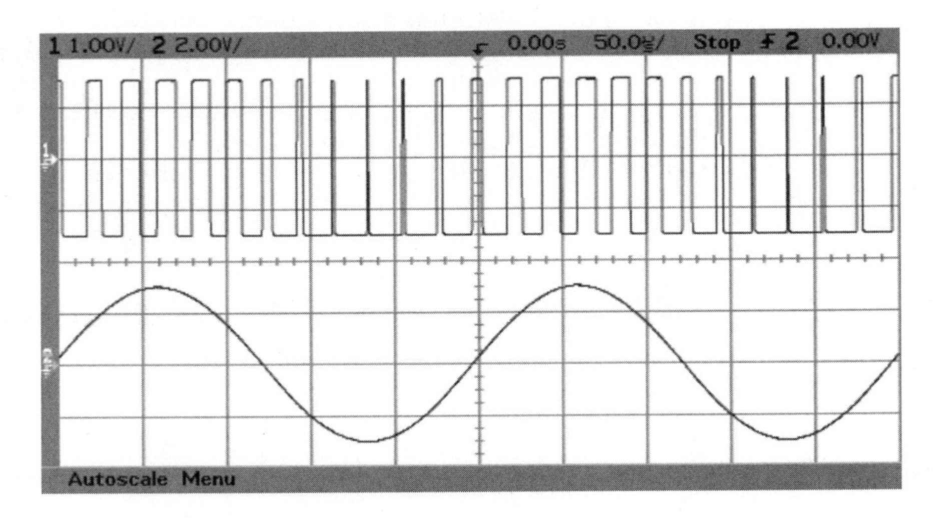

FIGURE 5-7 A pulse-width modulated signal

As can be seen from the figure, this form of pulse modulation varies the width of a pulse in proportion to the signal level at the time of the sample (in this case, the sampling occurs at the same time as the leading edge of the periodic pulse train). Note the relatively large pulse width when the signal has its maximum positive value, and the narrow pulse width when the signal has its largest negative value. A simple 555-timer IC can be used to produce PWM. A variation upon this theme is known as pulse-position modulation (PPM). For PPM, we modify a PWM signal by transmitting constant width pulses, which have leading edges that will coincide with the trailing edge of the PWM signal. Of what use is PWM or PPM? The answer is the same as before: they can be used in TDM schemes. They can also have a noise advantage over PAM because their noise-immunity characteristics are the same as standard frequency modulation.

At this time let us summarize our knowledge of analog pulse-modulation schemes. PAM, PWM, and PPM are all useful for time-division multiplexing schemes; however, in practice they are not used in practical telecommunications systems. PAM and PWM both suffer from the same noise limitations as AM and FM, respectively—there is no improvement in noise immunity. Recently, hybrid digital versions of PAM (2B1Q line code) are finding their way into high-speed digital-transmission systems. Interestingly, PWM, although not usually used in practical telecommunications applications, is extensively used in modern automobile ignition, fuel and braking systems, personal-computer (PC) audio amplifiers, and high-power AM transmitters! In all of these applications PWM is used to improve system efficiency.

5.3 Pulse-Code Modulation

At this point we want to discuss modern methods of pulse modulation that use digital encoding techniques. What is different about this form of pulse modulation is that the signal samples are not used to continuously vary some parameter of a pulse signal. Instead, the value of the signal sample is encoded into a digital code and sent as a burst of 1s and 0s at the sample rate.

Recall our example of the PSTN and T-carrier system pulse-code modulation (PCM). In this system every 125 μsec a burst of 1s and 0s are sent (8,000 bursts per second). The 1s and 0s represent the encoded value of the sample. At the receiver the 1s and 0s are decoded and converted back to the sample voltage. The advantage of this method is that the receiver is only tasked with determining whether the received signal is a 1 or a 0. It can be shown that this is the easiest decision a receiver can make. It also seems intuitively obvious that it would take a large amount of noise to cause the receiver to mistake a 1 for a 0 and vice versa; therefore, pulse-code modulation has noise immunity far superior to any of the other modulation schemes mentioned so far.

Let us now examine the details of PCM more closely. Figure 5-8 on page 182 shows how a signal might be quantized for PCM. The range of values of the signal to be transmitted is usually restricted in both voltage and frequency content. For this example, the amplitude range has been restricted to a 7-V span, corresponding to the bottom seven blocks on the oscilloscope display and to the binary values 000 to 111, as shown on the diagram. The signal shown has been sampled and assigned the nearest level. The need to encode a strictly band-limited signal comes from the Nyquist criterion mentioned earlier; the voltage restrictions come from the limitations of the ICs used to perform the encoding.

As shown by the figure, the input voltage range is divided into a number of standard levels. Because the encoding is binary the number of levels is equal to a power of two. A system with ten bits for encoding would have $2^{10} = 1024$ standard levels. As can be seen in the figure, at each sample time the actual signal level is encoded into the nearest standard level that can be represented by three binary bits (only eight levels are used in this depiction for emphasis). This process is called quantization and it has the

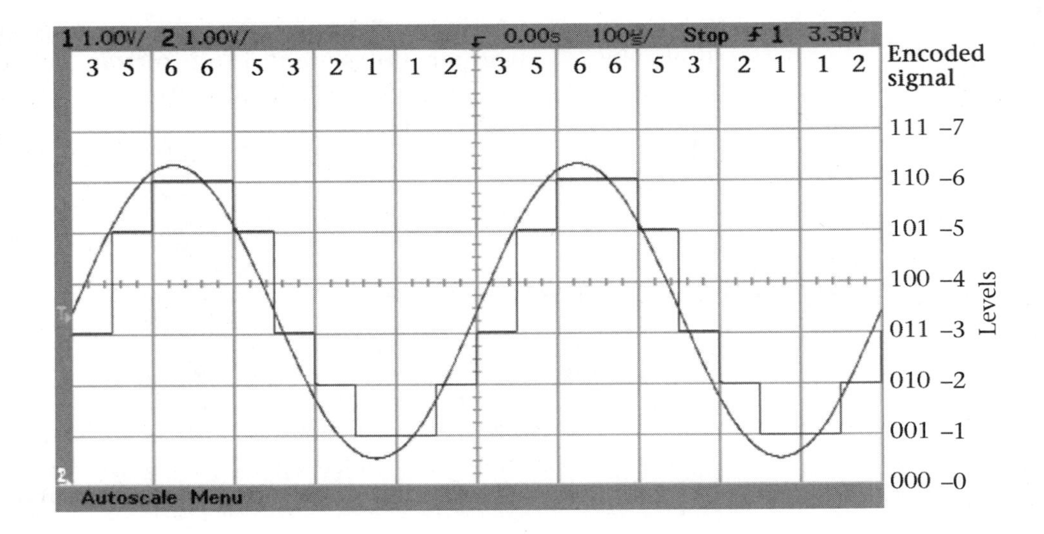

FIGURE 5-8 Typical signal quantizing scheme. A sample is taken every 50 μsec. Full scale is 7 volts.

effect of introducing quantizing noise into the system. For a 3-bit (or 8-level) system, the value of the sample might be quite different from the available levels. It should be obvious that the larger the number of levels the easier it is to match the sampled signal to a level; hence, there is less quantizing noise. The quantizing process is usually accomplished with an ADC IC.

Under-Sampling Problems

Although it has been alluded to as a problem, we have not explained the price paid for under-sampling. Before we go any further, let us consider the consequences of under-sampling a single modulating tone and then of under-sampling voice or music. Sampling a 10-kHz signal at 15,000 times per second or anything less than 20,000 times per second causes problems! For this example, we would discover that our attempts to recover the original tone from the transmitted samples would yield a distorted output consisting of additional frequencies that have been generated by the under-sampling process. These new frequencies are real; therefore, we cannot distinguish the original modulating tone at the receiver. What if we under-sample a voice or music signal? In this case we would observe that numerous new frequencies will mix in with the original information and they cannot be removed or separated from the desired signal. This is called either **spectral fold-over** distortion or **aliasing**. At the end of this chapter we will return to this topic and examine several examples of this problem in the frequency domain.

Sampling Details

At the sample time, the sample-and-hold subsystem of the ADC converter will lock on to the signal voltage while the ADC converter does the digital encoding. ADC converters are characterized by the number of bits of encoding, their input-voltage range, and their conversion speed. The number of **quantizing levels** that an ADC can divide a signal into is given by:

$$\# \text{ of levels} = 2^N \qquad \textbf{5.1}$$

where N is the number of bits. Present-day, fairly inexpensive ADC converters can sample at rates in excess of several 100s of megasamples per second (MSps) and can encode 16 to 20 bits of data.

EXAMPLE 5.1

An ADC IC encodes signals into 12 bits. Determine the number of possible levels.

Solution Using equation 5.1,

$$2^{12} = 4096 \text{ levels}$$

The ADC converter input voltage is usually restricted to a maximum input-voltage range (full-scale input or span). There are several possibilities of what the specific voltage range can be: a unipolar range of 0 to 2.5 V or –2.5 to 0 V, a bipolar range of –2.5 to +2.5 V, or some other variation upon this theme is typical. Early ADCs used to allow higher input-voltage ranges (10 V, full scale), but the most recent ADC ICs, with their lower power-supply voltages, have tended to have lower allowable input-voltage ranges. The **resolution** or quantization uncertainty of an ADC is equal to the smallest step size. This corresponds to the input voltage span divided by the number of quantizing levels or steps, as seen in equation 5.2.

EXAMPLE 5.2

Determine the resolution (smallest step size) for a 12-bit ADC with a 2.5-V input range.

Solution Using the same example values as above, the 12-bit ADC would quantize a 2.5-V signal range into 610.35-μV steps. Mathematically,

$$\text{resolution} = \frac{\text{input-voltage range}}{2^N} \qquad \textbf{5.2}$$

$$\text{resolution} = \frac{2.5\ \text{V}}{4096} = 610.35\ \mu\text{V}$$

As mentioned before, the input signal is usually sent through a low-pass filter to restrict the input-frequency range to less than the highest frequency that can be sampled successfully (½ the sample rate).

One might ask the question, How many bits (levels) of encoding are needed to successfully digitize a signal? The answer depends upon the amount of noise allowed. Practice has shown that seven to eight bits are sufficient for voice encoding and subsequent transmission over the PSTN.

Implementation of a Typical PCM System

Examine the block diagram of a typical PCM system shown in Figure 5-9. As can be seen, the transmitter section converts the input signal into a PCM signal. If the transmission path or channel is very long, PCM lends itself to the use of regenerative repeaters located in the transmission path. Finally, the receiver portion of the system converts the PCM signal back to an analog signal by using a digital-to-analog converter (DAC), the complementary function to the ADC used at the transmitter.

FIGURE 5-9 Typical PCM system

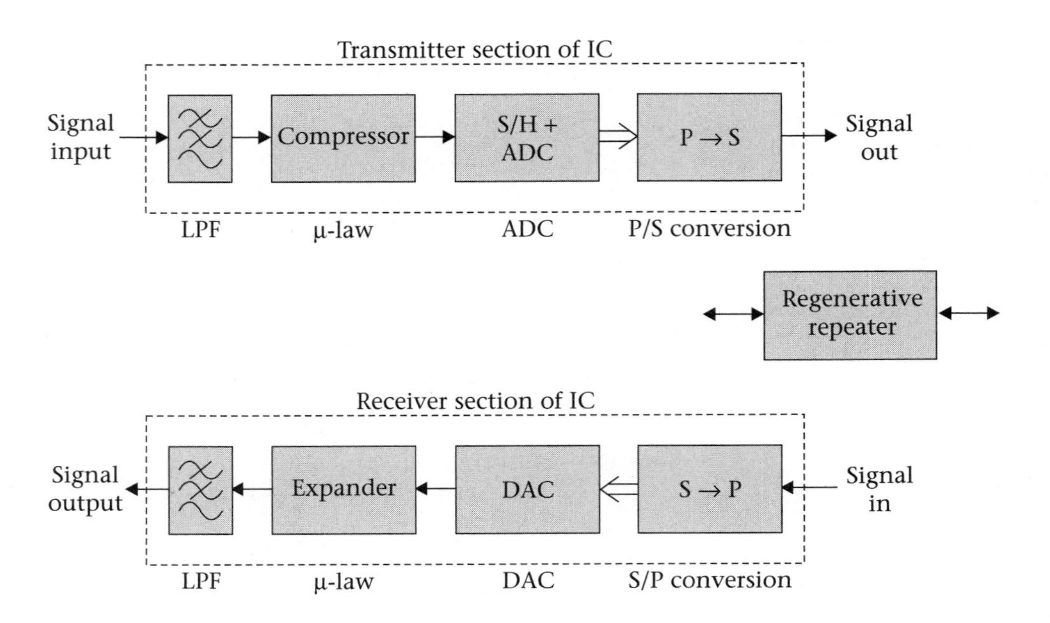

Quantizing Noise Revisited

Returning to the PCM transmitter, how bad is the quantizing noise problem? A measure of how well the system performs is given by the **signal-to-quantizing-noise ratio** (SQNR). This value depends upon how many bits are used and can be calculated by the following equation, which assumes that a full-scale signal is applied to the ADC. If the actual signal level is less than full scale, the SQNR will be less.

$$\text{SQNR} = 1.8 + 6N \text{ (in dB)} \tag{5.3}$$

EXAMPLE 5.3

Determine the SQNR for a 10-bit ADC with a full-scale signal applied.

Solution A 10-bit ADC gives 2^{10}, or 1024 levels. Therefore, from equation 5.3, with N = 10

$$\text{SQNR} = 1.8 + 6N = 1.8 + (6 \times 10) = 61.8 \text{ dB}$$

Companding Schemes

Modern PCM system design recognizes the fact that most speech signals tend to have an energy distribution that "hugs" the ground level with occasional bursts of energy having large voltage levels (see Figure 5-10). To make better use of the limited number of available signal levels, a form of nonuniform encoding is typically used. This nonuniform encoding scheme usually consists of smaller step sizes for low-level signals and increasing step sizes for larger-amplitude signals. The net effect is to

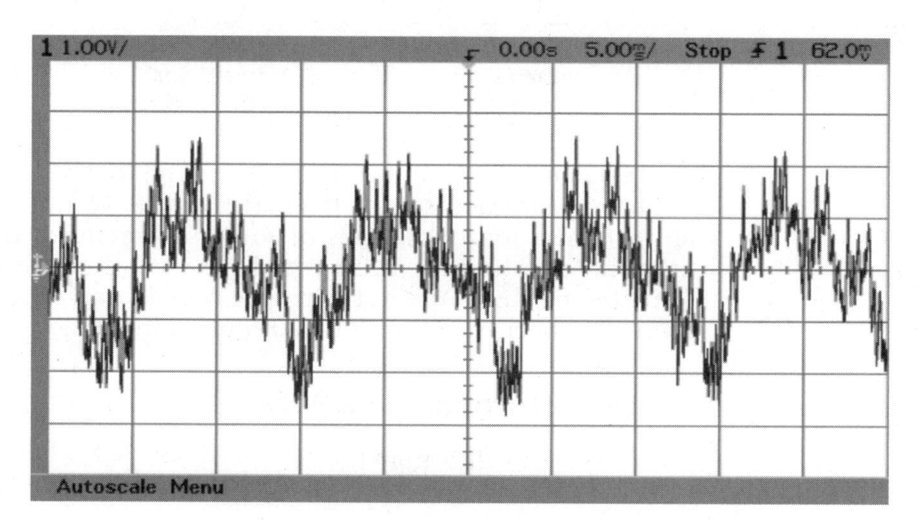

FIGURE 5-10 A typical voice signal

increase the SQNR for the low-level signals and decrease the SQNR for the high-level signals. This technique is known as **companding** a combination of compressing and expanding.

Companding can be accomplished in either of two ways—through analog or digital techniques. Historically, analog companding was implemented first. Two analog methods are currently in use. They are the **μ-law** system used in the United States and Japan and the **A-law** system used in Europe. These two methods tend to perform a logarithmic companding of the signal. The hardware necessary for analog companding is usually contained in what is called a CODEC or COMBO IC chip. Each of these chips contains the circuitry to both receive and transmit (code and decode) a PCM signal and process the audio signals as necessary.

The digital companding technique is performed by the PCM transmitter after a linear ADC conversion is performed on the signal. At the transmitter a compression algorithm is implemented to produce the digitally compressed signal. At the receiver the signal undergoes the corresponding process of digital expansion. The net effect of using either of these companding techniques is to improve the system quality.

The fact that PCM can be regenerated is another advantage of this type of system. In an analog system, each **repeater** used will add noise to the system and eventually there will be a limit to the number of repeaters. As a result of this there will also be a limit to the maximum possible transmission distance of the system. With PCM, if the correct spacing is used a **regenerator** can be used to decode the signal and retransmit it as an exact copy of the last transmitter or regenerator signal. See Figure 5-11 for an example of this regeneration.

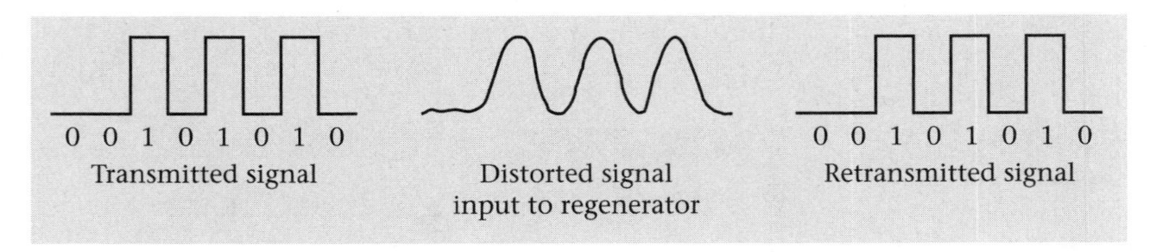

FIGURE 5-11
Regeneration of a PCM signal

See Figure 5-12 for a comparison of the forms of pulse modulation that we have examined so far. From the figure we note that PWM and PAM allow continuous variations of some parameter of the signal (width and amplitude), while the PCM signal consists of encoded bits used to represent the message. Again, the theme here is that PCM has improved noise immunity over the other forms of pulse modulation shown.

Noise Performance of PCM

What type of noise performance does PCM have? Before answering this question, it should be pointed out that PCM is influenced by two

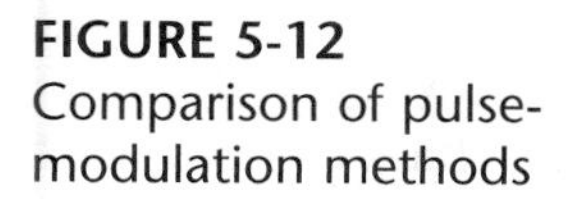
FIGURE 5-12
Comparison of pulse-modulation methods

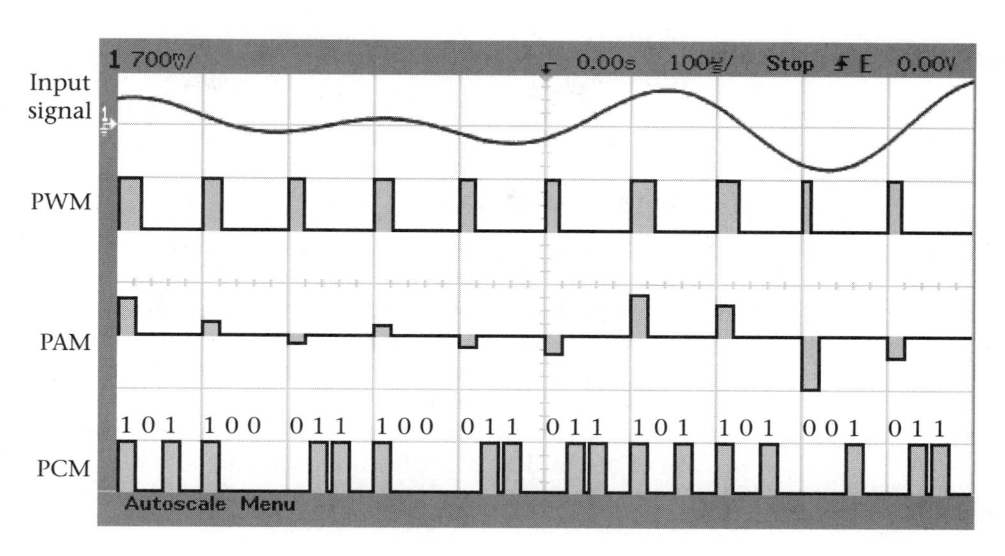

sources of noise: quantization noise that is introduced in the transmitter and present at the receiver output and channel noise. The effect of quantization noise can be reduced to a negligible level by the use of a sufficient number of encoding levels and a companding scheme that matches the type of signal being transmitted. The effect of channel noise is that it introduces bit errors into the received signal. The higher the bit error rate (BER), the more dissimilar the reconstructed signal is from the original signal.

Interestingly, it can be shown that there is an error threshold for PCM. What this means is that if the signal-to-noise ratio is above a certain level (the threshold) there are essentially no bit errors. However, below the error threshold the BER increases rapidly. For PCM this error threshold is approximately 11 dB. Because of this, the effect of channel noise can also be made negligible if the SNR is kept above the threshold value. Additionally, if regenerators are spaced so that the channel noise does not cause any errors, the overall BER of the system is basically independent of the physical length of the channel. Certainly, PCM is a much more robust modulation scheme than any other presented so far.

Other Forms of PCM

There are several variations of PCM used for speech transmission. These modified PCM systems are designed to use fewer bits than standard PCM. **Differential PCM** (DPCM) transmits fewer bits per sample by only transmitting the difference between successive samples. **Adaptive DPCM**

(ADPCM) is a much more complex scheme that uses adaptive quantization and adaptive prediction to encode speech information into a 4-bit code. These schemes allow for the implementation of systems that require lower data rates due to bandwidth limitations.

Delta Modulation

Let us turn our attention to another form of pulse modulation, **delta modulation** (DM). DM only encodes and transmits one bit per sample time. In a typical delta modulation scheme, shown in Figure 5-13a, the transmitter of the DM system transmits only the difference between successive samples. The DM output is therefore either a 0 or a 1. DM requires many more samples per second than PCM to be successful; however, it is a very simple scheme. The DM transmitter compares the new signal level with the signal level at the last sample time and outputs a 1 if the signal is larger, or a 0 if the signal is smaller. See Figure 5-13b for a more detailed look at the comparison of the two signals by the transmitter. See Figure 5-14 for a typical DM output signal shown along with the sampled message signal.

As can be seen in Figure 5-14, the DM output depends only on whether the signal voltage has increased or decreased since the last sample. Note the large number of 1s transmitted during the upsloping of the signal and the large number of 0s transmitted during the sharp downsloping of the signal. The DM shown here uses a constant step size regardless of the amount of voltage change in the signal.

A DM receiver is a fairly simple affair, as shown in Figure 5-15. The input DM signal causes the binary up/down counter to increment for a 1 input and decrement for a 0 input. The counter drives the DAC, and the output is proportional to the counter's value, which is tracking the input binary-pulse train (i.e., a series of 1s would cause the counter to count up and the output from the DAC would increase.)

Linear DM can experience problems with certain types of signals. If the signal voltage changes very rapidly or does not change at all, the linear

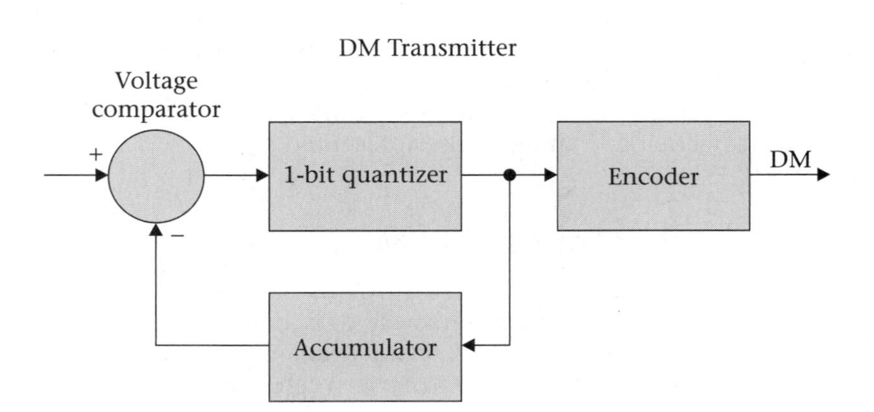

FIGURE 5-13a A delta-modulation transmitter

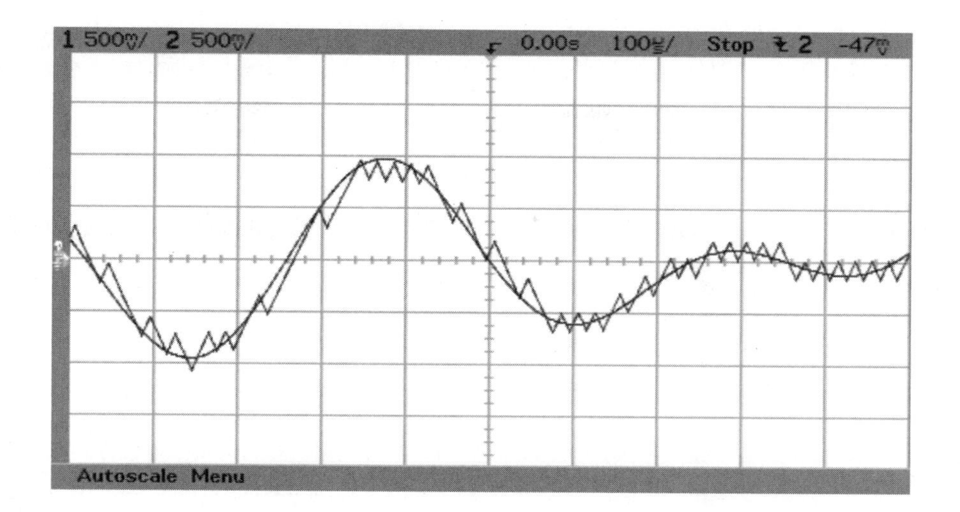

FIGURE 5-13b The two signals applied to the voltage comparator of the DM transmitter

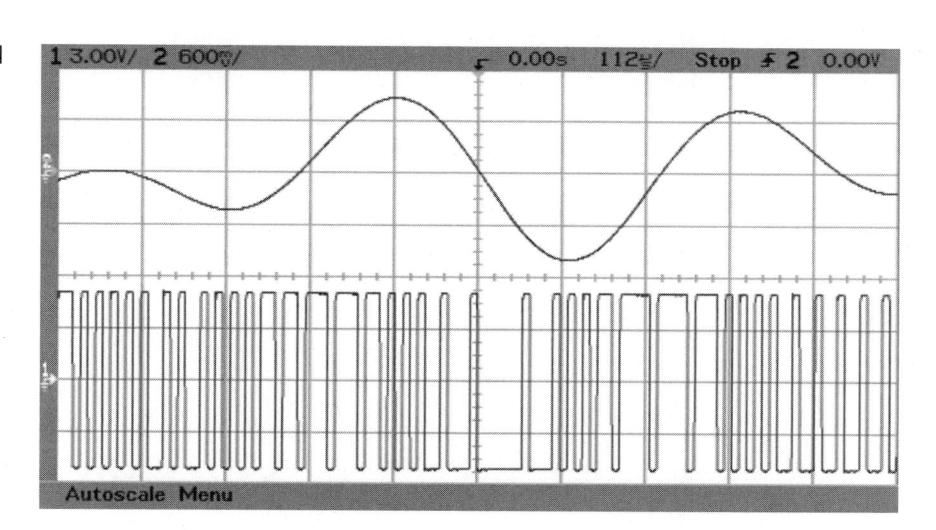

FIGURE 5-14 Typical DM transmitter output

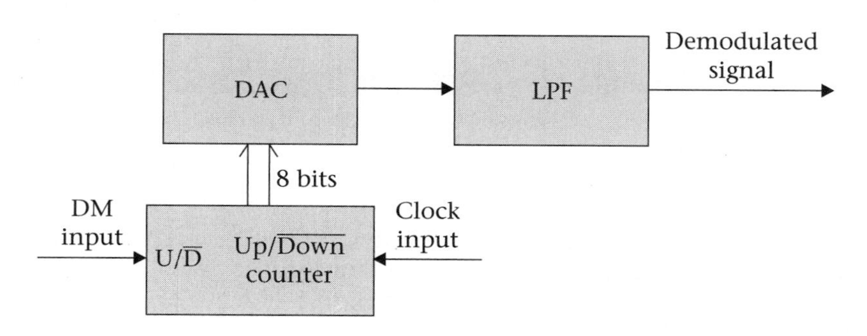

FIGURE 5-15 Typical DM receiver

delta modulator will experience slope overload (see Figure 5-16a) or granular noise (see Figure 5-16b). The cure for both of these problems is the use of adaptive DM, which uses varying step sizes triggered by the number of successive 0s or 1s in the case of slope overload, or alternating 0s and 1s in the case of granular noise.

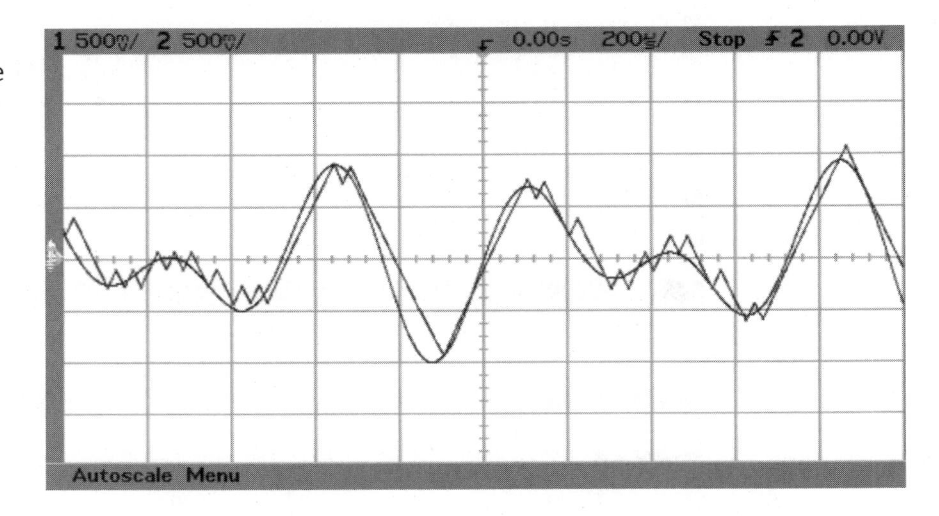

FIGURE 5-16a An example of DM slope overload

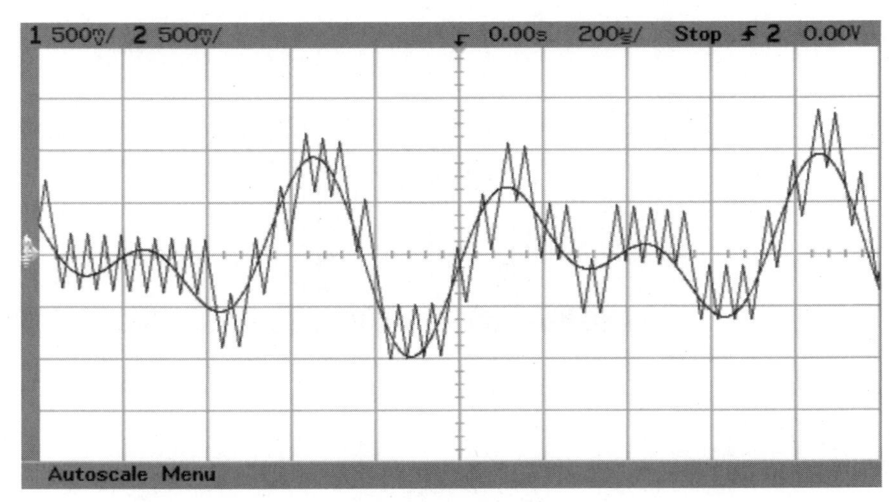

FIGURE 5-16b An example of DM granularity

DM Applications

Although DM systems appear to be simple to implement, to date they have had limited use in telecommunications systems. Popular applications include military uses and digital voice transmission over radio channels.

This concludes our overview of basic pulse-modulation techniques. We should note that up to this point we have only talked about the transmission of 0s and 1s to implement these modulation schemes. At this time, we should consider the transmission media and how best to encode these 0s and 1s onto the physical channel.

5.4 Line Codes and Encoding Techniques

As mentioned previously, PCM is used extensively over the PSTN on T-carriers. The question is: What type of electrical signal should be used to encode the 0s and 1s? The answer is that there are many possible ways to encode the 0s and 1s. The encoding scheme or **line code** used depends upon several factors. Different types of encoding schemes can be used by telecommunications providers to improve a signal's noise immunity (to lower the bit error rate) or to achieve higher data rates. Conversely, a certain line code can be used on lower bandwidth channels to achieve higher data rates or to increase the maximum transmission distance for a given data rate. Finally, a particular line code can be used to provide self-clocking for systems that require this feature.

Basic Definitions

Now let us look at basic properties of the most common types of line codes. First, some basic definitions:

- The length of time required to transmit a symbol or bit will be called a **bit time**.
- **Non-return-to-zero** (NRZ) encoding maintains the symbol for the entire bit time, while return-to-zero (RZ) encoding does not utilize the entire bit time.
- The voltage range used for encoding might be **unipolar** (either positive or negative) or **bipolar** (both positive and negative).

Non-Return-to-Zero (NRZ) Line Codes

Two common examples of NRZ line-code formats are shown in Figure 5-17 and Figure 5-18 on page 192. Figure 5-17 shows an NRZ-level bipolar line code. The code follows the TTL signal with a level shift. Figure 5-18 shows an NRZ-mark bipolar line code. There is a transition at the beginning of each 1 and no change for a 0.

Note that these two NRZ line codes experience the fastest change in voltage when there is a 0-to-1 or 1-to-0 transition. The fastest voltage changes in a signal correspond to the highest frequencies in a signal.

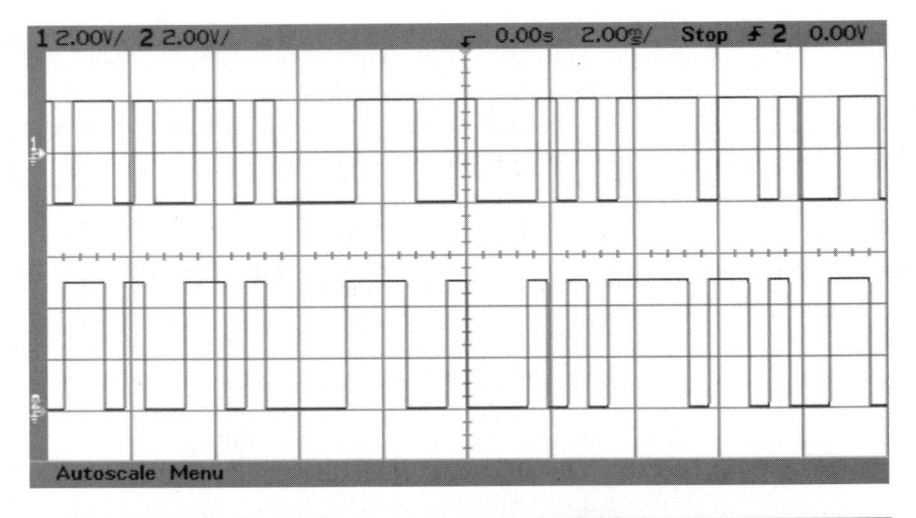

FIGURE 5-17 NRZ-L bipolar line code (Channel 1). Channel 2 is the TTL data stream.

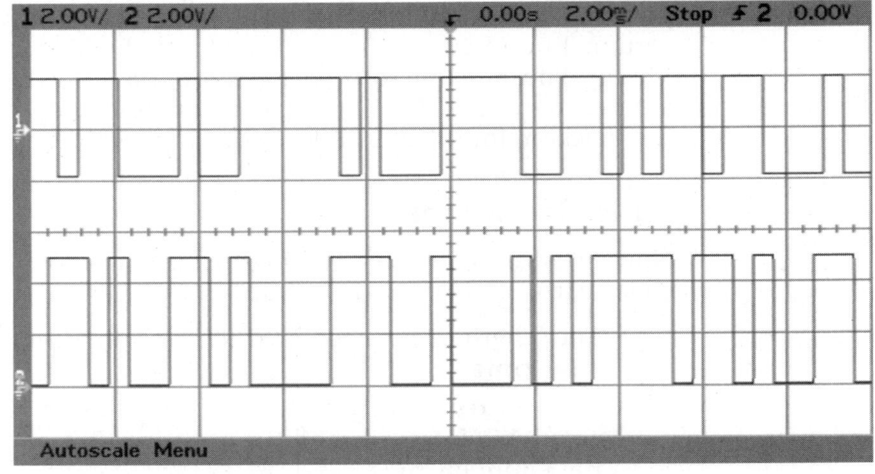

FIGURE 5-18 NRZ-M bipolar line code (Channel 1). Channel 2 is the TTL data stream.

Therefore, the highest fundamental frequency contained in NRZ line codes is equal to one half the bit rate, where the bit rate is the reciprocal of the bit time. In Figure 5-19, a sine wave of frequency, $f = 1/(2 \times \text{bit time})$ is overlaid on an NRZ line code to show the relationship between the bit time (480 μsec, which corresponds to a bit rate of $2083\frac{1}{3}$ Bps) and the signal frequency of $1041\frac{2}{3}$ Hz.

Return-to-Zero Line Codes

Figure 5-20 and Figure 5-21 on pages 193 and194 show two different RZ line codes. The unipolar RZ line code outputs a half-width output pulse for a 1 input and no output for a 0 input. This line code has a significant dc component. The bipolar RZ line code outputs a half-width positive-polarity

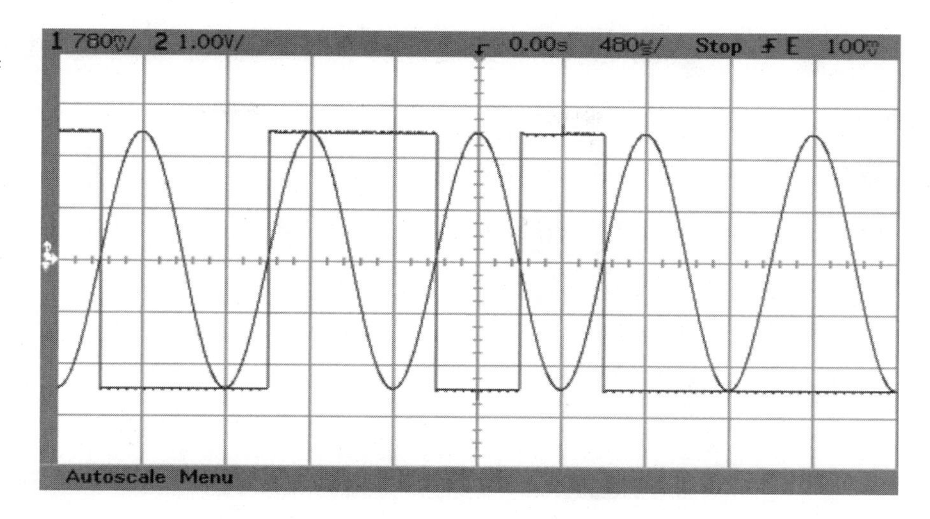

FIGURE 5-19
Frequency content of an NRZ line code

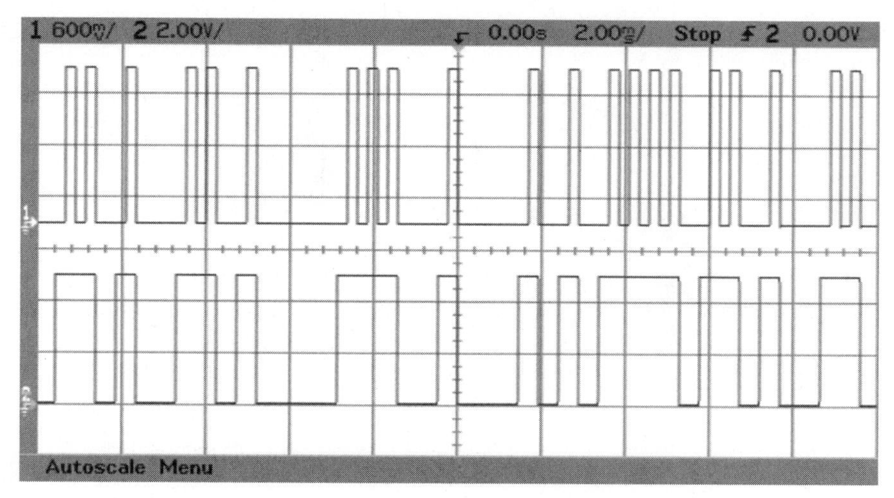

FIGURE 5-20
Unipolar RZ line code (Channel 1)

output pulse for a 1 input and a half-width negative-polarity pulse for a 0 input.

Figure 5-22 on page 194 shows the frequency content of a bipolar RZ line code. Again, the transitions from 0 to 1 or 1 to 0 contain the highest fundamental signal frequency. In this case, as can be seen from the figure, the frequency is equal to the bit time.

Comparison of Line-Code Formats

At this point we will compare these two different formats (NRZ and RZ). The unipolar line codes will have a dc component, while the bipolar line codes will have very little average value. This makes the unipolar codes less

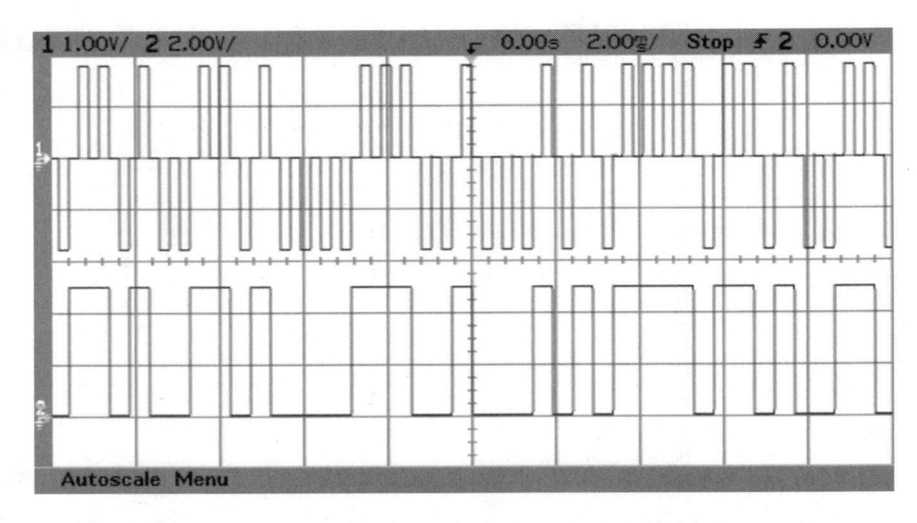

FIGURE 5-21 Bipolar RZ line code (Channel 1)

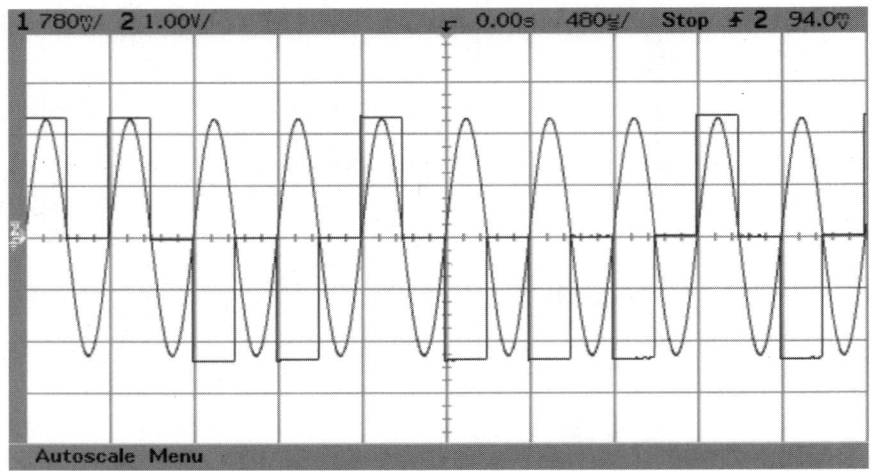

FIGURE 5-22 Frequency content of a bipolar RZ line code; $f = 1/(\text{bit time})$

power efficient than the bipolar codes. The NRZ line codes have a bandwidth of one-half the bit rate, while the RZ line codes have a bandwidth equal to the bit rate. However, the NRZ line codes will not have good clock-recovery characteristics, while the RZ line codes will. Neither type of line code will have any error-detection properties.

Other Line-Code Formats—AMI

Another line-coding scheme is shown in Figure 5-23. In this scheme, the two nonzero voltage levels both represent a logic 1, while the zero voltage level represents a logic 0. This type of line code is called return-to-zero alternate-mark inversion (RZ-AMI). There is a half-width output pulse if the

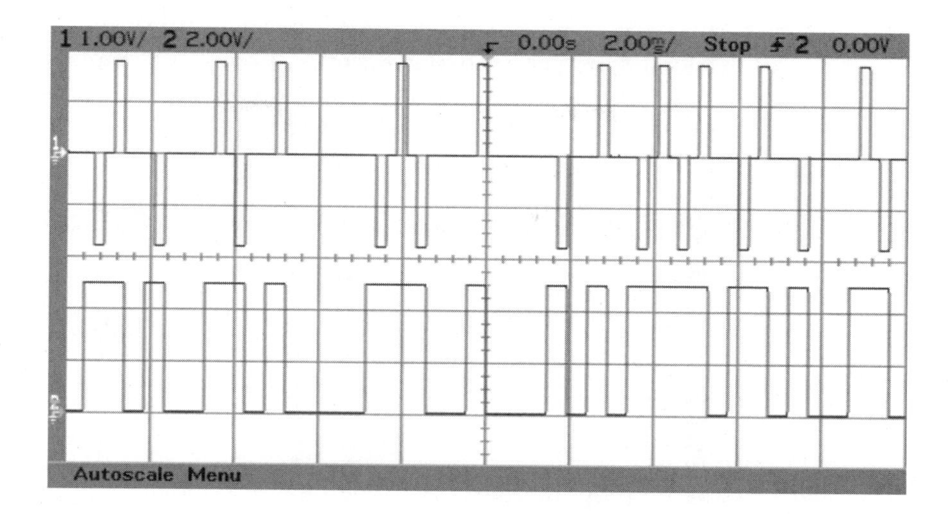

FIGURE 5-23
RZ-AMI line code

input is 1 and no output pulse if the input is 0. In addition, the polarity is inverted for every other output pulse. This line code will have a bandwidth equal to one-half the bit rate. This line code has error-detection properties, as a single error will cause a bipolar violation. However, with AMI coding it is possible for clock synchronization to be lost with a long sequence of consecutive 0s.

Table 5-1 compares these three basic types of line codes. As can be seen, the RZ-AMI line code has some advantages over the other coding schemes. Early T-1 carrier systems (twenty-four voice-band signals are transmitted using PCM and TDM) used AMI line coding with provisions to protect the system from the loss of clocking. Newer versions of the T-1 carrier system use a modified form of AMI called binary eight-zero substitution (B8ZS) coding, which ensures that sufficient voltage transitions occur so that clock synchronization is not lost.

TABLE 5-1
Line-coding summary

Encoding format	*Fundamental frequency*	*Ease of clock recovery*	*Error detection*
NRZ	$f = 1/(2 \times \text{bit time})$	Poor	None
RZ	$f = 1/(\text{bit time})$	Good	None
AMI	$f = 1/(2 \times \text{bit time})$	Good	Yes

Manchester Coding

Another type of line coding that provides for fairly trouble-free clock recovery is digital biphase, or Manchester code. This is a popular type of high-speed line coding used for Ethernet LANs. See Figure 5-24. Notice that transitions occur in the middle of each bit time. It is interesting to note that some of these biphase codes were developed for use in data recording on magnetic tape or on disk drives for video recording and digital computer mass-storage applications.

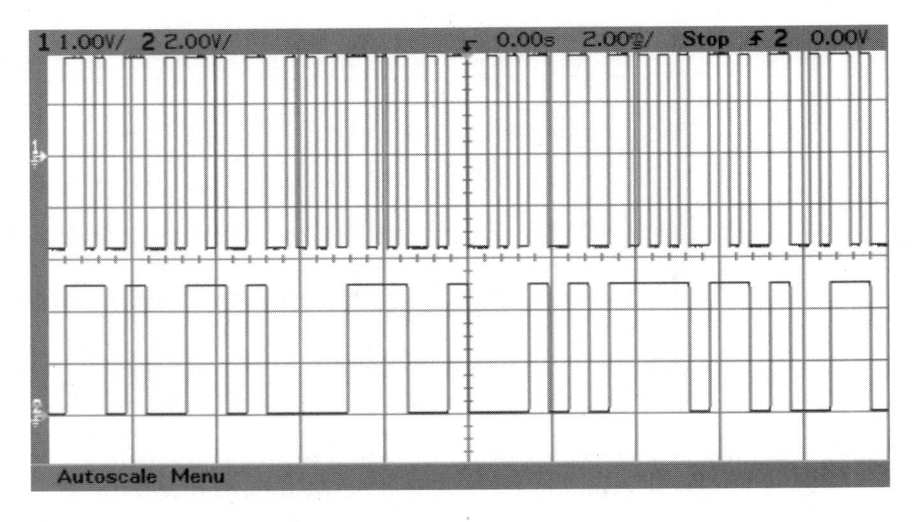

FIGURE 5-24 Digital biphase line coding (Manchester)

The 2B1Q Line Code

A relatively new line code used to implement digital subscriber lines (DSL) is the two-binary, one-quaternary (2B1Q) code shown in Figure 5-25. This line code uses 4-level PAM (digital PAM). Practical systems are implemented with four signal levels (+3, +1, –1, and –3), with each level representing two binary bits. The first bit of a binary pair determines the polarity of the

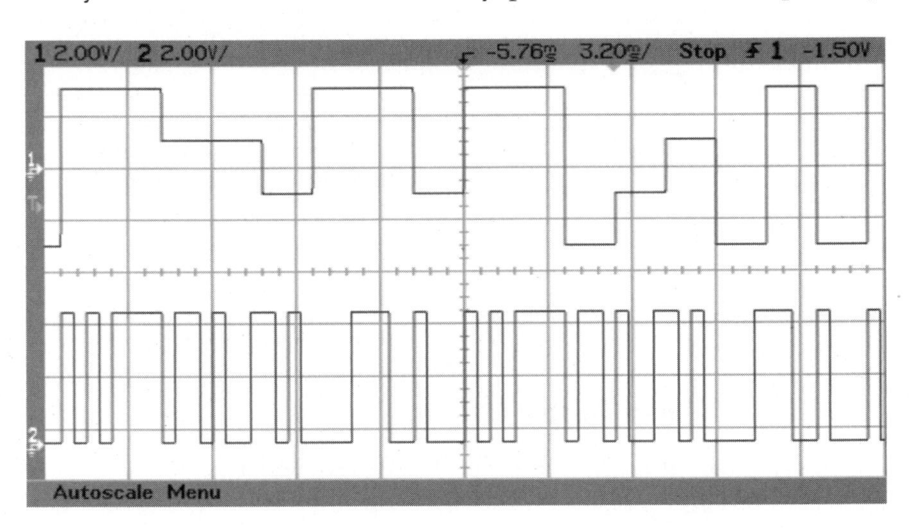

FIGURE 5-25 2B1Q line code

signal (1 = + and 0 = –) and the next bit determines the level (1 = ±1 and 0 = ±3). For example, 00 = –3, 01 = –1, 10 = +3, and 11 = +1. An advantage of this code is that the required bandwidth is one-half the bit rate. However, this code transmits two bits per bit time, so there is, in effect, an additional reduction in bandwidth by a factor of two. Note the slower rate of change in the 2B1Q code (Channel 1) in Figure 5-25.

Comparisons of Line-Code Bandwidths

To summarize the bandwidth needed for transmission of the various line codes presented here, Figure 5-26 shows their individual frequency spectra. This figure is based upon the standard T1-carrier data rate of 1.544 Mbps with a random sequence of symbols 1 and 0.

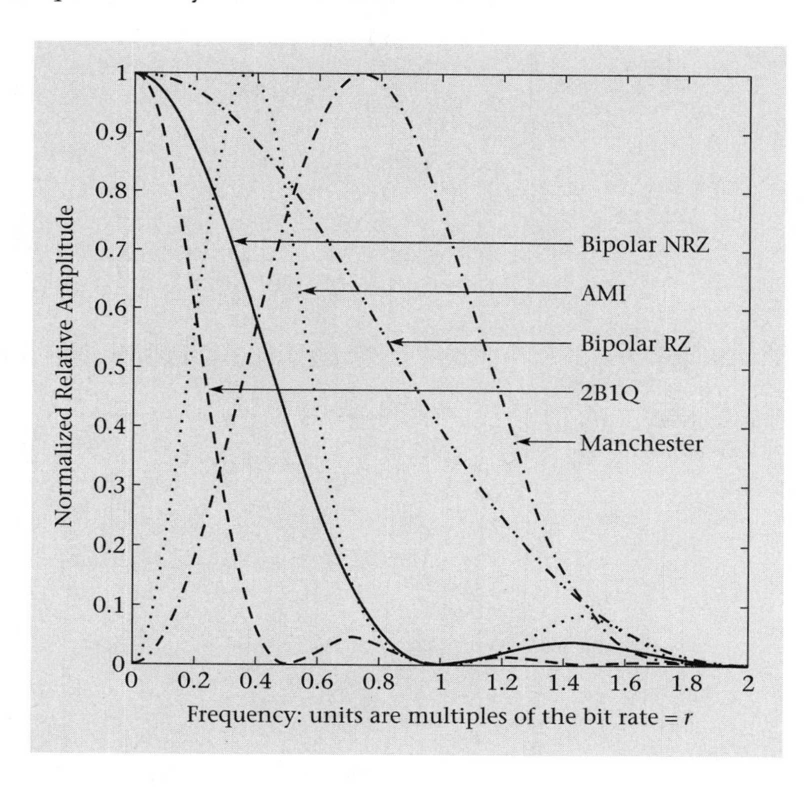

FIGURE 5-26 Frequency spectra of different line codes

From the figure it can be seen that the 2B1Q code has a much smaller bandwidth than the other line codes; hence the popularity of the use of the 2B1Q line code in the local loop. The use of 2B1Q line code can result in two scenarios, either a longer maximum distance of data transmission compared to the other transmission formats or a higher data rate over the same distance of operation compared to the other formats.

5.5 Intersymbol Interference

Now that we have discussed line codes, we can focus our attention on the limitations of the transmission media. The PSTN consists of copper-wire pairs that basically act like low-pass filters. As pulses travel down the copper pairs, and possibly through regenerators, they will encounter numerous different types of impairments to their transmission. The most significant problems that contribute to the improper interpretation of the received signal will be timing inaccuracies (jitter), amplitude and phase distortion, cross talk from other pairs, and insufficient bandwidth. Multiple twisted-pair Category-5 cable used for local-area networks (LANs) will also suffer from similar effects, as will coaxial cables. Optical signals used for the transmission of data through fiber-optic cables also suffer from signal dispersion and other impairments similar to wireline media.

Figure 5-27 shows the overall effect on the transmitted signal for any wireline or fiber-optic media. The top waveform is the transmitted signal, the middle waveform shows how the frequency response of the channel ideally rounds off the pulses, and the bottom waveform shows the combined effect of the various impairments. Note that when the received signal is sampled (usually at the midpoint of the bit time) to determine whether it is a one or a zero, it is no longer a sure thing that the correct determination will be made.

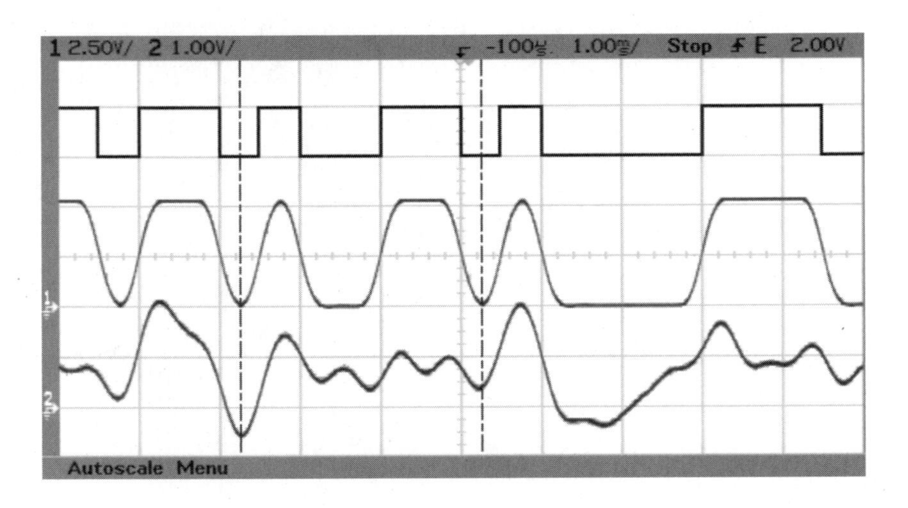

FIGURE 5-27 Pulse distortion from transmission on a copper pair

The term used to describe the deterioration of the signal due to the finite bandwidth of the channel and timing inaccuracies is called **intersymbol interference** (ISI). The channel's limited bandwidth will cause the energy in a pulse to be spread out over time, interfering with other pulses. For any type of channel, as one attempts to increase the data rate or the actual physical-channel span length, there eventually comes a point at which the pulses have been distorted so much that the bit error rate (BER) will become intolerable.

Eye Diagrams

We can measure the performance of a pulse transmission system by using an eye diagram to determine the amount of ISI. An **eye diagram** can be generated by displaying the received signal on an oscilloscope and triggering the trace at the signal data rate. Figure 5-28 shows a typical eye diagram.

FIGURE 5-28 A typical eye diagram

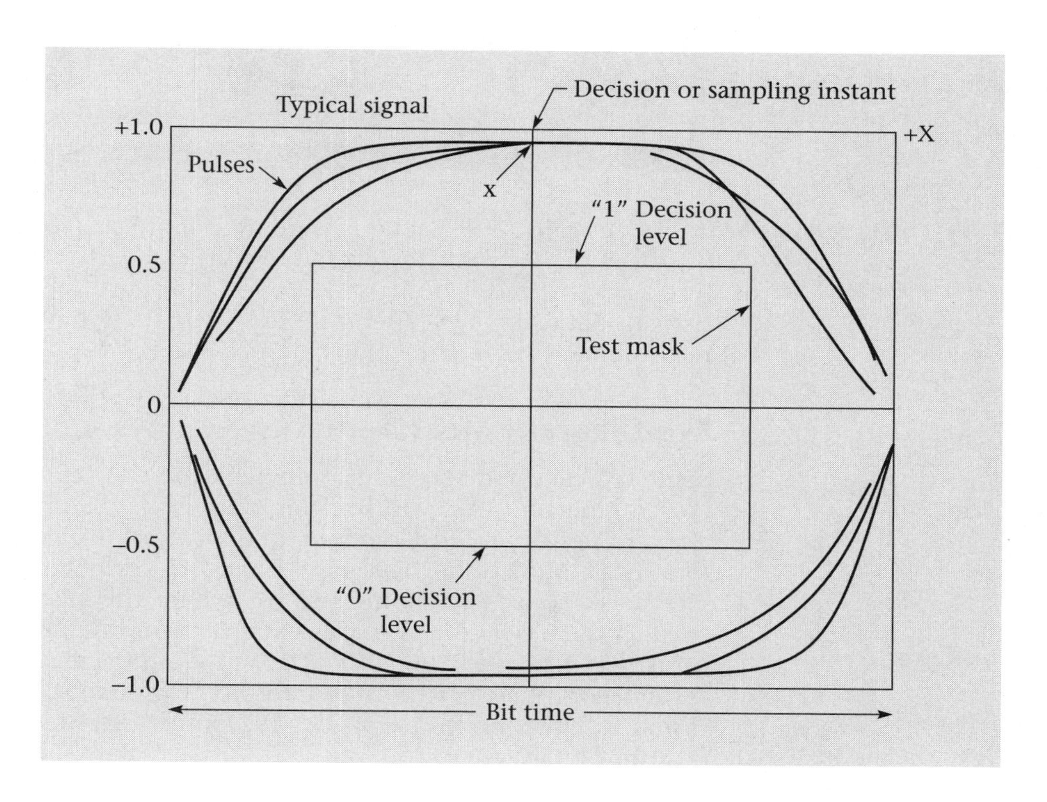

At one time, a useful measure or figure of merit for a pulse-modulation system was the amount of ISI degradation. Mathematically, it was equal to:

$$\text{ISI} = 20\log\left(\frac{x}{X}\right)\text{dB} \qquad \textbf{5.4}$$

where $\frac{x}{X}$ is the fractional value of eye opening.

An actual eye diagram for a baseband signal is shown in Figure 5-29. A mask has been included to illustrate present-day test-equipment capabilities.

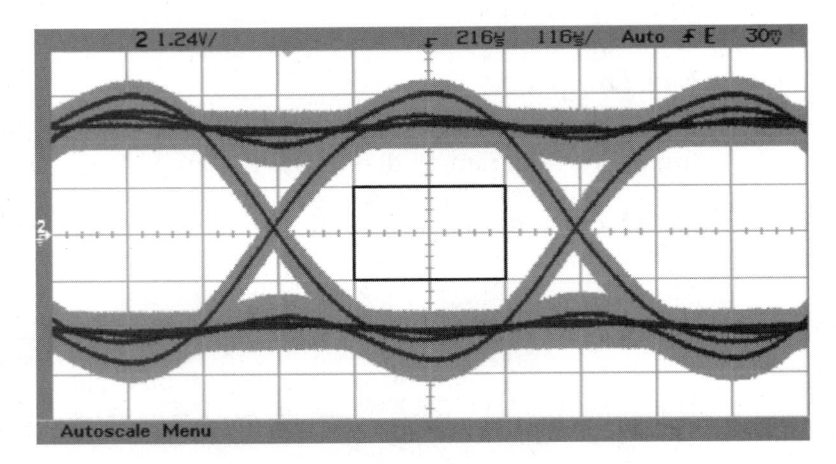

FIGURE 5-29 An eye diagram for a baseband signal

Historically, eye diagrams were used more as a qualitative measurement of system operation. However, due to the advent of communications analyzers (specialized digital oscilloscopes) that have the ability to generate statistical databases on the samples that they have taken, quantitative measurements of eye diagrams are possible.

Eye-Diagram Measurements

Some typical eye-diagram measurements are the extinction ratio, eye height, crossing percent, eye width, jitter, duty-cycle distortion, eye Q-factor, and rise and fall time. Most modern communications analyzers are able to test to a mask. *Mask testing* is a process used to verify correct system operation. The mask is used to define regions where the signal waveform may not exist. Industry-standard masks exist for wireline and fiber-optic standard transmission rates. A SONET/SDH eye-diagram mask allows one to check system compliance with standards for various OC-n signals. Figure 5-30 shows an eye diagram for an OC-192 optical signal generated by a communications analyzer.

5.6 Spectra of Pulse-Modulated Signals

The last topic in this chapter will be the spectra of pulse-modulated signals. Let us start by examining the spectra of a PWM or PAM signal. We must first determine the spectra of the low duty cycle unmodulated pulse waveform. The spectra for this type of signal is shown in Figure 5-31. As can be

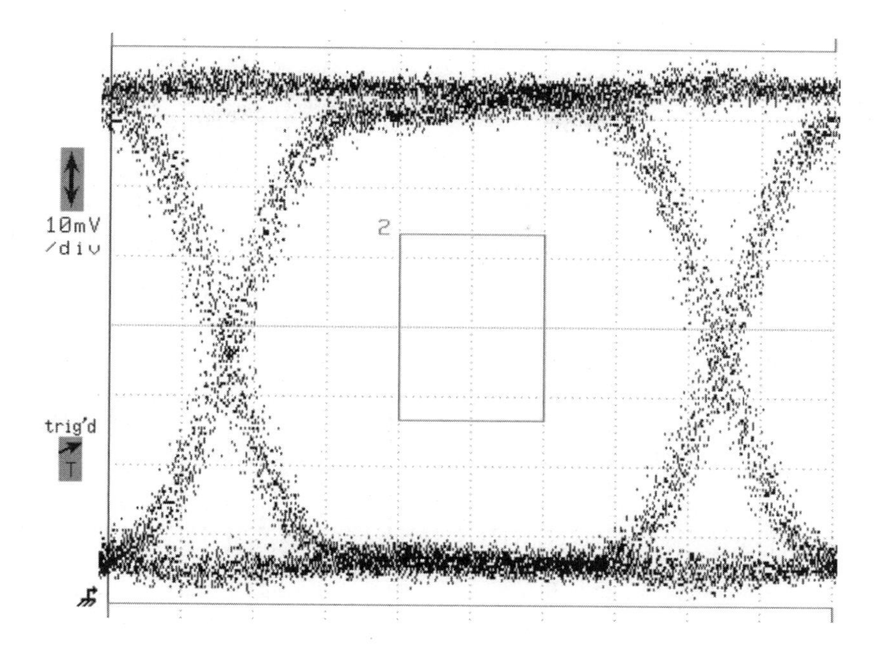

FIGURE 5-30 Eye diagram for an OC-192 (10 Gbps) optical signal

seen, this signal consists of a fundamental frequency equal to the pulse repetition rate and all the harmonics of this frequency that gradually decrease in amplitude.

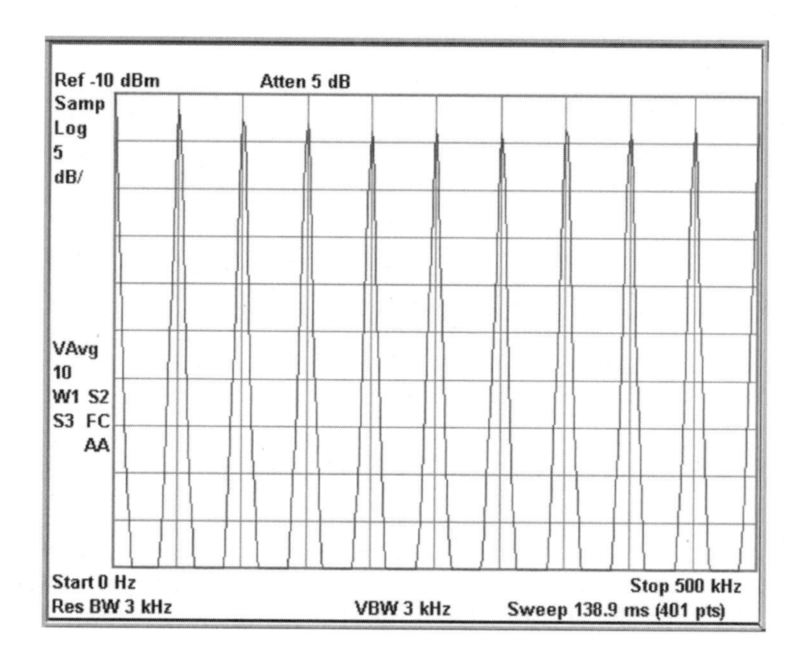

FIGURE 5-31 Spectra of a low duty cycle pulse waveform; from Figure 5-5 (100-kHz fundamental frequency)

Using a voice signal as the modulating signal, we note that it contains a range of frequencies, as shown in Figure 5-32a. The composite spectra of the resulting PAM or PWM signal is also shown (Figure 5-32b). The composite spectra indicates that the modulating signal's energy is carried by the pulses as a baseband signal and as sidebands located about the fundamental pulse-frequency component and the harmonic components of the pulse. These types of analog pulse modulation schemes produce baseband signals and are usually carried from point to point by transmission lines (copper-wire pairs, coaxial cables, and so forth).

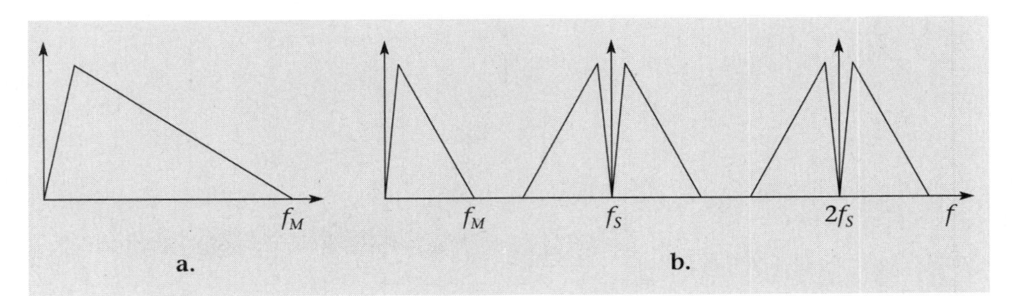

FIGURE 5-32 (a) A baseband modulating signal and (b) the resulting spectra for a PAM or PWM signal. The sample rate, f_S, is greater than $2 \times f_M$

If the signal has been sampled sufficiently, as shown in Figure 5-32, then the spectral components will be spaced far enough apart that they will not overlap. If the signal has not been sampled at a high enough rate, then aliasing will occur, as shown in Figure 5-33, and there will be distortion from the spectral overlap. In this case it will be impossible to recover the original modulating signal without distortion.

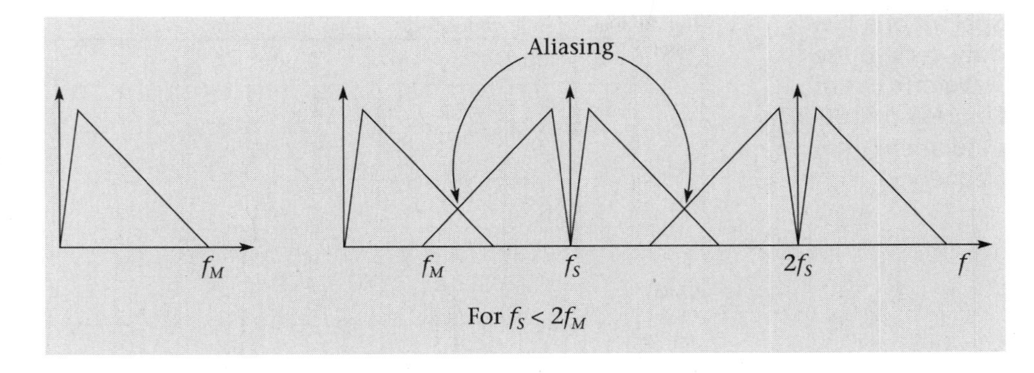

FIGURE 5-33 For this case the sample rate is not sufficient, $f_S \leq 2f_M$, and therefore aliasing occurs

Minimum Channel-Bandwidth Requirements

What amount of channel bandwidth is needed for pulse-modulated signals? Up until this time it has been assumed that these types of signals have been encoded in low duty cycle, ideal pulse-like signals. However, it has been found that some other alternate pulse shapes are more effective in terms of intersymbol interference and bandwidth efficiency and are therefore more commonly used to implement these modulation forms. Two of these alternate pulse shapes are the **sinc pulse** and the raised cosine. A typical sinc pulse is shown in the time domain in Figure 5-34.

FIGURE 5-34 A typical sinc pulse

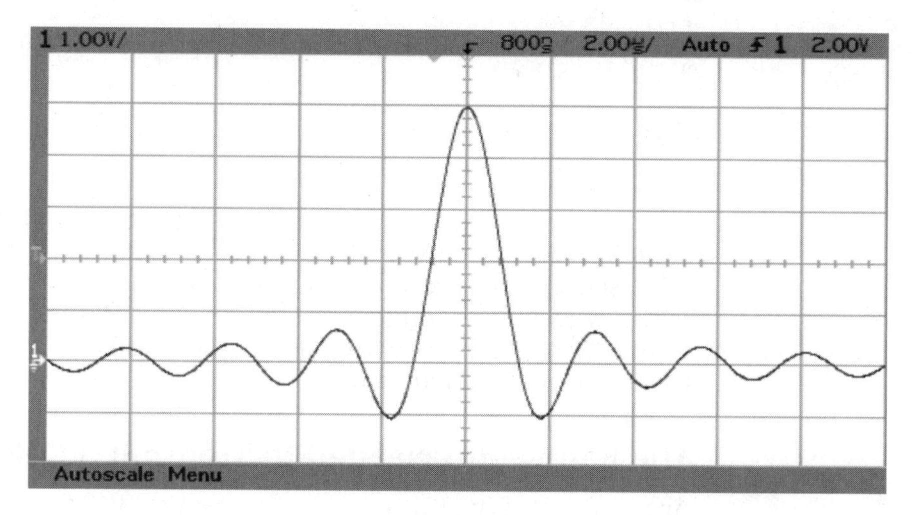

Using either low duty-cycle pulses or these alternate pulse forms, it turns out that most of the transmitted pulse's energy is contained in its fundamental frequency component. Therefore, a bandwidth equal to the reciprocal of the bit time is usually needed for the transmission of the pulses. Recall that the use of the various types of line codes discussed in section 5.4 will also yield improvements in bandwidth efficiency. Also recall Figure 5-26, shown earlier, which shows the 2B1Q line code with a required bandwidth equal to only one half of that required for a bipolar NRZ line code. Recall that this is because 2B1Q code transmits two bits per bit time.

Recovery of Analog Pulse Modulated Baseband Signals

Our last figure, Figure 5-35 on page 204, shows how the original modulating signal can be recovered. A simple low-pass filter will pick off the original signal while rejecting the higher-frequency components! Most pulse modulation systems will over-sample the signal. This has the effect of providing a guard band between the spectra components, thus easing the frequency response requirements (roll-off rate) for the low-pass filter.

The Spectra of a PCM Signal

In the case of a PCM signal sent over a wireline channel, and depending upon the type of line code used, the modulated signal will have a frequency spectra similar to that which was shown in Figure 5-32. There are

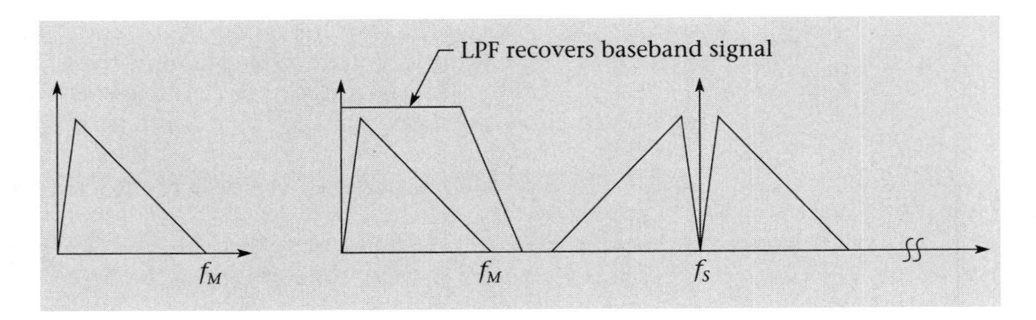

FIGURE 5-35 Baseband signal recovery can be achieved by using a simple LPF

frequency components at the bit-rate frequency and its harmonics that include information about the original modulating signal. In the case of the PCM signal, the receiver does a digital-to-analog conversion of the signal for each received sample, in effect converting the PCM signal into a PAM signal. This causes the spectra to be identical to Figure 5-32, only needing to be low-pass filtered to recover the information.

The Bandwidth Needed to Transmit a PCM Signal

Again, the bandwidth needed to transmit a PCM signal is the same as that needed for an analog pulse-modulation signal. In general, the bandwidth required is equal to the bit rate or the reciprocal of the bit time. The T-1 carrier system requires a bandwidth of 1.544 MHz, because it transmits bits at the rate of 1.544 Mbps using standard line encoding.

5.7 Overview of Digital-Compression Techniques

Up to this point we have spent a great deal of time discussing the most efficient use of the channel bandwidth and pointing out the fact that it is a precious resource. Our focus has been on examining various pulse-modulation schemes and the resulting adaptations of hardware implemented so as to make the most effective use of the channel in the transmission of analog signals.

MPEG Standards

It is now appropriate to introduce the topic of digital encoding and compression of audio (voice or music) and video signals. The transmission of either type of signal is extremely bandwidth consumptive and can easily exhaust the resources of a telecommunications network. A typical uncompressed NTSC television signal requires approximately 168 Mbps to be transmitted digitally. The Motion Pictures Expert Group (**MPEG**) has been working on signal-compression techniques since the late 1980s in an attempt to increase the efficiency of digital encoding. The MPEG-1 audio-coding standard was the first standard capable of transparent, lossless compression of stereophonic audio signals. MPEG-1 achieved its success by employing an encoder system that closely modeled the psychoacoustics of a human's auditory system. The MPEG-2 standard is designed for the compression of digital television and the MPEG-4 standard is for multimedia applications. An MPEG-7 standard is still under development, and is formally named "Multimedia Content Description Interface." It is a standard for describing multimedia content data and interpreting to some degree the information's meaning.

The MPEG-2 video-compression algorithm exploits both temporal and spatial redundancy present in video signals. MPEG-2 is used to encode NTSC television, resulting in 2- to 3-Mbps data rates (12- to 20-Mbps rates for HDTV). The success of the MPEG standards and their resulting compressed, more efficient data rates has been a driving force in the exponential expansion of the Internet and the quest to add broadband capability to the telecommunications infrastructure.

Summary

There are two basic types of pulse modulation. The first category is implemented by the continuous variation of some parameter of the transmitted pulse in accordance with the modulating signal. The second type of pulse modulation encodes the value of the modulating signal in a digital code and is transmitted as a burst of 1s and 0s at periodic intervals of time. Crucial to this entire process is the necessity of transmitting sufficient samples of the modulating signal per unit of time. The required rate of sampling is given by the sampling theorem, which states that the signal must be sampled at least twice as fast as the highest frequency in the signal.

The analog forms of pulse modulation—PAM, PWM, and PPM—facilitate the implementation of time-division multiplexing, but except for PAM are not widely used in today's telecommunications systems. The digital forms of pulse modulation—PCM and DM—are used extensively for the transmission of speech over telecommunications systems. PCM is used in conjunction with time-division multiplexing schemes to share channels with numerous other voice signals; DM is typically used for cordless telephones or other special applications. Both types of pulse modulation can be implemented with commonly available ICs.

The bandwidth required for each type of pulse modulation is essentially the same as the system bit rate. Exceptions to this occur when line codes that provide bandwidth reduction are used to transmit a baseband signal over wireline transmission media or special digital encoding schemes are used to transmit the signal at passband frequencies. Digital encoding schemes are addressed in chapter 6.

Questions and Problems

Section 5.1

1. What is the sampling theorem? How does it pertain to pulse modulation?
2. Determine the required sampling rate for a baseband video signal that contains frequency components up to 4.2 MHz.
3. What is the major advantage that pulse modulation offers?
4. Explain how sampling of a signal is accomplished.

Section 5.2

5. Describe either pulse-amplitude or pulse-width modulation.
6. What are the noise characteristics of analog pulse-modulation techniques?

Section 5.3

7. Describe the process of pulse-code modulation.
8. What advantages does PCM have over the analog forms of pulse modulation?
9. How many levels can one encode with 14 bits of data?
10. What is the resolution of a 14-bit ADC?
11. Define *quantizing noise*.
12. Explain the concept of companding.
13. Explain the basic concept of delta modulation.
14. If a video signal with a bandwidth of 4.2 MHz is to be digitized, what is the resulting data rate for 10-bit encoding, 12-bit encoding, and 16-bit encoding?

Section 5.4

15. What do all line codes have in common?
16. What is meant by the terms *unipolar* and *bipolar*?
17. What is meant by the term *non-return-to-zero*?
18. Why does the 2B1Q line code have a smaller bandwidth than the other line codes presented in this section?
19. What is meant by the term *clock recovery*?

Section 5.5

20. Describe the cause of intersymbol interference.
21. What is an eye diagram?
22. What is meant by the phrase *testing to a mask*?

Section 5.6

23. In general, what type of frequency spectra does a pulse-modulated signal have?
24. What is meant by the term *aliasing*?
25. How can aliasing be prevented?
26. Explain why the bandwidth required for a T-carrier system using 2B1Q line coding is equal to one half the bandwidth required for bipolar NRZ line code.

27. Explain why an LPF can recover the information contained in a pulse-modulated signal.

Section 5.7

28. Do an Internet search on MPEG compression techniques. Define the present standards. This could be a team project.

29. What is the basic principle that allows for video compression?

30. Compare the data rate for compressed HDTV with that for standard NTSC TV.

Chapter Equations

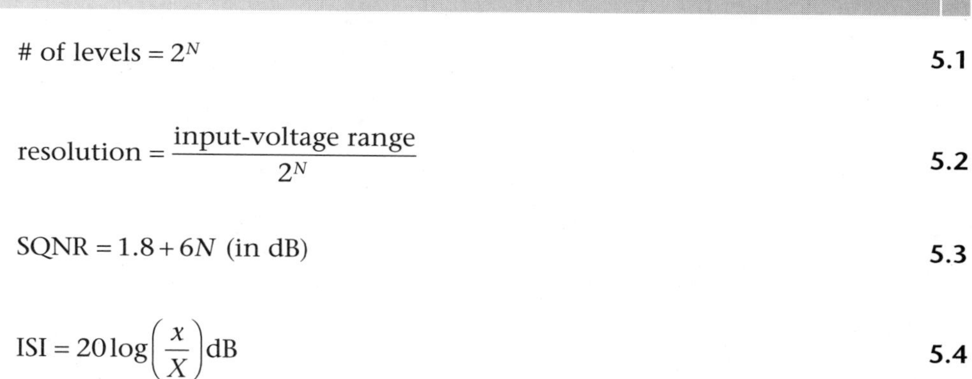

$$\#\text{ of levels} = 2^N \qquad (5.1)$$

$$\text{resolution} = \frac{\text{input-voltage range}}{2^N} \qquad (5.2)$$

$$\text{SQNR} = 1.8 + 6N \text{ (in dB)} \qquad (5.3)$$

$$\text{ISI} = 20\log\left(\frac{x}{X}\right)\text{dB} \qquad (5.4)$$

Digital Modulation

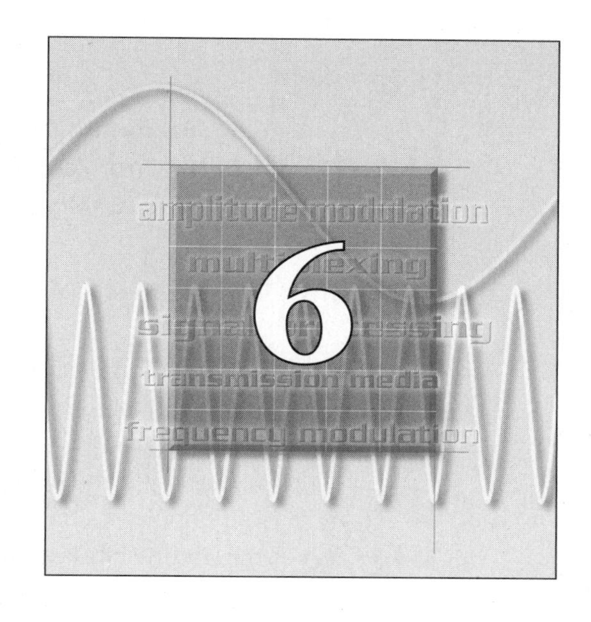

Objectives Upon completion of this chapter, the student should be able to:

- Discuss the relative merits of digital-modulation techniques versus analog-modulation forms.
- Discuss how channel capacity can be increased by a trade-off of more complex hardware for less spectrum.
- Discuss forms of frequency-shift keying (FSK).
- Discuss forms of phase-shift keying (PSK).
- Describe quadrature amplitude-modulation (QAM) techniques.
- Discuss the present uses of the modulation schemes presented.
- Discuss the type of hardware used to encode and decode the different digital-modulation schemes.
- Understand the information represented by a constellation diagram.
- Understand the concept of bandwidth efficiency.
- Understand the concept of bit error rate (BER).
- Discuss the use of time-division multiple-access (TDMA) and code-division multiple-access (CDMA) multiplexing schemes.

Outline

6.1 The Evolution of Digital Modulation
6.2 The Information Capacity of a Band-limited Channel
6.3 The First Generation of Digital Modulation
6.4 Second Generation Digital-Modulation Techniques
6.5 Other Second Generation Digital-Modulation Schemes

Key Terms

bandwidth efficiency
binary PSK
constellation diagrams
direct digital-frequency synthesis
frequency-shift keying
I Channel
minimum-shift keying
offset QPSK
passband
phase-shift keying
phasor
Q Channel
quadrature
quadrature amplitude modulation
quaternary PSK
scalar signals
signal-space diagram
symbol time
time-varient signals
Trellis coding
Trellis code-modulation
vector signals

Introduction

As the telecommunications industry strives to increase data rates and channel capacity, traditional methods of modulation are being replaced by sophisticated digital-modulation techniques. This transition from simple analog AM and FM transmission schemes has occurred because digital techniques offer greater information capacity and provide better quality communications. Furthermore, digital modulation provides compatibility with digital data services and higher data security through encryption. Some of the topics covered in this chapter include:

- forms of frequency-shift keying (FSK)
- forms of phase-shift keying (PSK)
- quadrature amplitude-modulation (QAM) techniques

Each discussion of a new modulation scheme includes an introduction to the necessary transmitting and receiving hardware. Also included is an introduction to I/Q modulators, constellation diagrams, bandwidth efficiency, signal spectra, bit error rate (BER) comparisons, and the use of new multiplexing schemes to increase channel capacity.

The entire thrust of this chapter is to show how digital-modulation techniques can provide enhanced data rates over limited bandwidth channels. The student should complete this chapter with a sense of appreciation for the benefits that digital-modulation techniques have brought to users of the telecommunications infrastructure.

6.1 The Evolution of Digital Modulation

During the 1960s, with the development of computer operating systems and high-level languages, digital computers became popular. Many businesses and most large universities and colleges began to introduce mainframe computers into their daily operation. Methods of sharing access to these mainframe computers from remote locations on the academic or business campus were developed, initially utilizing existing data-transmission technology and making use of the teletype machine as a replacement for the computer keyboard. Connected through modems over dedicated telephone lines, these electromechanical noisemakers achieved data-transmission rates of ten characters per second. The only telecommunications network available at the time was the PSTN; therefore, the transmission channel was basically a voice-grade circuit with very limited bandwidth. Data-transmission technology improved with the introduction of the Bell modems, which were designed around the characteristics of the PSTN channel. The Bell modems set many de-facto standards that have remained until today. A popular early model was the Bell System-103 modem. It was capable of full-duplex operation (simultaneous transmission in both directions) at speeds of up to 300 bits/sec over standard dial-up PSTN facilities. Over the span of approximately three decades, advances in digital-modulation technology have allowed us to increase the modem data rate by a factor greater than one hundred.

Because the modem was used to interface computers, early transmission schemes used standard computer-industry codes (ASCII or IBM) that were made up of combinations of 0s and 1s that stood for the alphanumeric character set. The early modems sent digital signals instead of analog signals, and traditional analog-modulation schemes were replaced by digital-modulation systems. A major focus of this chapter will be to see how these more complex digital-modulation schemes have increased channel efficiency.

6.2 The Information Capacity of a Band-Limited Channel

We will start our discussion of digital modulation by recalling Hartley's law and Shannon's theorem for information capacity. Hartley's law states that a channel's information capacity is proportional to bandwidth and transmission time:

$$I \propto B \times T \tag{6.1}$$

Shannon's theorem states that given a band-limited channel (e.g., the PSTN) the information capacity is given by:

$$I = 3.32BW\log(1 + S/N) \text{ bps} \qquad \textbf{6.2}$$

where, I is the information capacity in bits per second, BW is the bandwidth in Hz, and S/N is the signal-to-noise ratio (unitless) over the transmission path. Shannon's relationship gives us an upper limit for information capacity over a band-limited channel. It should be noted right away that to achieve this data rate requires the use of more than two symbols. As can be seen from the equation, the noisier the channel the less the amount of information that can be transmitted per unit time and, conversely, the cleaner the channel the greater the amount of information that can be transmitted per unit time.

From Early Modems to Today

The first modems used two symbols to represent 0s and 1s. As systems have become ever more sophisticated, more complex digital-modulation schemes have evolved. There has been a fundamental engineering trade-off in the design of these systems. As Figure 6-1 shows, the industry trend has been a switch from simple analog modulation to digital modulation. As shown in the diagram, as modulation schemes have increased in complexity, the size of the required "pipe" needed to carry the information has been reduced. The change from simple-modulation to complex-modulation schemes is somewhat misleading in terms of relative cost and size. The evolution of the IC has allowed for complete systems on a chip, with a corresponding reduction in cost, without any sacrifice of complexity or functionality. In practice, it is the transmission medium (the pipe) that is most costly in terms of both installation and maintenance.

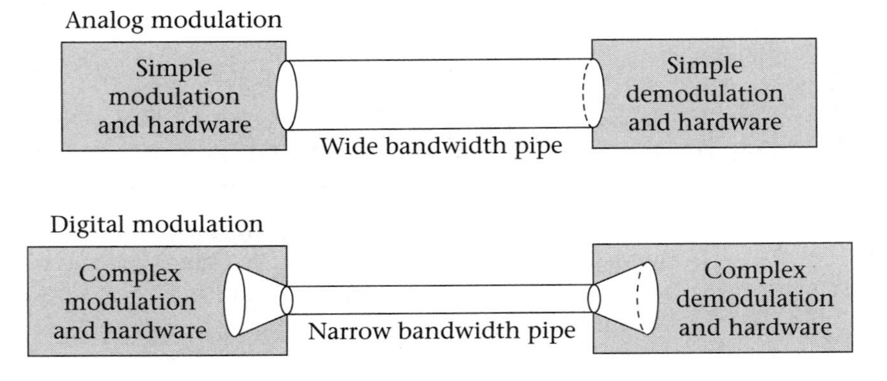

FIGURE 6-1 The trade-off of complex modulation and hardware for reduced bandwidth

Figure 6-2 illustrates this continuing evolution with three generations of modulated-signal types. Starting with the legacy analog-modulation schemes of AM and FM, these modulated signals can be considered **scalar signals** in that they are characterized simply by their amplitude and frequency. Information is embedded in variations in these two parameters. The next generation of modulation schemes includes frequency-shift keying (FSK), quadrature phase-shift keying (QPSK), quadrature amplitude modulation (QAM), and variations on these themes. These modulation schemes produce what we will call **vector signals** because they are characterized by combinations of both amplitude and phase or frequency and phase information. Finally, we have multiplexing schemes like time-division and code-division multiplexing that produce **time-variant signals.** These signals are characterized by vector signals that also have embedded time and/or code variations that are needed to gain access to the transmission channel.

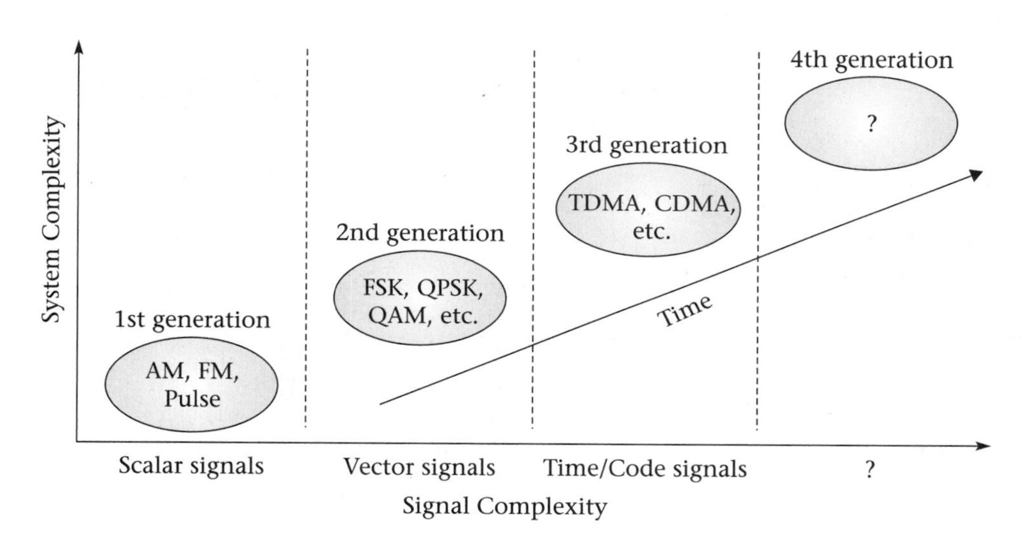

FIGURE 6-2 The evolution of signal/systems complexity needed to achieve bandwidth efficiency

Let us turn our attention to the different forms of digital modulation.

6.3 The First Generation of Digital Modulation

We start with the simplest form of digital modulation, which is on-off Keying (OOK). This form of modulation has many names (Morse code was a form of OOK), including binary amplitude-shift keying (BASK). Figure 6-3 on page 214 shows a typical OOK signal. It is produced by simply turning on and off the output of the transmitter. The bandwidth required for this signal (an alternating sequence of 1s and 0s) is shown in Figure 6-4.

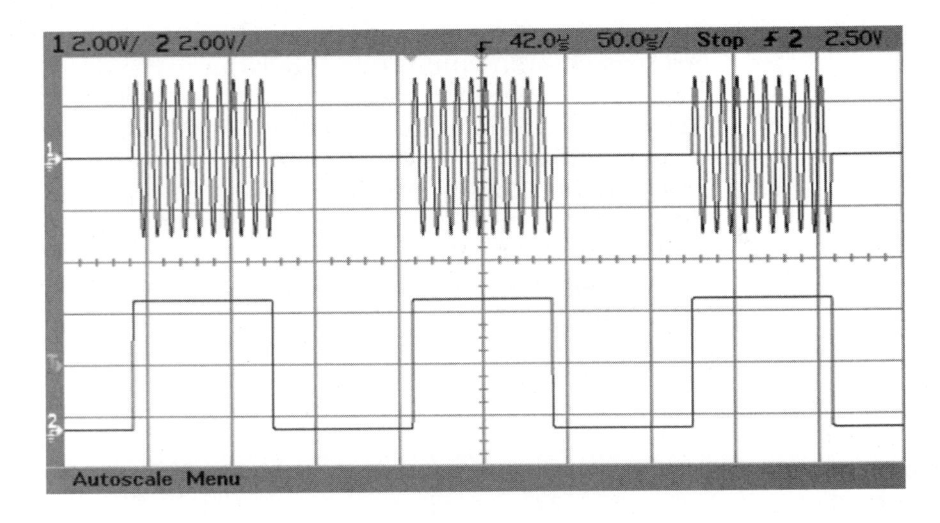

FIGURE 6-3 A typical BASK or OOK signal

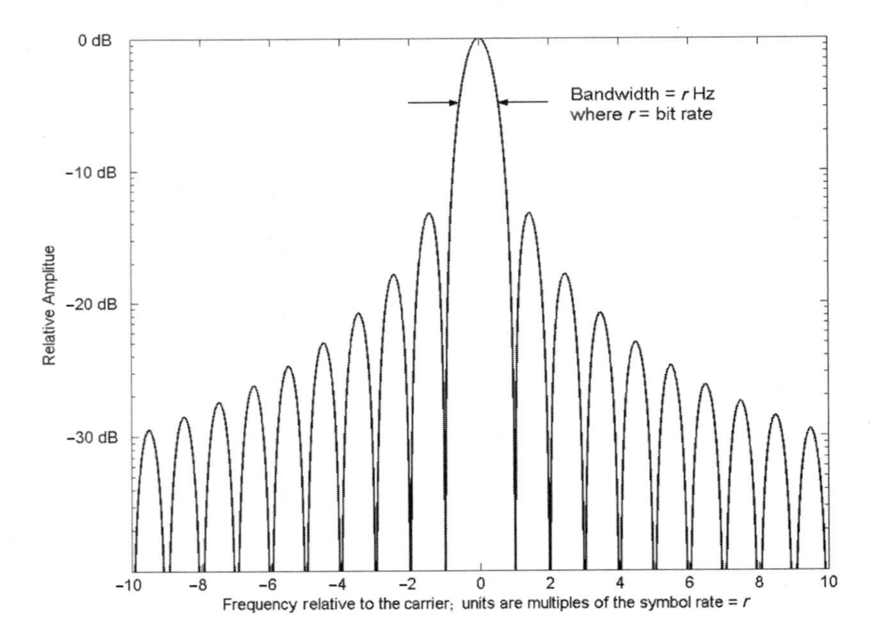

FIGURE 6-4 Frequency spectra of a BASK signal

The output signal spectra shows something interesting. The output signal does not merely consist of the transmitter frequency; it also contains frequency components related to the rate of signal transmission. These sideband components fall off rather rapidly and it appears that the majority of the signal power is included in a bandwidth equal to the bit rate. This form of digital modulation is seldom used today in high-performance communications systems.

Frequency-Shift Keying

The next form of digital modulation to consider is **frequency-shift keying** (FSK). This is the simplest form of FM and was mentioned earlier in the chapter. It was first used in the late 1800s and early 1900s for a particular type of wireless telegraph system. Figure 6-5 shows a typical FSK signal. As can be seen, the frequency of the transmitter (f_S or f_M) encodes the value of 0 or 1. Again, as with BASK, we should next consider the bandwidth and the data rate of an FSK signal.

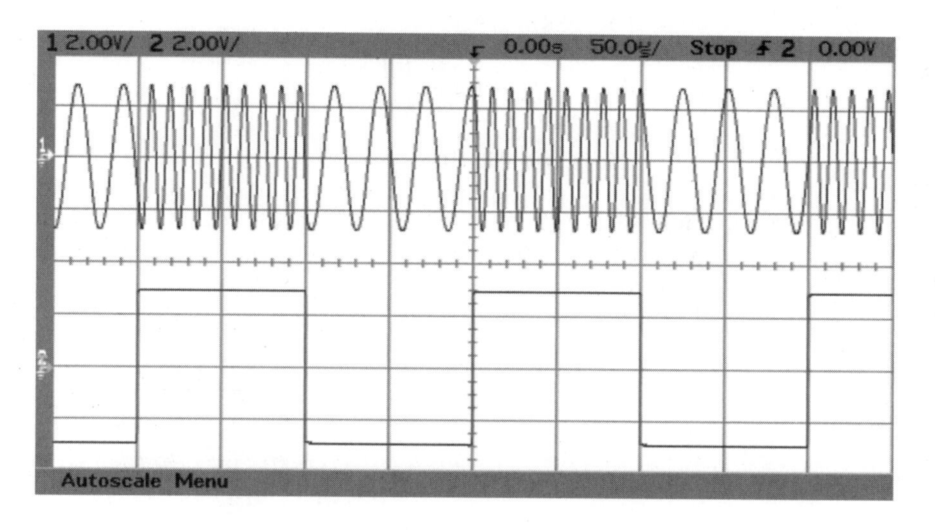

FIGURE 6-5 A typical frequency-shift keying (FSK) signal

The FSK transmitter is a simple voltage-controlled oscillator (VCO) and the input signal consists of two different voltage levels. The system can therefore be considered to be an FM system with a rest frequency, f_C. The input consists of a binary 1 or 0, and the output frequency of the VCO is shifted between a mark frequency, $f_M = f_C + \Delta f$, and a space frequency, $f_S = f_C - \Delta f$. Looking at a time display of an FSK signal (Figure 6-6 on page 216), it can be seen that one bit of information (one symbol) is sent per bit time period and that the highest fundamental frequency in the output signal is equal to one-half the bit rate. A signal consisting of alternating 0s and 1s will have the fastest signal transitions and therefore the highest frequency components.

Figure 6-7 shows the theoretical frequency spectra of an FSK signal. Note that the signal spectra consist of the mark and space frequencies and

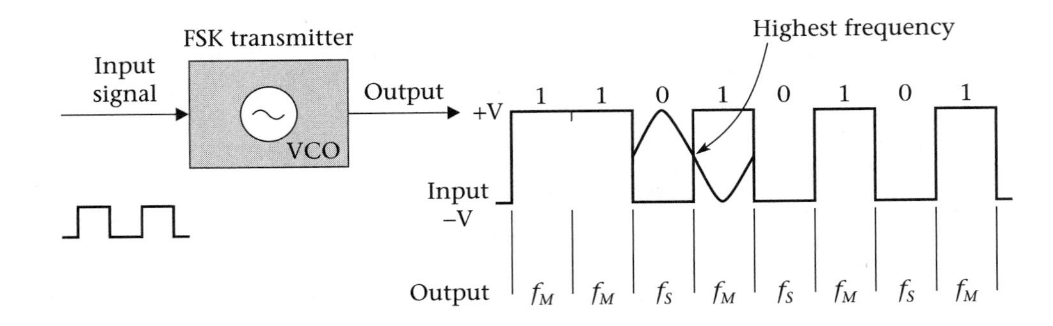

FIGURE 6-6 FSK transmitter and data rate

sidebands around each of these frequencies. In practice, the total bandwidth (BW) is equal to the difference in frequency between the two output frequencies and twice the bit-rate frequency. This can be written mathematically as:

$$\mathrm{FSK_{BW}} = (f_M - f_S) + 2\left(\frac{1}{t_b}\right)\ \mathrm{Hz} \tag{6.3}$$

where t_b is the bit time.

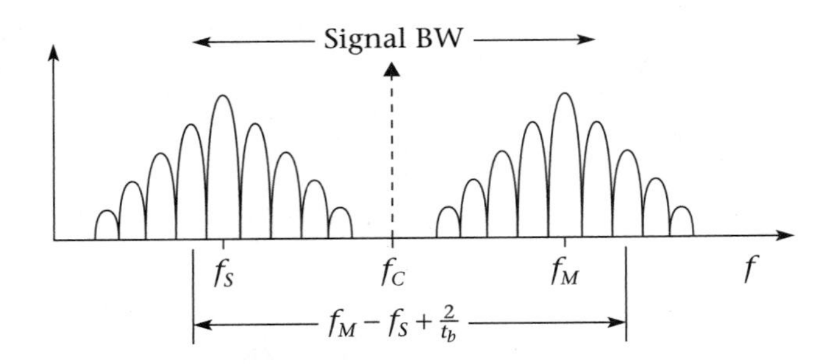

FIGURE 6-7 A typical FSK frequency spectra

The frequency spectra of an FSK signal, with f_M = 960 kHz, f_S = 1040 kHz, and f_C = 1000 kHz, is shown in Figure 6-8. The bit-rate frequency is 5 kHz.

EXAMPLE 6.1

Determine the bandwidth requirements for the early Bell 103 modem. This early dial-up modem was used over the voice-grade circuits of the PSTN to provide remote access to computers. To achieve full-duplex operation (transmission in both directions at the same time) it divided the available voice-grade circuit bandwidth into two channels. The low band was from 300 to 1650 Hz and the upper band was from 1650 to 3000 Hz. The low

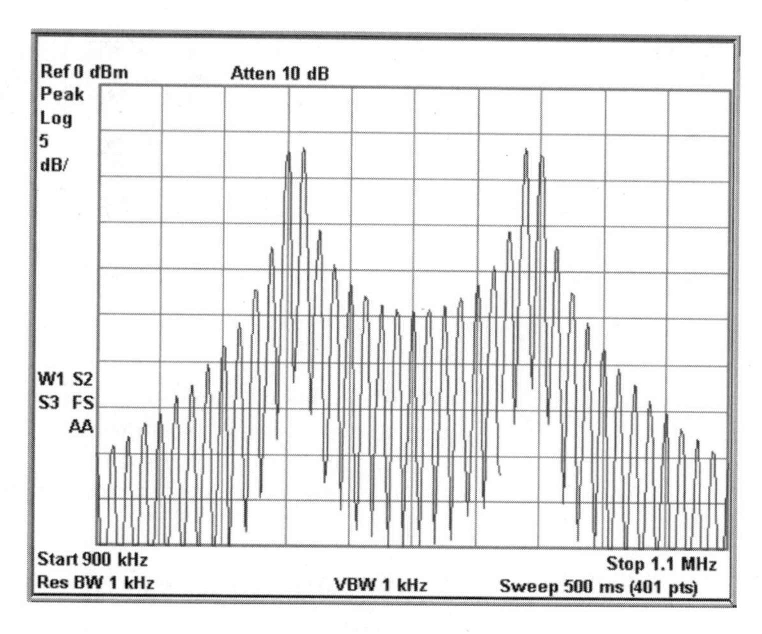

FIGURE 6-8 The frequency spectra of an FSK signal with a bit rate of 5 Kbps

band used mark and space frequencies of 1070 Hz and 1270 Hz and a data rate of 300 bps.

Solution The theoretical bandwidth of the lower channel was:

$$BW = 1270 - 1070 + 2\left(\frac{1}{t_b}\right)$$

and $t_b = \frac{1}{300}$; therefore,

$$BW = 200 + 2(300) = 800 \text{ Hz}$$

A similar calculation would yield the same results for the upper channel.

As can be seen by example 6.1, the limited bandwidth of the available channel was the limiting factor and set the maximum data rate. The same limitation in available bandwidth exists today if one talks about the local loop of the PSTN.

FSK Demodulation

How is FSK detected? Presently, an IC phase-locked loop (PLL) subsystem is used for this type of receiver. Figure 6-9 on page 218 shows a typical FSK

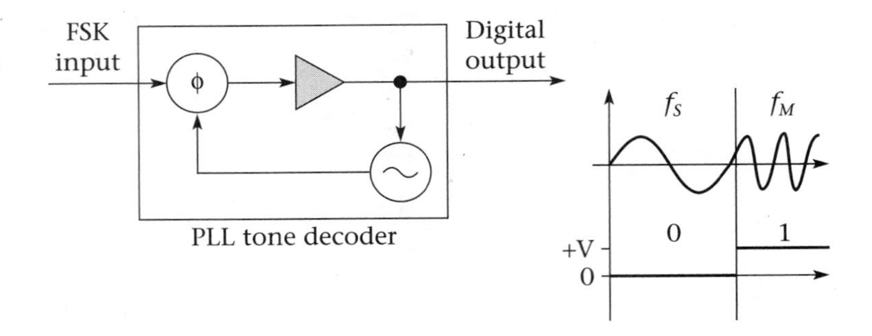

FIGURE 6-9 Typical FSK receiver function using a PLL

receiver. The voltage-controlled oscillator (VCO) of the PLL is set to either the space or mark frequency. When the signal's input frequency is equal to the frequency of the PLL's VCO the output is a digital 1; if the input frequency is at the other frequency, the output is a digital 0.

Minimum-Shift Keying

A variation on FSK is called minimum-shift keying. Standard FSK allows for abrupt phase changes when the output frequency changes due to an input-logic change (i.e., 0 to 1 or 1 to 0). **Minimum-shift keying** (MSK) synchronizes the mark and space frequencies with the bit rate (see Figure 6-10). The advantage of doing this is that frequency changes are continuous, yielding a narrower bandwidth signal. MSK requires that the mark and space frequencies be related to the center frequency in the following manner:

$$f_S \text{ and } f_M = n(f_b/2), \text{ Where } n = 1, 3, 5, 7, \ldots \quad \textbf{6.4}$$

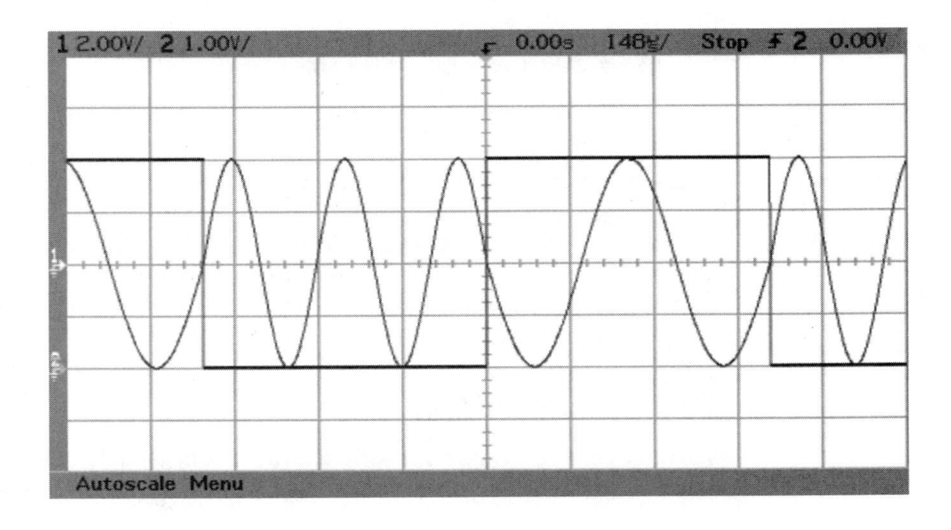

FIGURE 6-10 A typical MSK system with f_S = 3 kHz and f_M = 5 kHz

If we examine the output frequency spectrum of an MSK signal, we see a bandwidth that is equal to one-half the bit rate (see Figure 6-11). Again, with a limited bandwidth channel this is very important because it improves the channel efficiency. MSK is used extensively in cellular telephone systems.

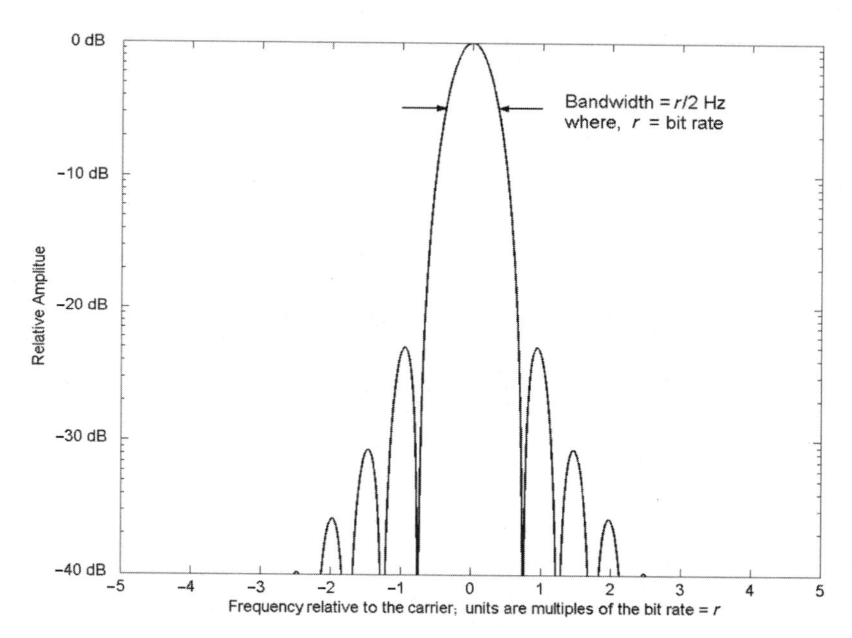

FIGURE 6-11 Typical frequency spectra of an MSK system

6.4 Second Generation Digital-Modulation Schemes

To this point, the digital-modulation schemes presented have only used either the signal amplitude or frequency to encode 0s and 1s. Furthermore, these schemes only transmit one bit per bit time. More complex second generation digital-modulation schemes allow for bandwidth efficiency because they can transmit multiple bits per bit time. In general, these newer schemes allow several signal parameters to be varied simultaneously. This section will examine several of these systems in detail and point out the corresponding bandwidth efficiencies.

Phase-Shift Keying

Phase-shift keying is a form of angle modulation. However, instead of allowing a continuous range of input voltages and corresponding output phase changes like FM or PM, phase-shift keying is limited to a finite number of output phases and the input signal is a binary digital signal. A typical **binary PSK** (BPSK) transmitter is shown in Figure 6-12. A BPSK transmitter has two possible output states that represent either logic 1 or 0.

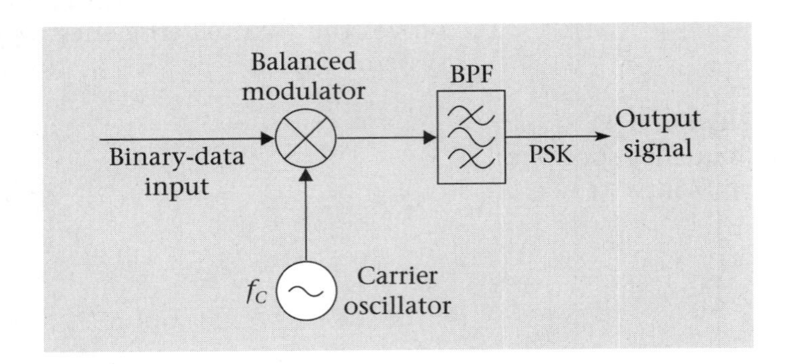

FIGURE 6-12 A typical BPSK transmitter

The Generation of Phase-Shift Keying

The heart of this BPSK transmitter is the balanced modulator introduced in chapter 3. Depending upon the polarity of the signal applied to this subsystem, the output phase of the signal will take on one of two values. It is instructive to see how this happens; therefore, a more detailed view of a typical balanced modulator is shown in Figure 6-13.

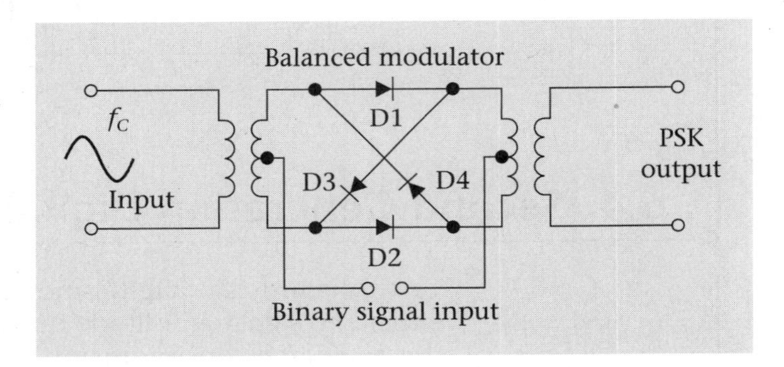

FIGURE 6-13 A typical balanced diode-ring modulator

As one can see from the figure, depending upon the polarity of the input signal, the two pairs of diodes (diode pair 1 and 2 or diode pair 3 and 4) will be either forward or reverse biased. These two possibilities are shown in Figure 6-14. As can be seen, if the binary input signal is positive (a logic 1) there is no signal phase change between the input and the output. If the binary input is negative, the output becomes inverted, or 180 degrees out of phase.

There are several commonly used graphical ways to represent the digital-modulation process: with a truth table, a **phasor** or **signal-space diagram**, or a **constellation diagram**. All three of these options are shown in Table 6-1 and Figures 6-15a and 6-15b, respectively.

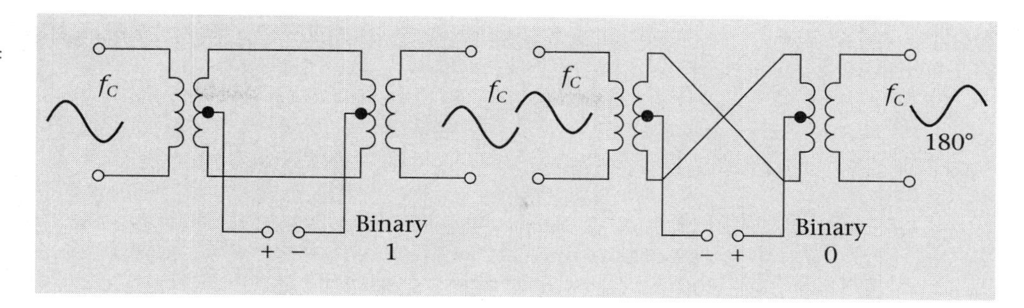

FIGURE 6-14 The two possible states of the balanced modulator

TABLE 6-1 A BPSK truth table

Binary input	*Signal output phase*
Logic 0	180°
Logic 1	0°

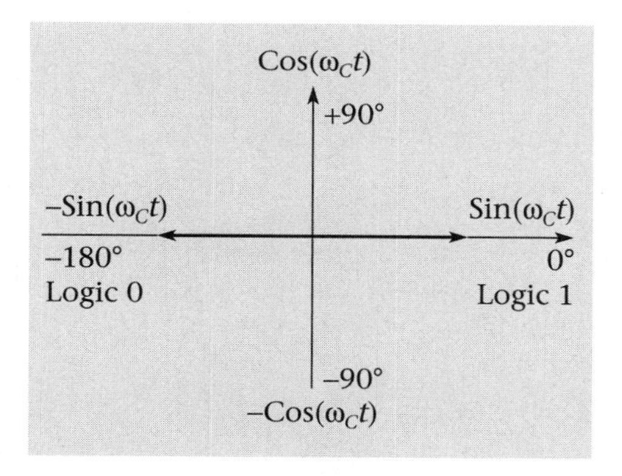

FIGURE 6-15a A phasor or signal-space diagram for BPSK

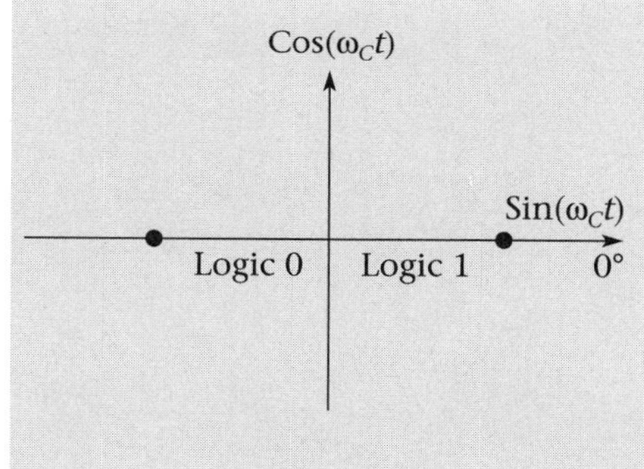

FIGURE 6-15b A constellation diagram for BPSK

The phasor diagram shows the relationship of the output signals to the reference phase of the signal. The reference lies on the *x*-axis, the horizontal line pointing to the right. The constellation diagram (named for similarities to the stars in the sky) shows the two output states in point form (the tips of the phasors in the phasor diagram).

What does the BPSK signal look like in the time and frequency domain? The time-domain display of a coherent BPSK signal is shown in Figure 6-16 on page 222. Coherent BPSK consists of an integral number of cycles of the carrier wave transmitted per bit time. The bandwidth of a BPSK signal (worst case) is equal to approximately 1.5 times the bit rate. Figure 6-17 shows the theoretical spectra of a BPSK signal produced by a random

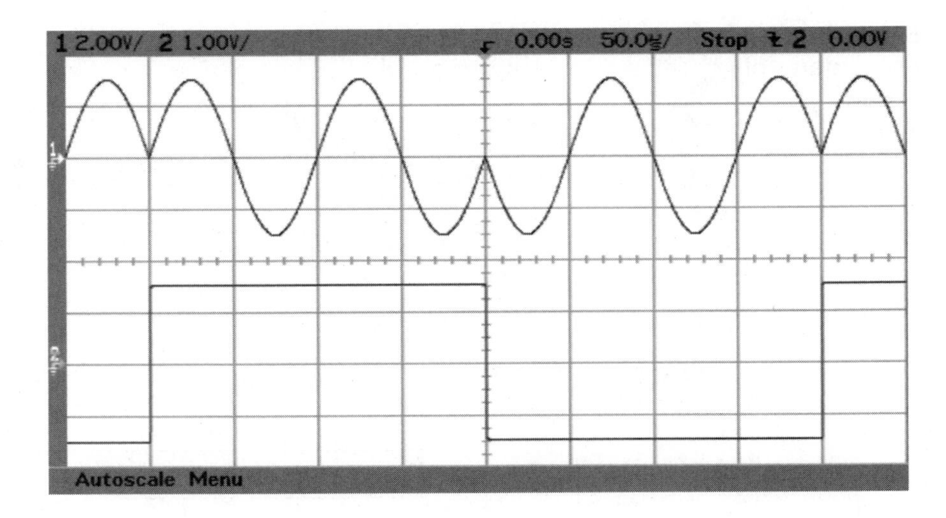

FIGURE 6-16 Time display of BPSK

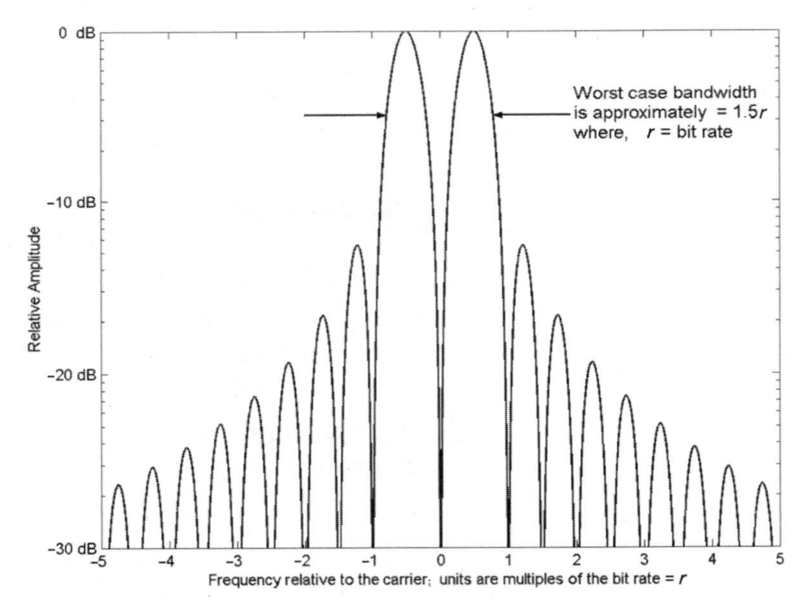

FIGURE 6-17 BPSK spectra

sequence of 1s and 0s. The frequency spectra of an actual BPSK signal is shown in Figure 6-18 on page 223. The center frequency is 400 kHz and the bit rate of an alternating sequence of 0s and 1s is 10.0 Kbps. Note the many sidebands produced by BPSK.

Phase-Shift Keying Receiver

To complete our coverage of BPSK we need to examine a typical BPSK receiver, depicted in Figure 6-19. As has been the case several times before,

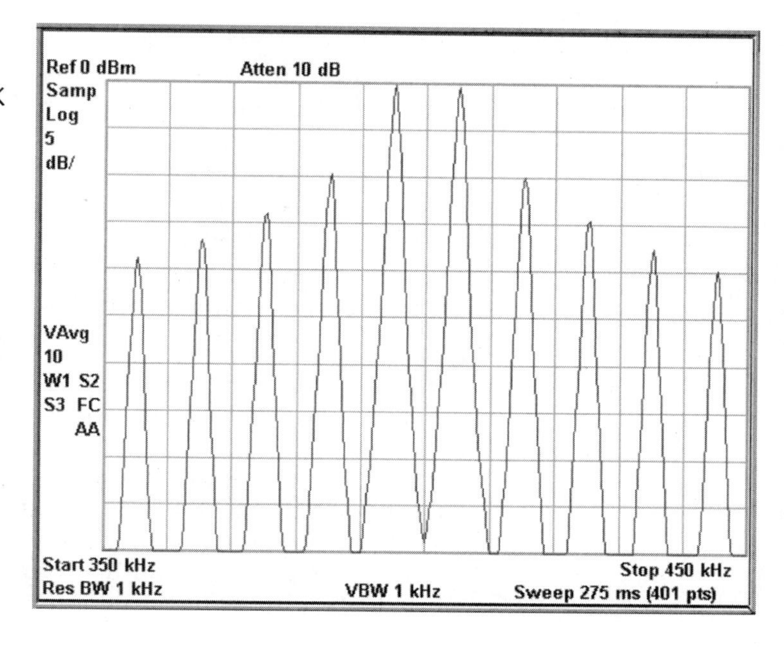

FIGURE 6-18 The frequency spectra of a BPSK signal

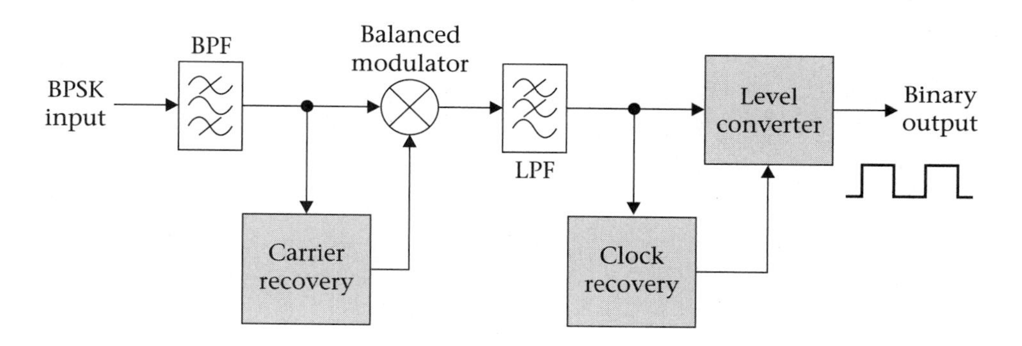

FIGURE 6-19 A typical BPSK receiver

the balanced modulator is the heart of this receiver. The BPSK input signal is selected with a BPF. A subsystem of the receiver recovers the carrier frequency; this signal plus the BPSK signal are applied to the balanced modulator. The balanced modulator effectively multiplies the two signals together. Mathematically, one input signal to the balanced modulator is either $\pm\sin(\omega_C t)$ and the other input to the balanced modulator is always $\sin(\omega_C t)$. Therefore, the two outputs will be either $\pm\sin^2(\omega_C t)$. Using trigonometric identities, the output for $\sin^2(\omega_C t) = \frac{1}{2} - \frac{1}{2}\cos(2\omega_C t)$ and the output for $-\sin^2(\omega_C t) = -\frac{1}{2} + \frac{1}{2}\cos(2\omega_C t)$.

The LPF at the output of the balanced modulator will filter out the $\cos(2\omega_C t)$ terms leaving $\pm\frac{1}{2}$. This signal will be interpreted as logic 1 or 0, as the case may be.

At this point we note that both FSK and BPSK have not yielded any significant gains in our attempt to make the channel as efficient as possible. However, this is about to change. There is no reason why one cannot encode more than one bit per symbol time.

Quaternary Phase-Shift Keying

Quaternary phase-shift keying (QPSK) uses four different output phases with each phase representing two binary bits. What is so important about this fact is that during the same bit time we can send twice the amount of digital information. At this time, we will stop referring to the basic time interval as a bit time and instead refer to it as a **symbol time**. This is because we can now transmit a symbol during the basic time interval that can represent n bits of data (where $n = 1, 2, 3, 4 \ldots$). So, restating what has just been said, we can transmit more than one binary bit per symbol time. Mathematically, we have the following:

$$BW = \frac{\text{bit rate}}{N} \qquad \textbf{6.5}$$

where N is the number of bits transmitted per symbol time. It can be seen that as N increases the required bandwidth decreases. This is referred to as **bandwidth efficiency**. Look at the QPSK truth table in Table 6-2 and the constellation diagram in Figure 6-20.

TABLE 6-2 A QPSK truth table

Binary input I and Q bits	*QPSK phase*
10	$\frac{\pi}{4}$ or +45°
00	$\frac{3\pi}{4}$ or +135°
01	$\frac{5\pi}{4}$ or −135°
11	$\frac{7\pi}{4}$ or −45°

As can be seen from the truth table and the constellation diagram, each signal output phase encodes two binary bits. How is QPSK generated? A typical transmitter is shown in Figure 6-21.

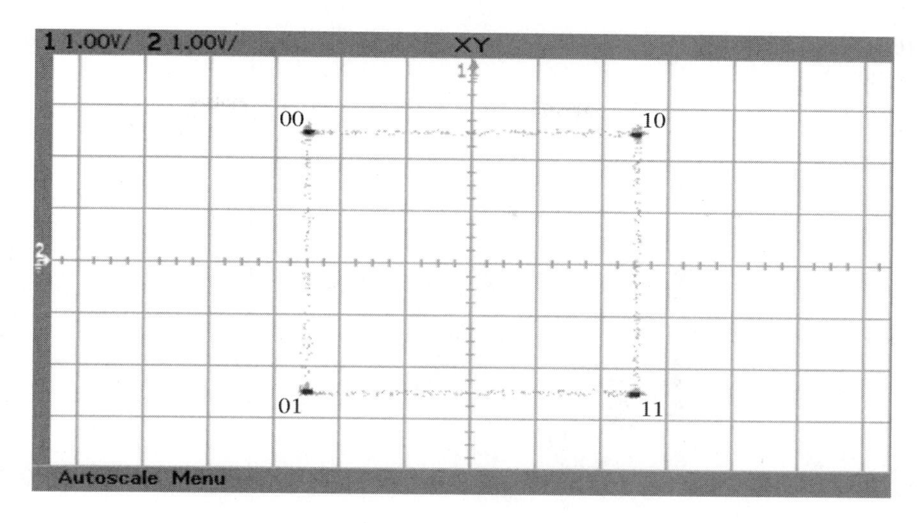

FIGURE 6-20 A QPSK constellation diagram displayed on an oscilloscope. From top right, in a counter-clockwise direction, the points represent 10, 00, 01, and 11.

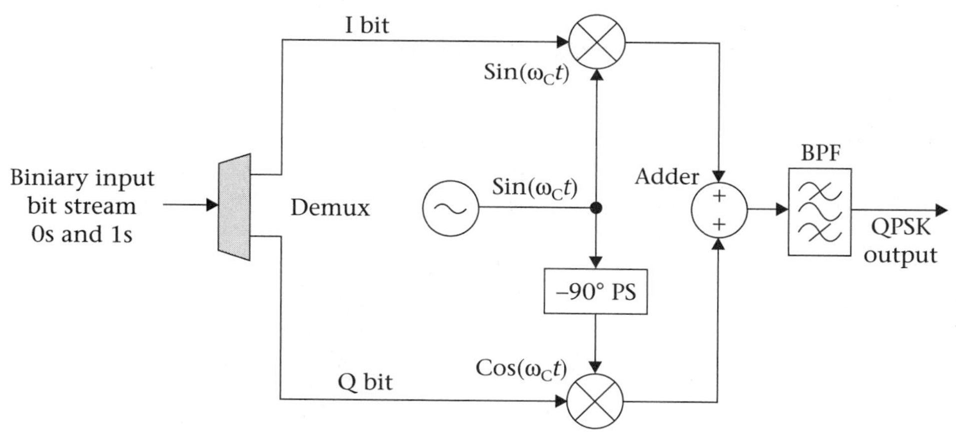

FIGURE 6-21 A typical QPSK transmitter

Notice that the binary-input bit stream is clocked into a demultiplexer and then two bits are simultaneously output to separate balanced modulators. The two separate channels have been named the I channel and the Q channel. The **I channel** feeds the balanced modulator that receives the in-phase signal from the reference carrier oscillator; the **Q channel** feeds the balanced modulator that receives the same signal 90 degrees out of phase, or in **quadrature**, with the reference signal. The I-channel balanced modulator will output a signal of $\pm\sin(\omega_C t)$, while the Q-channel balanced modulator will output a signal of $\pm\cos(\omega_C t)$. These two signals are linearly added and bandpass filtered to generate the QPSK output. A time display of the four possible outputs is shown in Figure 6-22 on page 226. The reader should also refer back to the constellation diagram in Figure 6-20 for a comparison of the time display and the phase angles of the constellation points.

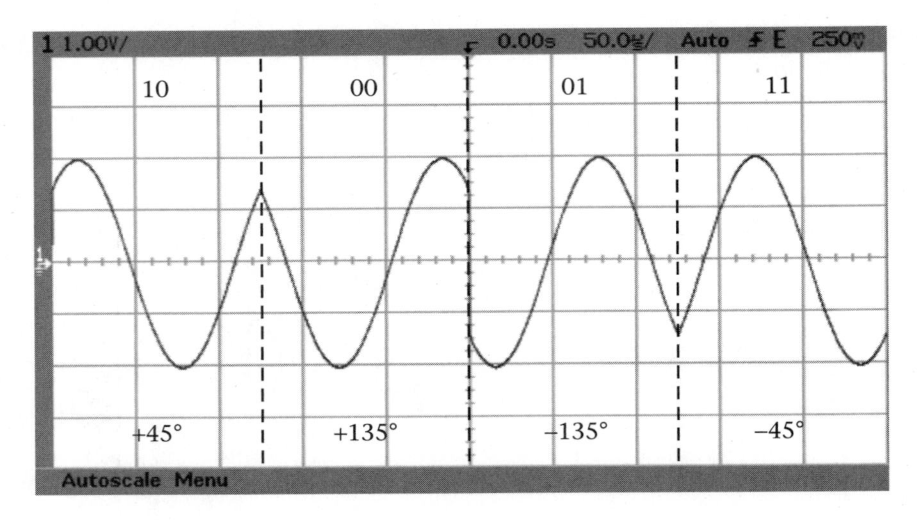

FIGURE 6-22 QPSK transmitter output

How is QPSK received? A QPSK receiver is shown in Figure 6-23. The QPSK receiver must recover the carrier signal and then use this signal to drive the I-channel balanced modulator and the Q-channel balanced modulator after a 90-degree phase shift. The output from the I-channel balanced modulator feeds the I bit to an output multiplexer, while the Q-channel balanced-modulator output feeds the Q bit to the same multiplexer. The two bits are clocked out of the multiplexer after one symbol time.

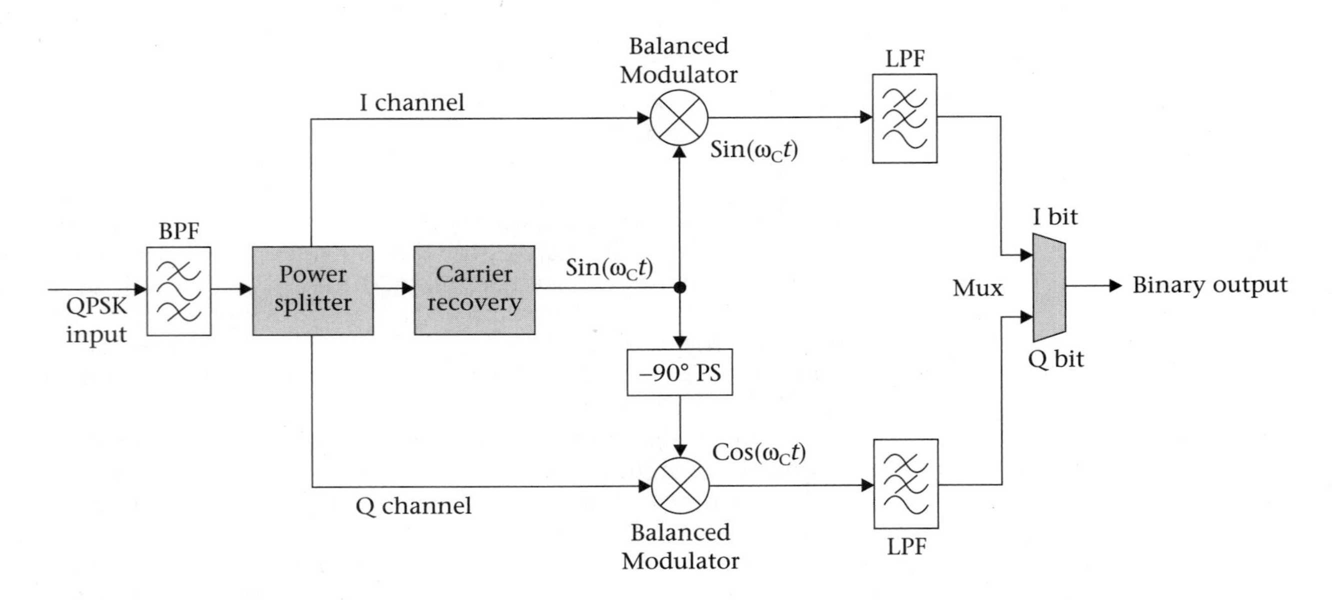

FIGURE 6-23 A typical QPSK receiver

Baseband versus Passband Systems

It should be pointed out that digital-modulation techniques can be used for data transmission over baseband channels as well as at radio frequencies or as **passband** signals. The transmitters and receivers for QPSK shown in previous figures will work just as well for both applications. For baseband operation, the transmitter hardware can be replaced by a read-only memory (ROM), DACs, and a low-pass filter to generate the signal through **direct digital-frequency synthesis** (DDFS). One might consider DDFS as being the opposite of signal sampling, which was described in chapter 5. A large number of data points are stored in memory and then sequentially accessed and applied to a DAC. The output of the DAC is passed through an LPF or processed by a digital filter implemented with a DSP to eliminate the harmonic energy contained in the resulting signal. This is the same type of technology used to implement the present-day arbitrary waveform generator, the successor to the laboratory-function generator.

QPSK Advantages and Alternative Implementations

Let us pause to review what has been gained with a QPSK system. For each symbol time, two binary bits are transmitted. The output-frequency spectrum for this system is identical to the BPSK system. Therefore, twice the information can be sent over the same bandwidth or, conversely, one-half the bandwidth is needed to send the same amount of information.

Several variations of QPSK exist. One is called **offset QPSK** (OQPSK). It is used primarily in cellular-telephone applications and has the advantage of placing less-stringent linearity requirements upon the active device used as the transmitter's final-power amplifier (FPA) due to smaller phase shifts (90 versus 180 degrees) between outputs. Another variation is $\frac{\pi}{4}$ differential QPSK ($\frac{\pi}{4}$ DQPSK). This type of modulation is also used for cellular-telephone applications for the same reason just stated for OQPSK. Still another variation is minimum-shift keying (MSK) QPSK. In this system, which is used by the GSM cellular system, there are four output phases, but the transmitter power remains constant when transitioning from one phase angle to another.

Phase-Shift Keying Limits

Can we keep increasing the number of PSK outputs to generate *m*-PSK, where $m = 2^n$? The answer is *no* and the reason is simple. The phase spacing (recall the constellation diagram) between the possible symbols will eventually become too close and noise (phase jitter) will degrade the system to the point at which the bit error rate becomes intolerable.

Presently, 8-PSK systems are deployed for various services. An 8-PSK transmitter will encode three binary bits at a time and transmit them during one symbol time. This means that 8 output phases are possible, each separated by 45 degrees. The hardware used to implement 8-PSK becomes more complex; one popular implementation actually utilizes a digital form of pulse amplitude modulation (PAM) as part of the encoding process. It should be pointed out that any of the systems discussed so far are implemented in ASIC form. Also, recall that 8-PSK affords a threefold increase in bandwidth efficiency because three binary bits are sent per symbol time. 16-PSK systems are certainly possible; however, 16-PSK can only be used over relatively noise-free channels because the phase difference between output symbols shrinks to 22.5 degrees and correct decoding becomes problematic. With 16-PSK, four binary bits can be sent per symbol time, causing the bandwidth efficiency to become fourfold.

6.5 Other Second Generation Digital-Modulation Schemes

The *m*-PSK systems examined to this point have used discrete phase shifts to encode a 0 or 1 or combinations of 0s and 1s. At this time we are going to introduce another digital-modulation scheme that uses discrete values of several signal parameters to encode several to many binary bits per symbol time.

Quadrature Amplitude Modulation

Our next topic is **quadrature amplitude modulation** (QAM). In this form of digital modulation, information resides in both the phase and amplitude of the transmitted signal! Let us start our discussion of QAM by examining the truth table and constellation diagrams of an 8-QAM system. As can be seen from any one of these representations in Table 6-3 and Figure 6-24, QAM is a hybrid form of analog AM and PM. The lone restriction is that only finite values of amplitude and phase are allowed. For 8-QAM, three binary bits are encoded and transmitted per symbol time. 8-QAM is identical to 8-PSK in terms of bandwidth reduction or efficiency, for each system yields a threefold increase in bandwidth efficiency.

What does 8-QAM look like in the time domain? Figure 6-25 on page 230 shows an 8-QAM signal.

We will now examine a possible implementation for a 16-QAM transmitter. As can be seen in Figure 6-26, this 16-QAM transmitter takes four binary bits at a time and encodes them. The four bits are broken up into two pairs of two bits each. One pair of bits drives a digital pulse-amplitude modulator (4-level PAM), which in turn feeds the I-channel balanced

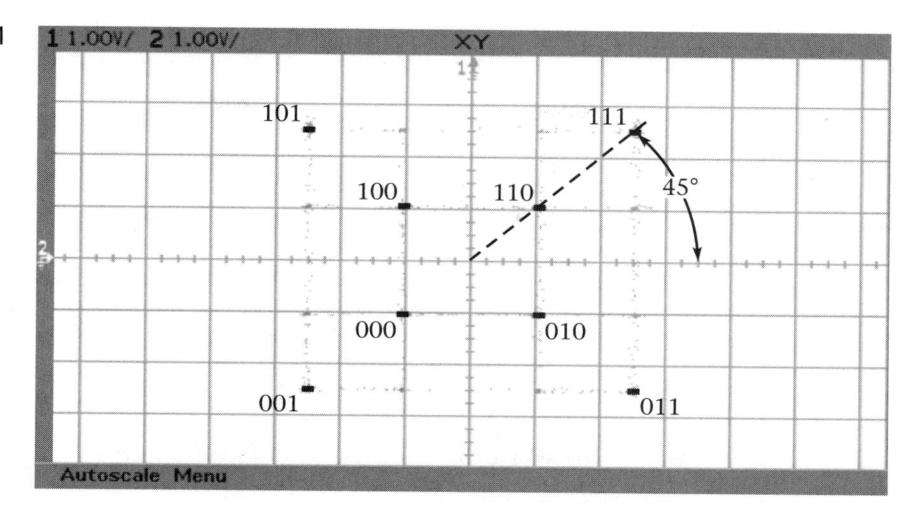

FIGURE 6-24 8-QAM constellation diagram

TABLE 6-3 An 8-QAM truth table

Binary input	*8-QAM output amplitude and phase*
000	1.0 V −135°
001	3.0 V −135°
010	1.0 V −45°
011	3.0 V −45°
100	1.0 V +135°
101	3.0 V +135°
110	1.0 V +45°
111	3.0 V +45°

modulator. The other pair of bits performs the same operations for the Q channel. The two balanced-modulator outputs are summed to form the 16-QAM signal. As mentioned before, a baseband QAM transmitter can be implemented with direct digital-synthesis technology, or a passband system can be implemented with a system like that shown in Figure 6-26 on page 230.

The question again arises, Are we limited in the number of bits we can encode per symbol time with QAM? Again, the answer is that there is a practical limit to the number of bits encoded because of noise considerations. Presently, 64-QAM and 256-QAM systems are operational. With 256-QAM, 8 bits (a byte) are transmitted per symbol time, thus yielding an eightfold increase in bandwidth efficiency! This is a rather impressive gain.

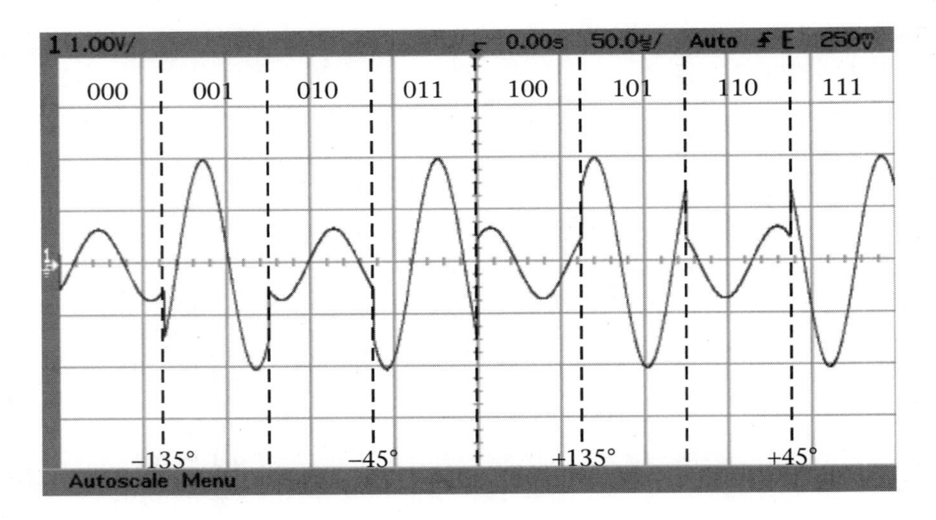

FIGURE 6-25 A time display of 8-QAM

It should also be pointed out that the SNR required to have low bit error rates increases as the number of QAM levels increases.

An Example of a Modern Modem

Earlier in the chapter example 6.1 was used to illustrate signal bandwidth for the Bell 103 modem. Let us refamiliarize ourselves with a look at the

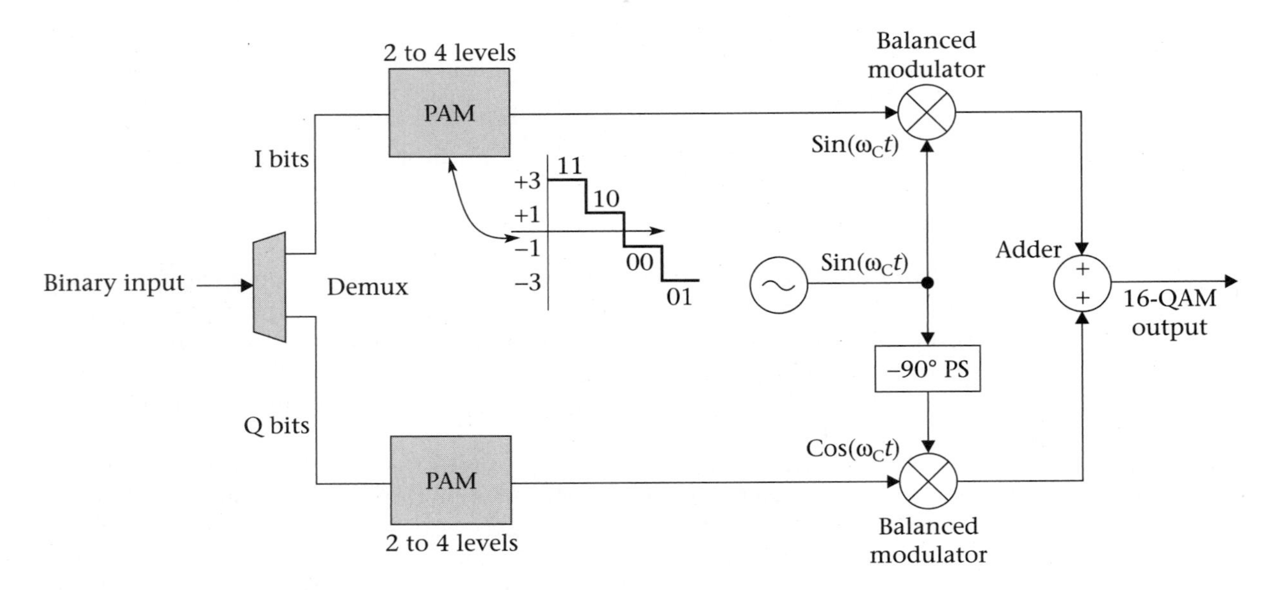

FIGURE 6-26 A typical 16-QAM transmitter

standard V.34 modem. A V.34 modem operates at data rates up to 33.6 Kbps. The design of this modem incorporates several distinctive features:

1. It detects the frequency response of the channel by sending test tones. This information is then used to set an appropriate carrier frequency and bandwidth.
2. The modem sets the bit rate according to the receiver's estimation of the maximum possible bit rate in conjunction with a maximum allowable BER.
3. The modem uses a form of **Trellis coding**. Trellis-coded modulation (TCM) is an error-control coding technique that provides an effective coding gain in the SNR of the channel.
4. A form of decision-feedback equalization is used to extend the channel bandwidth near the band edges.
5. It uses a form of 960-QAM (a super constellation).

The V.34 modem utilizes as much of the typical PSTN channel (typically about 3.5 kHz) as it possibly can to achieve the highest possible data rate. This is certainly an improvement upon the early Bell 103 modem!

6.6 Bandwidth Efficiency

Table 6-4 on page 232 is presented to summarize our coverage of digital-modulation techniques by looking at the bandwidth efficiency of the various schemes.

Trellis-Code Modulation

Let us return to equation 6.2, Shannon's theorem for information capacity. If we desire a certain data-transmission rate over a band-limited channel, Shannon's theorem can predict the required signal-to-noise ratio (SNR). Assume that we desire a 56-Kbps data rate over a channel with 3100 Hz of bandwidth. From,

$$I = 3.32BW\log(1 + S/N)\ \text{bps}$$

we can calculate that the necessary SNR is 54.4 dB.

As mentioned briefly in our discussion of the V.34 modem, recent developments in **Trellis-code modulation** have, in effect, increased the channel's signal-to-noise ratio, and thus increased the maximum data-transmission rates available over the PSTN. Trellis coding implements a form of controlled redundancy that effectively reduces the probability of transmission errors.

TABLE 6-4 Digital-modulation bandwidth efficiencies

Modulation Type	*Encoding (bits)*	*Bandwidth (Hz)*	*Bandwidth efficiency*
FSK	1	$\geq f_b = \frac{1}{t_b}$	≤ 1
BPSK	1	$\geq f_b$	≤ 1
QPSK	2	$\frac{f_b}{2}$	2
8-PSK	3	$\frac{f_b}{3}$	3
8-QAM	3	$\frac{f_b}{3}$	3
16-QAM	4	$\frac{f_b}{4}$	4
64-QAM	6	$\frac{f_b}{6}$	6
256-QAM	8	$\frac{f_b}{8}$	8

6.7 Bit Error Rates for Digital-Modulation Schemes

As a practical matter, all of the digital-modulation schemes presented so far will have a finite probability of error. Mathematically, this probability of error can be predicted, and is usually stated as one bit error per some number of transmitted bits. It is possible to measure the BER, and the measurement is usually stated as BER = 10^{-x}, where x is an integer.

For analog-modulation schemes we talked about the signal-to-noise ratio as a measure of the quality of the received signal. For digital-modulation schemes the probability of an error is a function of the carrier-to-noise (C/N) power ratio and the energy per bit-to-noise power ratio (E_b/N_0). These measures are analogous to the SNR for analog modulation. The mathematical analysis of the probability of error, P_{error}, is rather complex and will not be shown here. However, some error performance curves will be shown for some of the schemes discussed so far.

PSK Bit Error Rates

Figure 6-27 shows the bit-error performance curves for various levels of PSK. From these plots one can see that the probability of error increases as the number of levels increases (phase spacing decreases) and the value of SNR (E_b/N_0) decreases. These are not unexpected results.

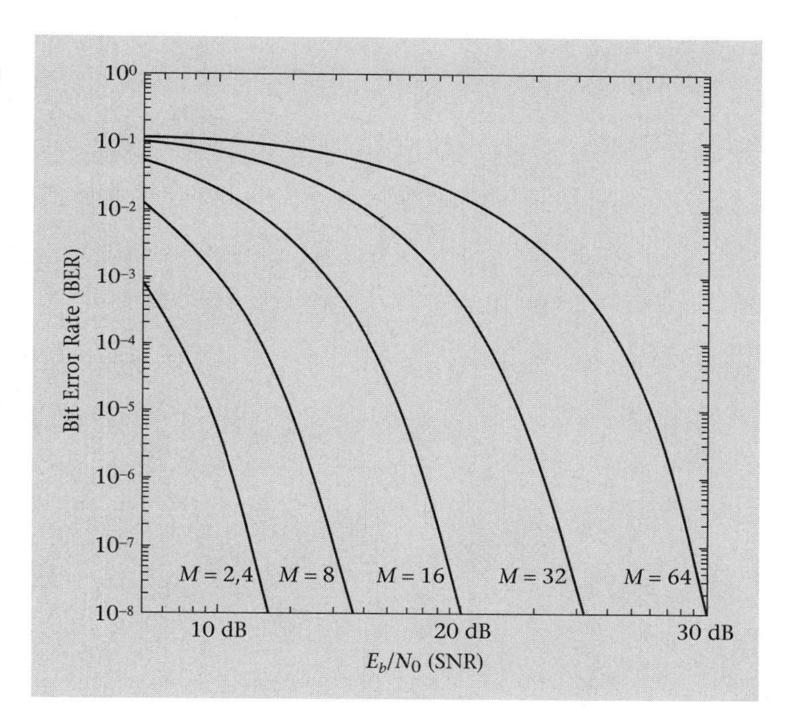

FIGURE 6-27
Bit error rates of PSK modulation

QAM Bit-Error-Rate Performance

Figure 6-28 on page 234 shows the same type of error-rate information for QAM. Again, it can be seen that the probability of error increases for increasing numbers of levels and decreasing values of (E_b/N_0).

6.8 Third Generation Digital-Modulation Schemes

As discussed earlier, the third generation of digital-modulation schemes includes vector-type signals that also incorporate time variance or complex coding to gain access to a channel. Time-division multiple access (TDMA) or code-division multiple access (CDMA) schemes would be examples of

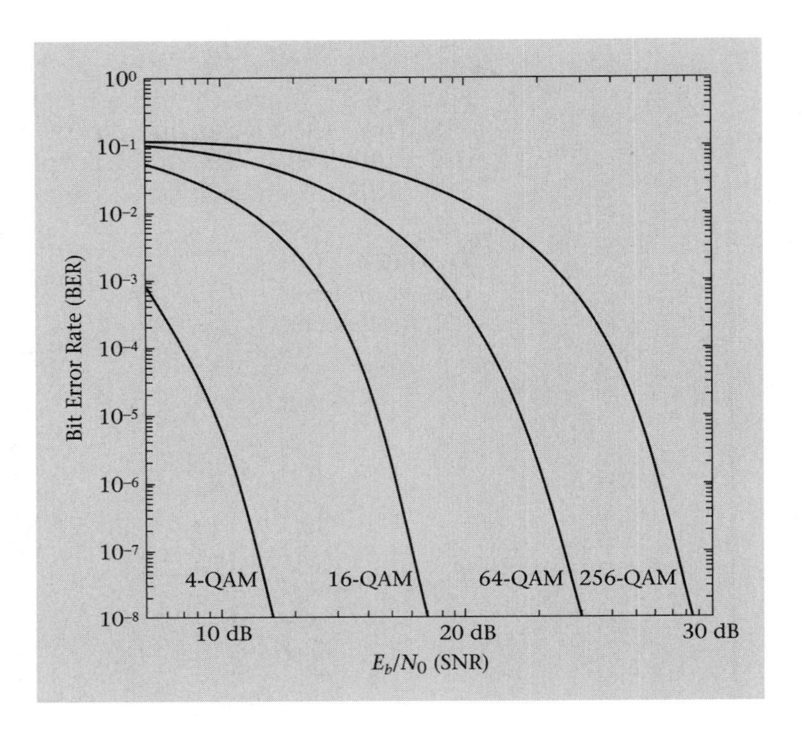

FIGURE 6-28 Bit error rates of QAM modulation

third generation digital modulation. CDMA, a form of spread-spectrum transmission, allows multiple signals to use the same channel simultaneously. These topics will be covered in more detail in chapter 7.

Summary

Let us conclude our study of digital-modulation techniques by restating the basic use of this technology: Digital modulation is primarily used for the transmission of digital data. This digital data can simply be computer data or it can be multimedia content that has been digitally encoded and is being transmitted from computer to computer. Pulse-modulation techniques, described in chapter 5, tend to be used for the transmission of speech. Digital modulation can be summarized fairly easily. Digital modulation uses more complicated combinations of traditional modulation schemes to achieve increased information capacity over a limited bandwidth channel. Recall our trade-off of increased complexity for reduced bandwidth! The two most common digital techniques are m-PSK and m-QAM, where m is a power of 2.

In m-PSK, a transmitted symbol (a certain phase angle) can encode one or more bits of information. When more than one bit is encoded per symbol, bandwidth efficiency is obtained. If one is dealing with a fixed bandwidth, the amount of information that can be transmitted over the channel can be increased. For wireline media, this could also translate into the same information rate, but over a longer distance due to a reduction in needed bandwidth.

In *m*-QAM, a transmitted symbol (a certain phase angle and amplitude) can encode several bits of information. 256-QAM transmits 8 bits per symbol time. This allows for either 8 times the channel capacity or one-eighth the required bandwidth. Both PSK and QAM techniques have been used in modern modems to extend the bit rate that is possible over the PSTN. They can also be used at RF. This makes them suitable for use by cable-TV providers, cellular operators, microwave radio, high-definition and digital TV, direct-satellite broadcasts, and so forth.

Digital modulation also provides compatibility with digital data services, higher data security, and has better noise immunity than traditional modulation schemes.

Questions and Problems

Section 6.1

1. What has been the driving force behind the conversion from analog to digital-modulation techniques?

Section 6.2

2. What does Hartley's law state about bandwidth and transmission time?
3. What is Shannon's theorem?
4. What effect does noise have on the maximum information rate of a band-limited channel?
5. What has been taking place with the hardware requirements for the transmission of digital signals?

Section 6.3

6. Describe amplitude-shift keying (ASK).
7. Describe the basic operation of frequency-shift keying (FSK).
8. In theory, using FSK, what is the maximum duplex data rate that could be obtained over the PSTN? Consider it to have a bandwidth of 3.5 kHz.
9. Do an Internet search on the 567 Tone-Decoder IC. Write a short description of its operation.
10. What advantages does minimum-shift keying (MSK) have?

Section 6.4

11. Describe the process of phase-shift keying (PSK).
12. What is meant by the term *constellation diagram*, and what does it represent?
13. What is the bandwidth of a binary PSK signal with a bit time of 100 μsec and a random sequence of 0s and 1s?
14. Describe the process of quaternary PSK (QPSK).
15. Define a symbol time. How does it relate to the bit rate?
16. What are the I and Q channels?

17. Explain the difference between baseband and passband transmission of digital signals.
18. What is direct digital-frequency synthesis?
19. Do an Internet search to learn about offset QPSK or $\frac{\pi}{4}$ DQPSK operation.
20. What limits the number of PSK levels?

Section 6.5

21. Describe the process of quadrature amplitude modulation (QAM).
22. What advantages are there to QAM?
23. Do an Internet search on Trellis coding. Write a short report about its benefits, uses, and so forth.
24. Do an Internet search on modem standards. What is the newest standard? What are the types of digital techniques used by this standard?
25. What is the basic limitation to the number of QAM levels?

Section 6.6

26. Compare the bandwidth efficiencies of the first generation digital-modulation systems to those of the second generation systems.

Sections 6.7 and 8

27. Why do the bit error rates increase for a decrease in E_b/N_0?
28. Why do the bit error rates increase for an increase in the number of levels used to implement a digital-modulation system?
29. Do an Internet search to discover the type of modulation used by the Cable-TV industry to deliver digital-TV signals. What type of signals are these?
30. Name some of the third generation digital-modulation schemes. Do an Internet search on one of these schemes and report what you find.

Chapter Equations

$$I \propto B \times T \tag{6.1}$$

$$I = 3.32BW \log(1 + S/N) \text{ bps} \tag{6.2}$$

$$\text{FSK}_{\text{BW}} = (f_M - f_S) + 2\left(\frac{1}{t_b}\right) \text{Hz} \tag{6.3}$$

$$f_S \text{ and } f_M = n(f_b/2), \text{where } n = 1, 3, 5, 7, \ldots \tag{6.4}$$

$$\text{BW} = \frac{\text{bit rate}}{N} \tag{6.5}$$

Multiplexing and Access Technologies

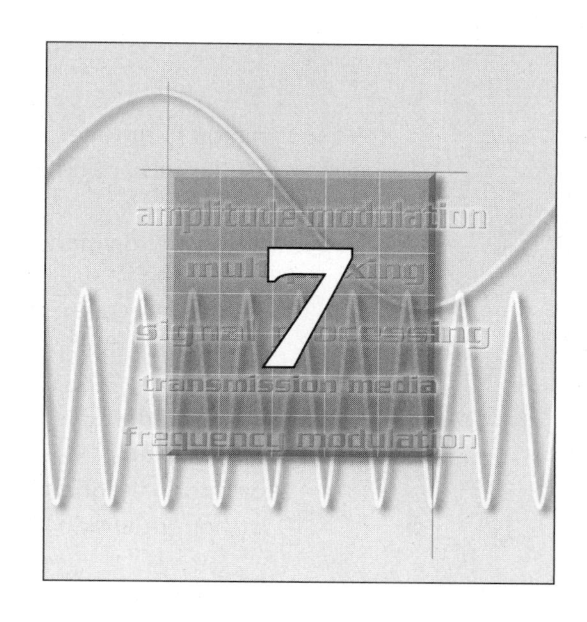

Objectives Upon completion of this chapter, the student should be able to:

- Discuss the need for multiplexing.
- Describe frequency-division multiplexing (FDM) and its application to wireless and wireline channels.
- Describe time-division multiplexing (TDM).
- Discuss the evolution of TDM to present-day standards.
- Describe wavelength- and dense wavelength-division multiplexing (WDM and DWDM, respectively) and its application to fiber-optic cables.
- Describe code-division multiple access (CDMA).
- Describe time-division multiple access (TDMA).
- Describe spatial or geographic multiplexing.

Outline

Key Terms

access technologies
ADSL
BlueTooth
circuit switching
code-division multiple access
dense WDM
digital cross-connect switches

- digital multitone
- digital subscriber line
- downstream
- frequency-division multiplexing
- frequency reuse group
- geographic multiplexing
- ISM bands
- MXY multiplexers
- orthogonal FDM
- orthogonal signals
- orthonormal signals
- personal-area networks
- party line
- phase-division multiplexing
- pseudonoise
- Quality of Service
- refarming
- space-division multiplexing
- split band
- spread spectrum
- super group
- super master group
- time-division duplex
- time-division multiple access
- time-division multiplexing
- time-slot interchange switch
- upstream
- Walsh codes
- wavelength-division multiplexing

Introduction

Multiplexing is used to separate different users of the channel/transmission medium or to allow more than one user over the same physical medium. During the late 1800s, Alexander G. Bell and Thomas Edison, amongst others, developed practical, albeit limited, electromechanical multiplexing schemes for the telegraph. Starting at the beginning of the twentieth century and continuing to today, frequency-division multiplexing (FDM) has been used by radio services and television broadcasting. In the United States the FCC assigns different frequencies to broadcasters and other radio-service users to prevent interference. The telecommunications industry first used multiplexing to reduce the number of facilities needed between cities or towns. Physical channels tend to be expensive to install and maintain; therefore, without multiplexing most telecommunications systems would tend to be very inefficient. This chapter describes the various forms of classic guided-wave multiplexing schemes with examples taken from the telecommunications industry. The student is introduced to the newest forms of multiplexing and accessing schemes that use digital-modulation techniques for data transmission over the local loop and through free space and fiber-optic channels.

7.1 Introduction to Multiplexing

This chapter is going to present an overview of multiplexing and access technologies. We start with a definition of multiplexing:

> Multiplexing is the technique used to separate different users of the electromagnetic spectrum or other types of transmission media.

We might further define the various transmission media or telecommunications channels by subdividing them into those that guide electromagnetic waves and electrical signals and those that do not. Examples of the first type of media are twisted-wire pairs, coaxial cables, and fiber-optic cables. With these media signals are physically guided from the source to the destination (a point-to-point type of operation). Free space is an example of the other type of transmission media in which the propagating electromagnetic signals are essentially unguided waves. This type of electromagnetic wave can be aimed at a destination via an antenna or can be launched into space for all to have access to, as typified in broadcasting applications. Unguided waves bring a whole new set of circumstances for efficiently getting energy from point A to point B and for sharing the channel. This chapter will not discuss these issues to any great extent, but one should be aware of them.

The Use of the Electromagnetic Spectrum

The electromagnetic spectrum has always been a precious and limited resource. The ability to generate electromagnetic signals and transmit them either through free space or over some type of conductor is a process that has been slow to evolve. Over the course of time, the type of signal that we have desired to transmit has generally increased in bandwidth as a result of an increase in information content and a desire for increased noise immunity. This need for increased bandwidth has usually resulted in a move to increasingly higher frequency bands that provide the necessary bandwidth. The current desire to transmit vast quantities of data from computer to computer has placed a strain on our telecommunications infrastructure. This has never been more apparent than with today's service providers scurrying to upgrade their facilities in an attempt to give consumers access to high data-rate services that require large bandwidth transmission facilities and the ever-increasing number of consumers demanding these services.

Early Multiplexing History

Historically, the first transmission medium for electrical signals was the telegraph cable. Many individuals like Elisha Gray, A.G. Bell, and Thomas Edison attempted to invent ways to send multiple independent signals over the cable. Various forms of what were known as *harmonic* or *multiplex telegraphs* were invented by Gray and Edison. These telegraph systems used a crude form of multiplexing in which different electromechanically generated currents or frequencies were transmitted that would only activate a particular receiver at the other end. Edison's quadruplex telegraph system relied on different current values to activate certain electromechanical receivers to the exclusion of other receivers. However, these systems were complex and, given the technology of the day, were difficult to adjust and operate. They were never successfully implemented as transparent systems.

The next advancement in electrical communications technology was the telephone, a by-product of Bell's efforts to improve the telegraph. The early telephone system used point-to-point transmission with a manual switchboard that was used to determine the final connection of the two end points. However, the **party line** was introduced, another idea that enhanced the sharing of facilities. With a party line, as many as ten subscribers could share the same wire cable. For this system to be successful (could this be the earliest definition of **Quality of Service**?), all the users would have to observe time constraints. That is, only one user at a time could use the wire connection unless otherwise agreed upon; therefore, this was an early form of time-division multiplexing. When the first wireless telegraph transmitters were put into operation, they also used a form of frequency-division multiplexing by employing unsophisticated LC (inductance and capacitance) tuning circuits to set their output frequency. With the advent of high-frequency alternators and then the introduction of reliable vacuum tubes, signals could be generated at set frequencies with more accuracy and with a stability determined by high Q-resonant circuits. In the early days of radio broadcasting, prior to government regulation, a new station could select whatever frequency they desired and just start transmitting.

The FCC's Role in the Implementation of Multiplexing

In the United States, as radio-transmission technologies improved and radio broadcasting became popular, the FCC was given the responsibility of allocating radio bands for certain services and of making channel assignments to broadcasting stations that applied for licenses to operate. Again, the primary form of multiplexing used to prevent interstation interference was frequency-division multiplexing (FDM). Another form of multiplexing known as space or geographic multiplexing was also employed. Because of signal-propagation limitations the FCC could assign the same channel or frequency to many stations in different geographical areas of the country. These stations were usually limited to low output powers. In the 1940s AT&T became the first telecommunications service provider to use FDM over a coaxial cable. As FM broadcasting and television technology were introduced, FDM was again used to allocate the radio resource. Later, as digital technology matured, it was again AT&T that was the first to use time-division multiplexing (TDM) with a point-to-point system. The first commercial system was known as a *T1-carrier* and it could transmit twenty-four simultaneous telephone calls over the same cable or wire pair.

With the advent of wireless cellular technology in the early 1980s and its immediate and immense popularity, the FCC reallocated radio spectrum to the wireless service providers from the upper part of the UHF television band (channels 70 to 83). As the number of cellular subscribers grew at an exponential pace, the service providers adopted new access technologies to share the limited radio-spectrum resources. The European GSM system was one of the first major systems to implement time-division multiple access (TDMA) technology and North America quickly followed suit in the early 1990s with the implementation of digital AMPS service. In 1995, code-division multiple access (CDMA) was introduced worldwide as a new access technology. With the continuing technological evolution toward third generation (3G) cellular technology (high-speed data service), numerous new **access technologies** have been suggested.

In a recent development, the FCC has refarmed the spectrum used by the Private Land Mobile Radio Services to allow for more efficient use of the available spectrum. This **refarming** will change the rules of how the spectrum is allocated to Public Safety and industrial/business users of this radio service. The refarming effort was driven by a demand from these users for more mobile radio access and was made possible by technological improvements in modulation techniques and in the hardware available for this frequency range.

Final Thoughts on Multiplexing

One should not overlook the use of fiber-optic cables with their ability to transmit vast quantities of high-speed data. They have not been left out of the recent onrush of technology. Advances in fiber-optic cable technology and laser-diode sources have resulted in commercial deployment of **wavelength-division multiplexing** (WDM) and **dense WDM** (DWDM) systems. This technology can yield almost limitless bandwidth from the fiber-optic cables already in place.

Lastly, the World Administrative Radio Conference (WARC) now meets regularly to coordinate frequency allocations on a worldwide basis as the pace of reallocation of radio resources has started to escalate.

7.2 Traditional Forms of Multiplexing

Let us recap the traditional types of multiplexing available and then take a closer look at each type of technology:

- frequency-division multiplexing (FDM)
- phase-division multiplexing (PDM)
- time-division multiplexing (TDM, TDD, and TDMA)

Frequency-Division Multiplexing

Frequency-division multiplexing splits the available frequency bands into smaller fixed-frequency channels. For the free-space channel this is easily implemented by channel assignment and transmitter hardware; however, it requires national and international agreement and administration. In the United States, new channel assignments for broadcasting are made on the basis of "no interference to existing facilities" via published FCC engineering guidelines. To promote new technologies there are several international unlicensed bands available. The instrumentation, scientific, and medical (**ISM**) **bands** at 2.4 GHz and 5.5 GHz are two such unlicensed bands. Recently, these bands have been used for wireless LANs and will be used for the new **Bluetooth™** technology that holds the promise of **personal-area networks** (PANs). However, operation in these bands does not rule out interference from other radio services or RF applications (microwave ovens operate in the 2.4-GHz frequency band also). In an effort to add additional radio spectrum for emerging high-speed second generation wireless LAN technology, several bands in the 5-GHz frequency range have been recently reassigned and are now known as unlicensed national information infrastructure (U-NII) bands. Figures 7-1 and 7-2 show some examples of wireless radio services using FDM.

The over-the-air spectrum-analyzer display in Figure 7-1 shows the FM broadcast band (88 to 108 MHz) as received in the author's home area. Note the marker on the strong signal at 102.1 MHz. The author can see this station's antenna from his bedroom window. Upon application for a

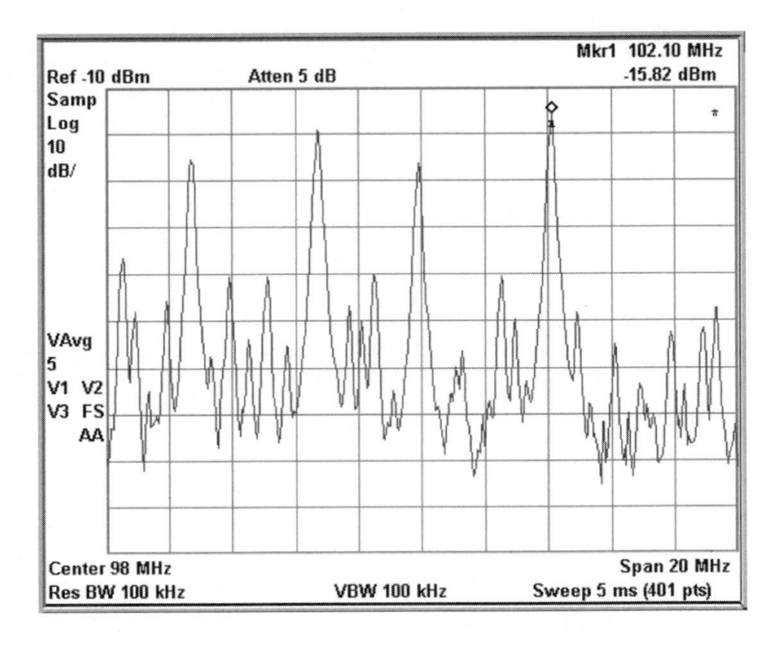

FIGURE 7-1
FM broadcast-band spectrum received off air

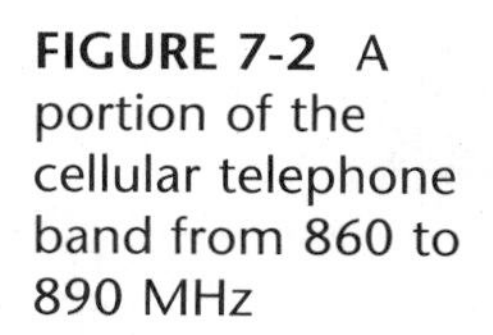

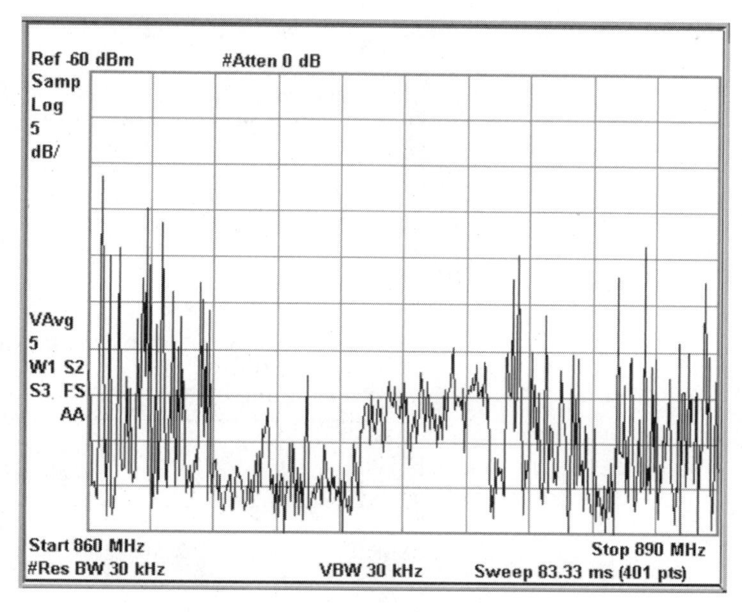

FIGURE 7-2 A portion of the cellular telephone band from 860 to 890 MHz

license, the FCC assigns channels to FM broadcast stations if there are any frequencies available that do not cause interference to already existing stations.

The over-the-air spectrum-analyzer display in Figure 7-2 shows part of one of the cellular-telephone bands (860 to 890 MHz) as received in the author's home area. The SA display shows two different kinds of signals within the same band—standard FM signals (the narrow spikes) and wide-band CDMA signals (the mesa-shaped areas in the middle of the display). The service provider is licensed by the FCC to use a particular block of frequency assignments. Multiple channels are located in the licensed frequency range and the local cellular providers assign them to the system's users on an "as needed" basis.

Cable-Television FDM Methods

Newly upgraded cable-TV systems provide upwards of 750 MHz of frequency spectrum and are utilizing split-band plans to transmit standard analog TV signals as well as digital TV signals based upon MPEG-2 standards. Using modern digital-modulation techniques, cable operators are increasing the efficiency of the bandwidth that they presently have available. Figure 7-3 on page 244 shows both types of split-band plans popular in the United States. The figure also indicates the **upstream** frequencies available to the cable operator for delivery of data services, Internet access, and telephone service. In all cases, FDM is used to separate the uses of the available spectrum.

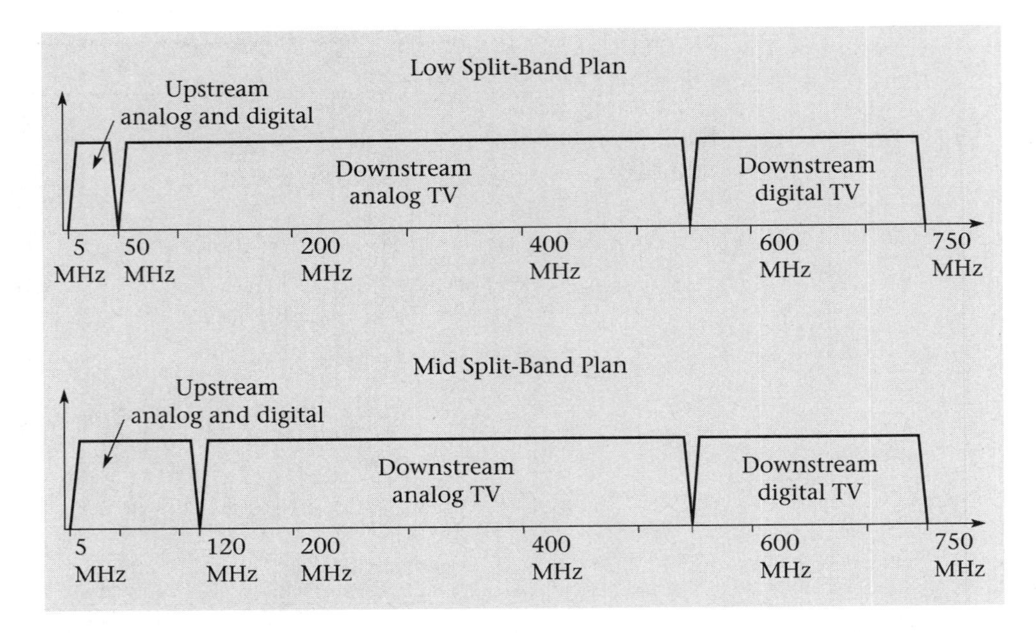

FIGURE 7-3 Cable TV split-band plans

Figure 7-4a shows an SA display of the signals on the local cable TV system in the author's home area. This is a **split-band** system. Analog TV signals, each with a 6-MHz bandwidth, are frequency-division multiplexed on the lower portion of the RF spectrum available on the cable system. Digital TV signals are located at the upper end of the spectrum.

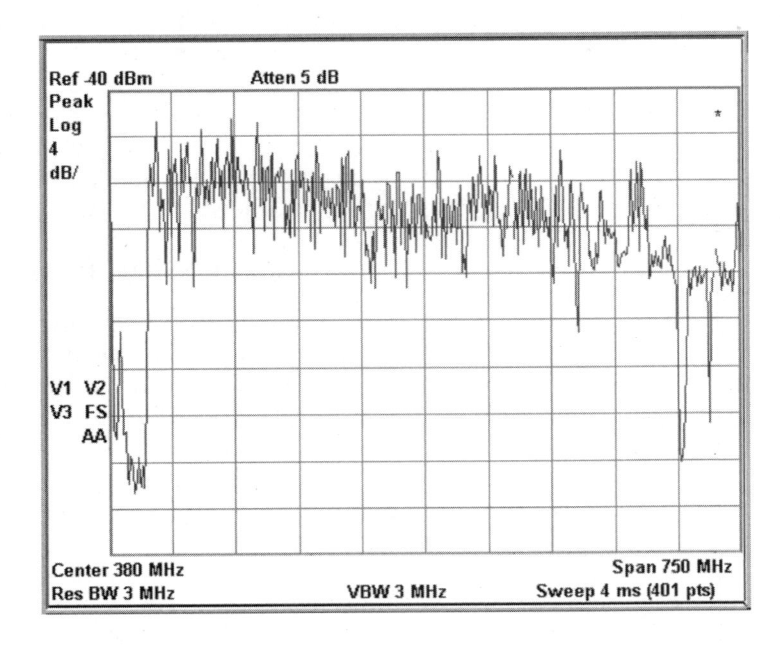

FIGURE 7-4a A local cable TV system spectrum

Figure 7-4b shows an expanded view of the analog portion of the band. One can see three NTSC analog TV signals in this 18-MHz frequency span. The large peaks are the VSB video carriers and the shorter peaks, located 4.5-MHz above the video signals, are the FM-modulated sound carriers. See Figure 7-4c for a display of a single analog TV signal.

FIGURE 7-4b Three mid-band analog cable TV signals

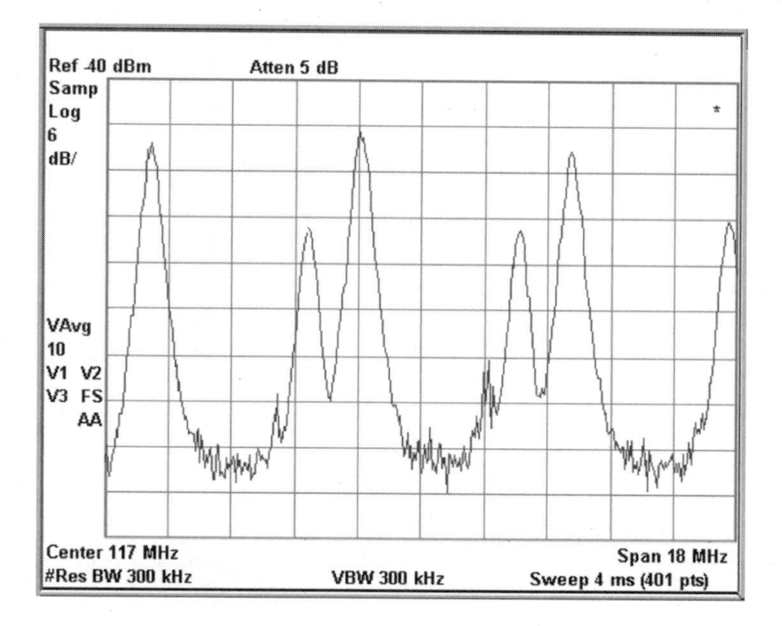

FIGURE 7-4c A single analog TV signal

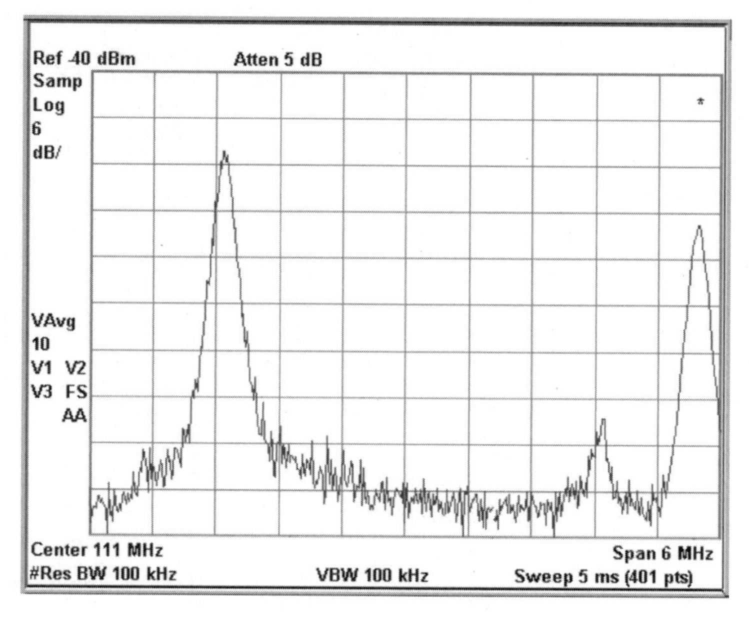

At frequencies between 550 MHz and 750 MHz, the cable system is using digital modulation and MPEG-2 compression to reduce signal-bandwidth requirements. Figure 7-4d shows this span of signals on a typical system. The frequency spectra of two channels of these types of signals are shown in Figure 7-4e (note the mesa shape of the signal spectrum).

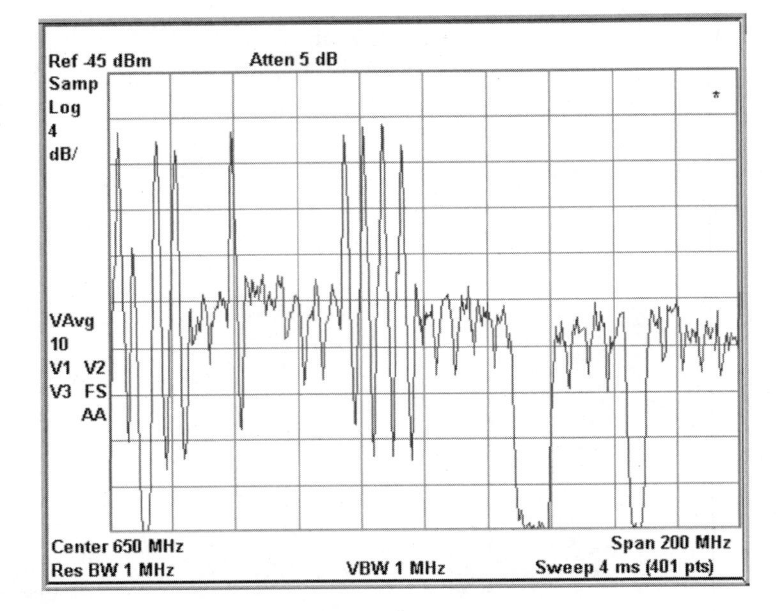

FIGURE 7-4d The digital downstream portion of a cable TV system

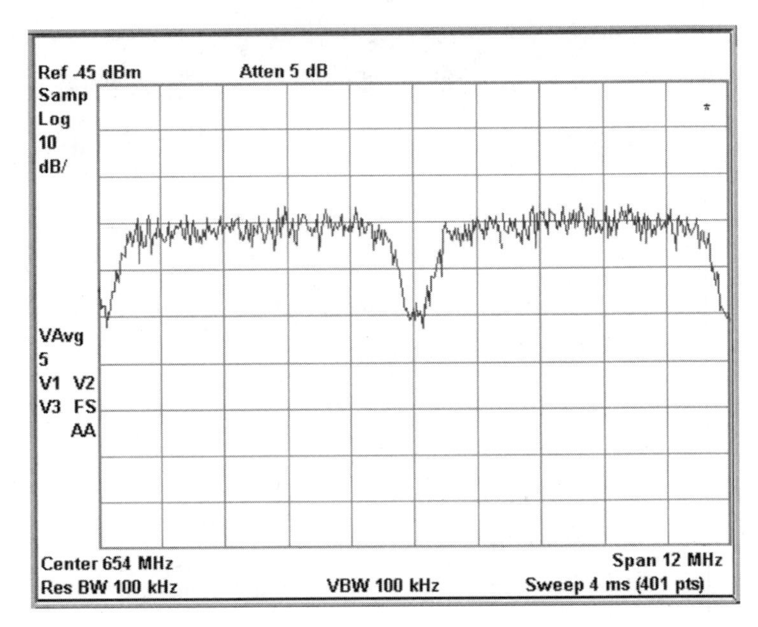

FIGURE 7-4e Two digitally modulated cable TV signals

The channels that are normally broadcasted over the air can be placed in any **downstream** analog channel slot on the cable. For this system the frequency spectra below 42 MHz is shared upstream spectra.

Legacy-Telephone Coaxial-Cable FDM Systems

At one time the PSTN (Bell System) and the long-distance service provider (AT&T) used baseband FDM schemes to deliver numerous simultaneous analog telephone calls over point-to-point coaxial cable or microwave facilities. These were known as *L-carrier* and *U-carrier* systems. These systems used internationally standardized frequencies and started with groups of 12 voice signals that were single-sideband suppressed-carrier (SSB-SC) modulated on to different carrier frequencies. The 12 channels formed a group and occupied 48 kHz of frequency spectrum. The group was then SSB-SC modulated on to another carrier frequency with 5 other groups to form a **super group** of 60 channels. The process was repeated with 5 super groups forming a master group. The process could be repeated further to form **super master groups** of 600, 900, and 1800 channels.

At the receiving end of the system the signals would have to be successively demodulated as many times as they had originally been modulated. In their heyday, both the L600 and U600 systems consisted of 600 voice channels and had bandwidths of 1.232 MHz. These signals had to be sent by coaxial cable with repeaters every mile or over point-to-point microwave-radio facilities in the 6-GHz frequency range. The final coaxial cable systems to be deployed were the Bell L5 (10,800 circuits) and the AT&T L5E (13,200 circuits). The very broadband signals generated by these systems were carried by extremely large-diameter multiple coaxial cables, as shown in Figure 7-5. These systems have been made obsolete by fiber-optic cables and digital-modulation schemes.

FIGURE 7-5 A typical multiple coaxial cable used in L-carrier FDM systems

A Summary of FDM Operation

As depicted in Figure 7-6, a transmitter typically employs a PLL frequency synthesizer and a narrowband filter to set its output to its assigned frequency or channel (the blip in the frequency spectrum). The receiver employs a narrow bandpass filter to pass the desired signal and to reject unwanted signals (the other blips). Although this figure shows free-space transmission, the principle is still the same if guided-wave point-to-point transmission media is used. For cordless- and cellular-telephone operation, the system determines the best frequencies of operation within an operational band and assigns them to the users of the service.

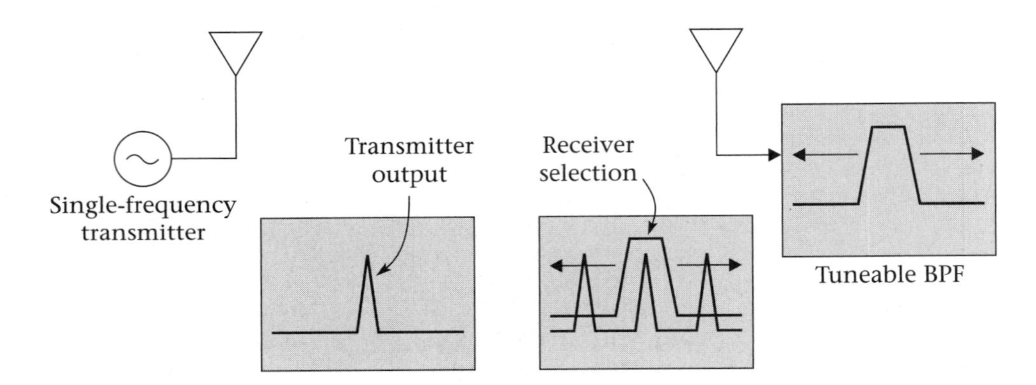

FIGURE 7-6 The FDM concept

Phase-Division Multiplexing

Phase-division multiplexing (PDM) is a fairly new technology and recently the wireless telecommunications industry has started to refer to it as **orthogonal FDM** (OFDM). High-speed wireless data transmission usually employs some form of advanced digital modulation to achieve the desired speeds. Quadriphase or quaternary-shift keying (QPSK) is a form of PDM. In this system, the in-phase and quadrature-phase (I and Q) channels of a transmitter are modulated by two bits at a time. The I-channel bit modulates a sine wave at the carrier frequency and the Q-bit modulates the same carrier frequency, but shifted by 90 degrees. The resultant two carriers are linearly added together and transmitted over the channel as a passband signal. Because the two carriers are orthogonal to one another, they do not interfere with each other; therefore, they can be separated at the receiver fairly easily.

Orthogonal Signals

Let us pause for a moment to discuss what we mean by **orthogonal** signals. In the prior example, it was stated that one carrier was a sine wave and the other carrier was phase-shifted by 90 degrees. That would indicate that the other carrier is a cosine wave. It turns out that these two signals have the special property of being orthogonal to one another. Note that the sine wave is at its maximum while the cosine wave is at its minimum and vice versa. Mathematically, it can be shown that there is no cross correlation or similarity between these two signals. This might come as a surprise to the reader, as sine waves and cosine waves are usually used interchangeably when discussing signals. Indeed, they certainly look the same, only shifted in time relative to one another. See Figure 7-7 for an example. However, if one considers the fact that these two signals are always doing exactly the opposite of one another (consider their slopes and zero crossings), then one might consider them as possessing no similarity. Furthermore, the mathematical definition of signal coherence or similarity can be used to prove this fact. This special property of orthogonality is important because the two signals will not interfere with one another at the receiver. Furthermore, if two or more orthogonal signals all have the same energy content, they are called orthonormal signals. There are many other examples of orthogonal signals that are employed in digital-modulation schemes to increase the effectiveness of the systems.

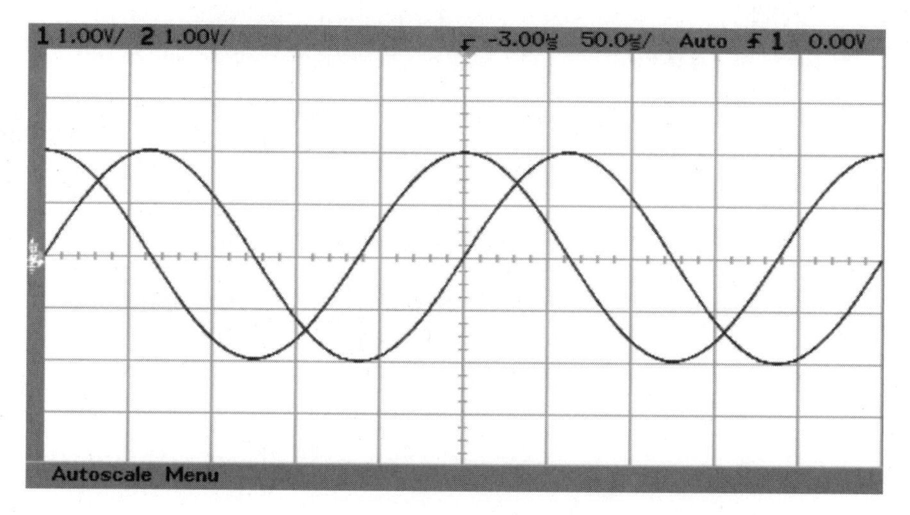

FIGURE 7-7 Orthogonal signals: a sine wave and a cosine wave of the same frequency

It is important to remember that there is an advantage to the use of this modulation technique. The net gain when using PDM or QPSK modulation is a doubling of the data rate in the same bandwidth channel or the halving of the required bandwidth for the same data rate. See chapter 6 for a more detailed explanation of QPSK.

Time-Division Multiplexing

The next form of multiplexing to consider is that of **time-division multiplexing** (TDM). TDM involves separating in time the outputs of the transmitters that share the channel or transmission media. The simplest and earliest form of TDM is called **time-division duplex** (TDD). Consider CB radio, FM walkie-talkies, the new Family Radio Service (FRS) walkie-talkies, or any radio service that requires the use of "push-to-talk" buttons with which only one person can talk at a time. TDD jargon—"over" and "over and out," for instance—is used to facilitate the exchange of information and the protocol of who speaks when. **Time-division multiple access** (TDMA) is used by GSM (Global System for Mobile Communication) digital cellular and the North America TDMA (NA-TDMA) system. As can be seen in Figure 7-8, each user of the channel is assigned a particular time slot for their use on a round-robin basis. Each GSM channel uses 8 time slots to share the radio resources and the NA-TDMA system uses 6 time slots.

Figure 7-8 shows how users are assigned to a time slot and how each signal is combined with the others to be transmitted over the channel as a continuous signal. TDMA is normally combined with FDM for use by the cellular-telephone system.

FIGURE 7-8 TDD and TDMA operation

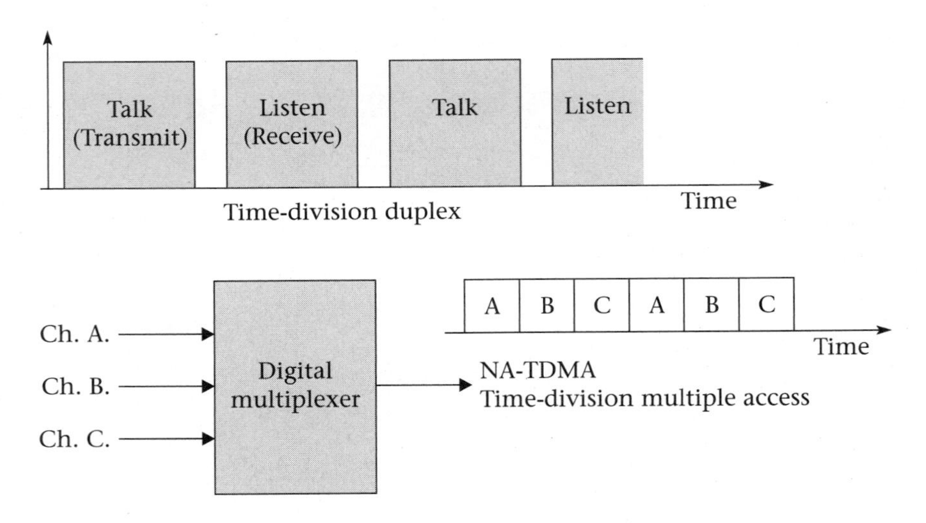

7.3 Time-Division Multiplexing and the PSTN

The PSTN uses TDM extensively in point-to-point guided-wave systems. Starting with the T1-carrier systems in the early 1960s, this type of technology has matured considerably, now including higher order T1- and E1-carrier systems, frame-relay and ATM (asynchronous transfer mode) systems, and SONET (synchronous optical network) technology. SONET systems have the ability to transmit 10s of Gbps over fiber-optic cables using TDM technology. We will not endeavor to explain all the different

TDM systems that exist at the present time. However, we will use the T1-carrier system to explain the basic operation of TDM. The TDM concept is shown in Figure 7-9 and Figure 7-10.

As shown in Figure 7-9, each telephone call is sampled and digitized and then encoded into pulse-code modulation (PCM) in a round-robin fashion (i.e., channel 1, channel 2, ..., channel 23, channel 24, channel 1, and so forth). In the T1 system this means 24 calls are grouped together. After the 24th call is sampled, the process repeats itself again starting with the 1st call and so on. In the figure, we see the encoded value of the sample for the 2nd telephone call and its resulting line code. Note that the least-significant bit (LSB) of the coding bits is used for signaling and control purposes at certain repeating time intervals. Note that one can also transmit digital data over a T1-carrier system.

FIGURE 7-9
T1-carrier system operation

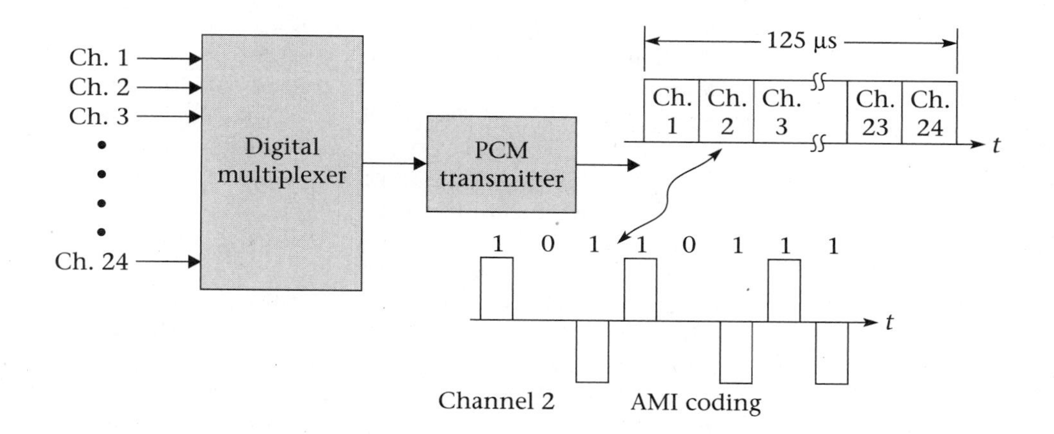

Figure 7-10 on page 252 shows the T1-carrier **frame**. The specifications for a T1-carrier system frame are as follows: 24 telephone calls at 8 bits per sample plus one framing bit, which equals 193 bits/sample every 125 μsec or 8,000 times per second:

$$24 \times 8 + 1 = 193 \quad \text{and} \quad 193 \times 8000 = 1.544 \text{ Mbps}$$

Therefore, the T1-carrier bit rate is equal to 1.544 Mbps. Note that this is also the nominal required signal bandwidth (1.544 MHz) depending upon the type of line code used.

In Europe, the E1 TDM system uses 32 time slots (30 calls and 2 signaling slots) with the basic frame consisting of 256 bits.

$$32 \times 8 = 256 \quad \text{and} \quad 256 \times 8000 = 2.048 \text{ Mbps}$$

Therefore, the E1-carrier bit rate is equal to 2.048 Mbps and the signal bandwidth is 2.048 MHz.

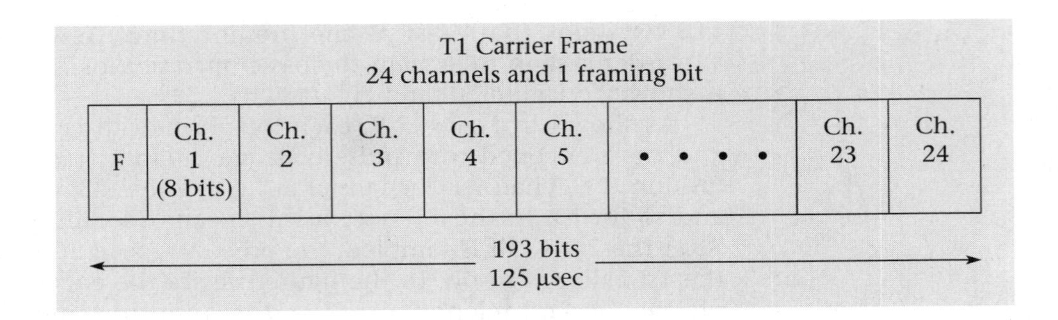

FIGURE 7-10 The T1-carrier frame

TDM Details: The Time-Slot Interchanger at the Central Office

One might question how a typical telephone call's time-slot assignment is determined. Recall that the vast majority of telephone calls start as analog signals being transmitted over the local loop to the telephone company's central office (CO). For a local call that can be serviced by a local telephone exchange, the call will be converted to PCM format at the CO. This happens in a "line card," which is part of the stored-program electronic switch (5ESS or DMS-100) located at the CO. At this point the switch will assign a vacant time slot to the call, which identifies the call. To complete the call it is necessary to connect the caller's signal to the correct wire pair for transmission to the desired destination. This function is provided by a **time-slot interchange** (TSI) **switch**. The purpose of the TSI switch is to convert the time-slot assignment to match the outgoing line. For the local call a line card will convert the PCM signal back to an analog signal to be transmitted by the local loop to its destination. Furthermore, if the call needs to be directed to another local exchange or is to be sent to a long-distance carrier, it must be allocated the correct time slot to match what is available on the outgoing T-carrier trunk line (interexchange circuit). In this case, the T1-carrier signal is translated into an appropriate line code such as B8ZS or 2B1Q and sent over a point-to-point transmission link to its destination.

This explanation is a simplified version of what takes place in a circuit-switched telecommunications channel. The most important point to grasp is that the network contains a switching fabric within the channel that is used to direct the information to the proper destination. This type of legacy telecommunications system is referred to as **circuit switching** because the switching resources are dedicated to the telephone call for its duration.

Higher Order T-Carrier Systems

The T1 system is the basic building block in North America. T1s can be multiplexed together to form higher order systems. Table 7-1 and Figure 7-11 show the digital-multiplexing hierarchy for North America.

TABLE 7-1 North American digital-multiplexing hierarchy (* denotes a nonstandard level)

Carrier system	*Digital signal #*	*Number of voice channels*	*Transmission rate*	*Transmission media*
T1	DS1	24	1.544 Mbps	copper pair
T1C	DS1C	48	3.152 Mbps	copper pair
T2	DS2	96	6.312 Mbps	copper pair
T3	DS3	672	44.736 Mbps	fiber, radio
	DS3C*	1,344	90.0 Mbps	fiber, radio
	DS3X*	2,016	135.0 Mbps	fiber, radio
T4	DS4	4,032	274.176 Mbps	fiber, radio
	DS4E*	6,048	405.0 Mbps	fiber
T5	DS5	8,064	560.160 Mbps	fiber

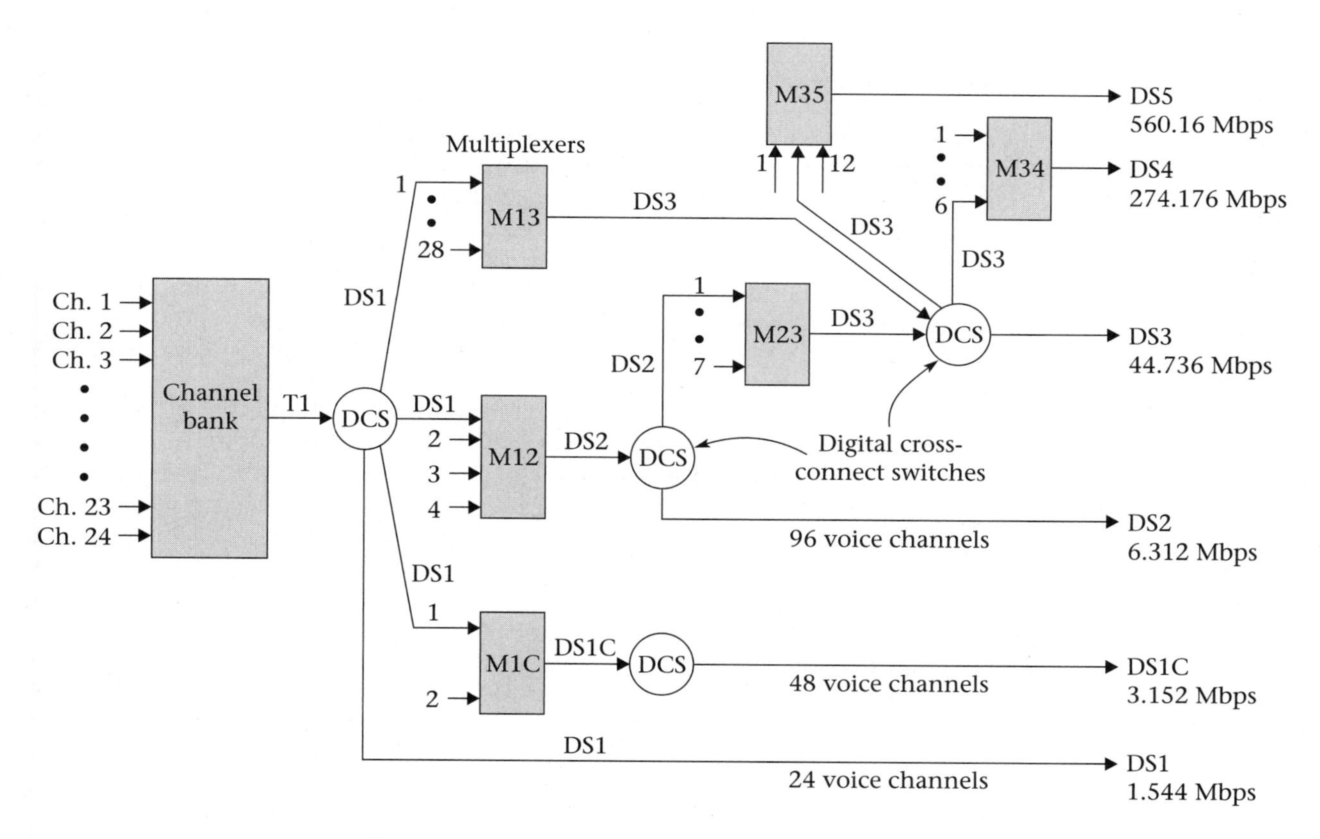

FIGURE 7-11 Block diagram of the North American digital-multiplexing hierarchy

Figure 7-11 shows various hardware components that facilitate the combining of lower order T1-carrier signals into higher order signals. The blocks labeled as **MXY** are multiplexers that combine *X*-level inputs into a *Y*-level output. For example, an M13 multiplexer will combine 28 T1/DS1 (level 1) signals into one T3/DS3 (level 3) signal. The circle blocks labeled as DCSs are **digital cross-connect switches**. Their function is simply to switch digital signals. Notice that in each case the bandwidth necessary to transmit the signals increases due to the increasing bit rate. At some point, the transmission media must become a fiber-optic cable to accommodate the wide bandwidth signal.

7.4 Access Technologies: Code-Division Multiple Access

Code-division multiple access (CDMA) is a relatively recent addition to the modulation techniques used by the telecommunications industry. This is a form of spread-spectrum modulation used by the wireless mobile industry (cellular telephone). Although spread-spectrum techniques have been used for secure communications by the military for years, it was not until recently that these techniques were adopted for use by the wireless industry. The adoption of CDMA is a direct response to the overwhelming popularity of cellular-telephone use. With a limited radio spectrum, the providers of cellular service first adopted TDMA to increase the capacity of their systems; they then quickly embraced CDMA for the same reason. For CDMA, a **pseudonoise** (pn) digital code is overlaid on the signal. Through system assignment the receiver and transmitter share this particular orthonormal code. With a CDMA system, multiple users are permitted to transmit simultaneously on the same frequency or channel. For other receiver and transmitter pairs, the orthonormal signal codes transmitted by other users appear as noise. The choice of coding (**Walsh codes**) places some limitations on the number of code channels possible. Figure 7-12 shows the CDMA/direct-sequence spread-spectrum operation in more detail. The spreading signal has a characteristic chip rate; CDMA uses 1.2288 megachips/sec (Mcps). It has been proposed that newer 3G systems will use multiple chip rates with bandwidths of 15 MHz and higher.

Figure 7-13 shows the spreading process in even finer detail. The two signals are "exclusive-OR'd" together to obtain the output. Note how in this example the original signal bandwidth will be increased (spread) by a factor of 8 because the bit time has been divided by 8. Recall that bandwidth is proportional to bit rate ($f_b = \frac{1}{t_b}$).

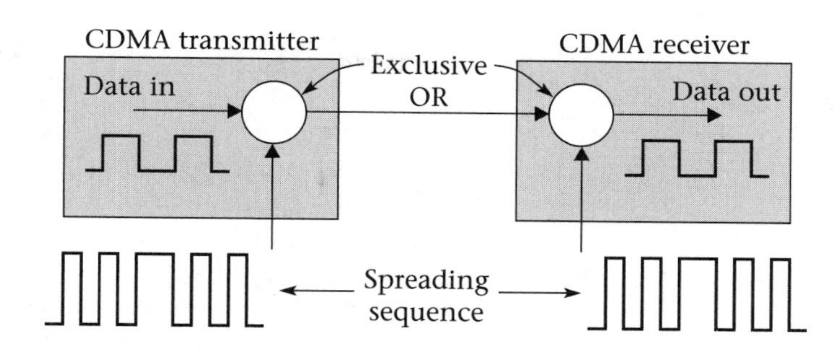

FIGURE 7-12 CDMA coding

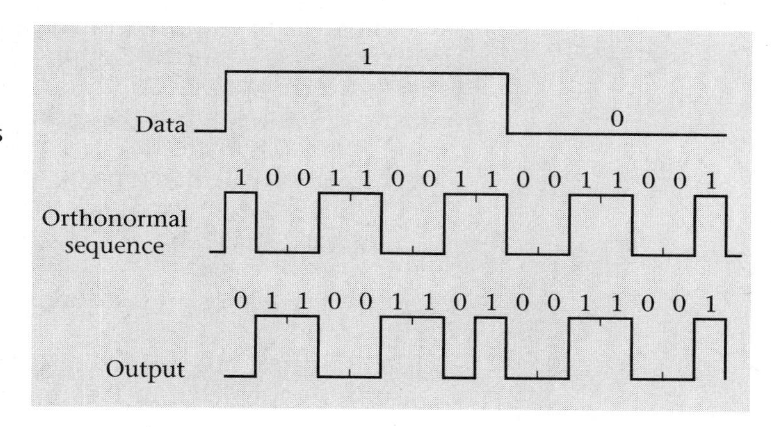

FIGURE 7-13 The CDMA spectrum-spreading process in more detail

Figure 7-14 shows CDMA combined with FDM. In this figure four CDMA channels are shown and each channel has multiple CDMA users active.

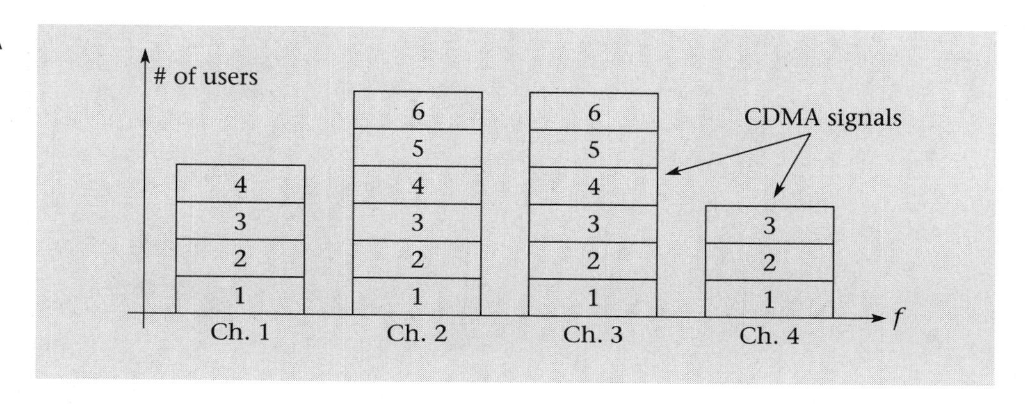

FIGURE 7-14 CDMA combined with FDM

7.5 Geographic or Space Multiplexing

In our short history of multiplexing technology we mentioned the use of **space-division multiplexing**, an easy concept to grasp. In short, stations using the air interface can use the same frequency if they are out of transmission range of one another. Cellular systems use space or **geographic multiplexing** extensively (see Figure 7-15).

The same frequencies can be reused in each cell, but they must be separated in space to prevent interference. In Figure 7-15 adjacent cells cannot utilize the same frequencies. Frequency reuse depends on several factors, including but not limited to transmitted power, terrain, antenna height, frequency, and so forth. Figure 7-16 shows a diagram of one **frequency-reuse pattern** with an indicated reuse distance. Each cell with the same numeric designation can reuse the same set of frequencies.

In Figure 7-16 the frequency reuse pattern is 7. The frequency-reuse distance can be calculated by:

$$D = R(3N)^{\frac{1}{2}} \quad \textbf{7.1}$$

where R = cell radius and N = reuse pattern.

Cell splitting can be used to increase the capacity of the system further. Cell splitting, depicted in Figure 7-17, is accomplished by using sector antennas. Each cell is served by a 120° (1/3 of a circle) antenna, as shown in Figure 7-18.

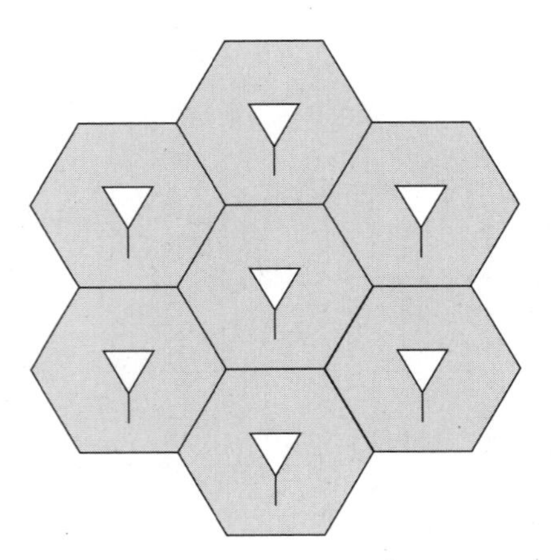

FIGURE 7-15 Typical cell structure: geographic multiplexing (cellular telephone)

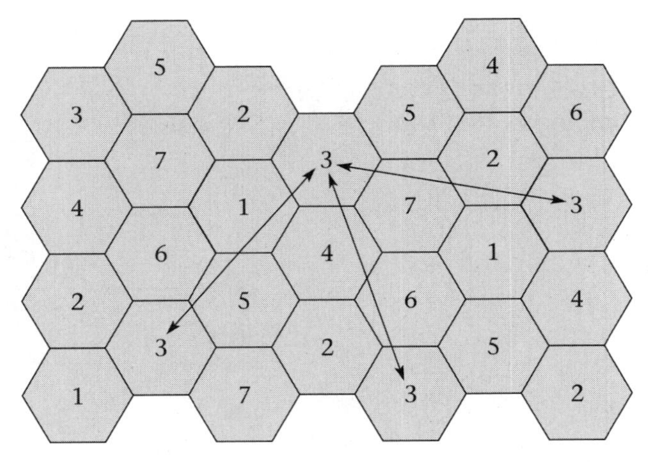

FIGURE 7-16 A frequency-reuse diagram with the reuse distance indicated

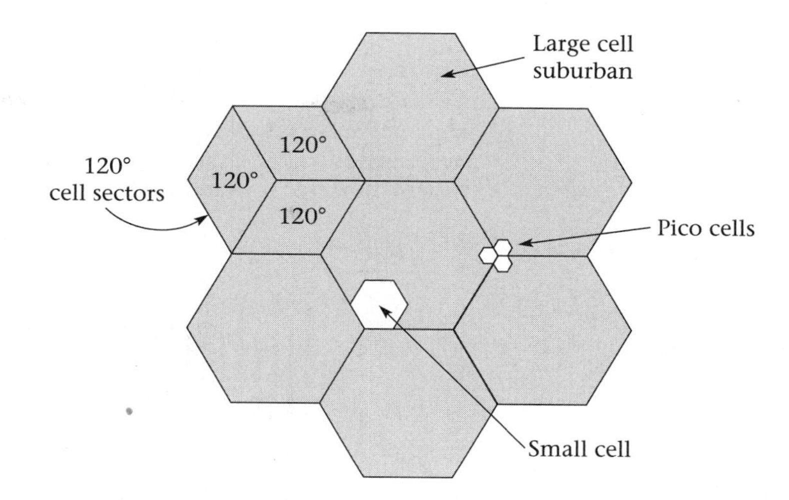

FIGURE 7-17
Cell splitting

FIGURE 7-18 This picture shows a typical cell tower with three sector antenna arrays per level (each level represents a different cellular-service provider)

7.6 Wavelength-Division Multiplexing

Wavelength-division multiplexing (WDM) is really the same as FDM, but with a different label. The users of fiber optics and laser technology use wavelength (λ) to represent the frequency of operation, much as the pioneers of wireless radio did at the end of the nineteenth century. Fiber-optic cables have a tremendous amount of available bandwidth that can be utilized. Advances in laser-diode sources and optical filters have allowed the number of possible wavelengths that can be transmitted simultaneously down a fiber to be increased dramatically. Present commercial systems can transmit 128 different wavelengths with a frequency spacing of 50 GHz. Recent reports indicate successful laboratory results for 1-GHz spacing! Figure 7-19 shows a typical optical spectrum of a WDM signal. These systems are capable of transmission rates equal to several to many terabits/sec (Tbps). The use of WDM or dense WDM (DWDM) technology allows already in-place fiber-optic infrastructure to be upgraded as needed.

FIGURE 7-19 A typical WDM signal (*Courtesy of JDS Uniphase Corporation*)

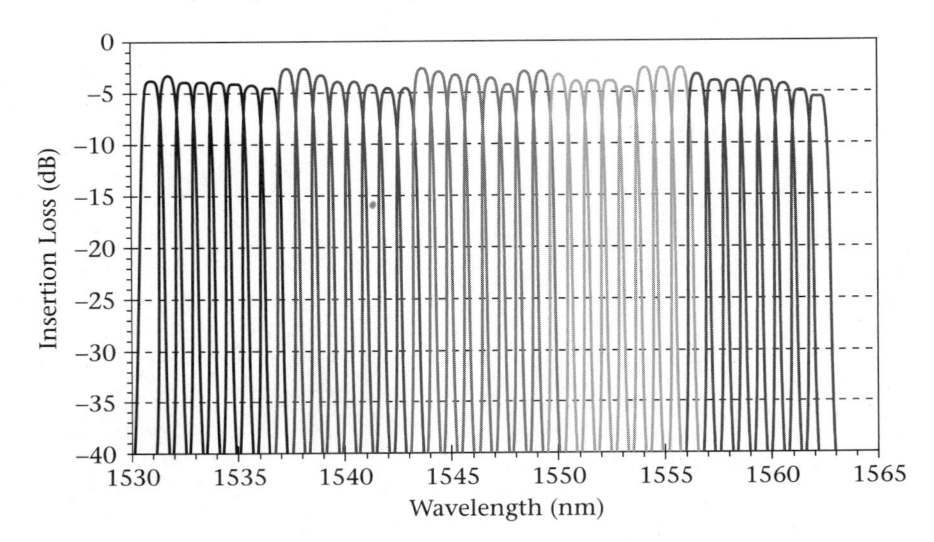

7.7 Multiplexing Technology for ADSL

Many of the incumbent telephone-service providers are in the process of rolling out what are known as **digital subscriber line** (DSL) technologies with the promise of high-speed data transmission/Internet access. These new technologies utilize the local loop (copper-wire pairs) with its limited bandwidth to provide high-speed data transmission. How is that possible? A short overview of the technology follows.

Two popular forms of DSL technology are asymmetrical digital subscriber line (**ADSL**) and rate-adaptive DSL (RADSL).

ADSL uses **digital-multitone** (DMT) technology (sometimes called OFDM). DMT ADSL uses the scheme shown in Figure 7-20 on page 259. This figure shows an FDM scheme with both upstream and downstream bands and a "plain-old telephone" (POTS) band that are all separated in the frequency spectrum. This system is designed to work over copper pairs, thus pushing the limits of the available bandwidth that can be utilized to transmit high-speed data. What is not obvious in the figure is that the frequency bands are broken up into a total of 256 frequency "bins" or "tones," and each bin of 4.3125 kHz (above 30 kHz) has its own QAM

FIGURE 7-20 Digital multitone ADSL

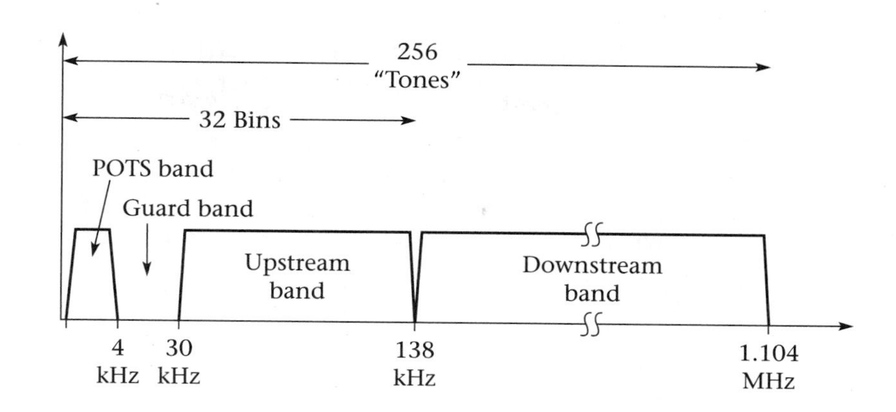

transmitter or receiver. Figure 7-21 shows the organization of the QAM transmitters (all 256 QAM transmitters are located on a single VLSI IC chip). A varying number of binary bits are encoded per QAM transmitter depending upon the channel-frequency response.

FIGURE 7-21 A DMT modulator with 256 bins with spacing of 4.3125 kHz. The first bin is at 4.3125 kHz and the last bin is centered at 1.104 MHz. Only certain ranges of bins are used for transmission or reception.

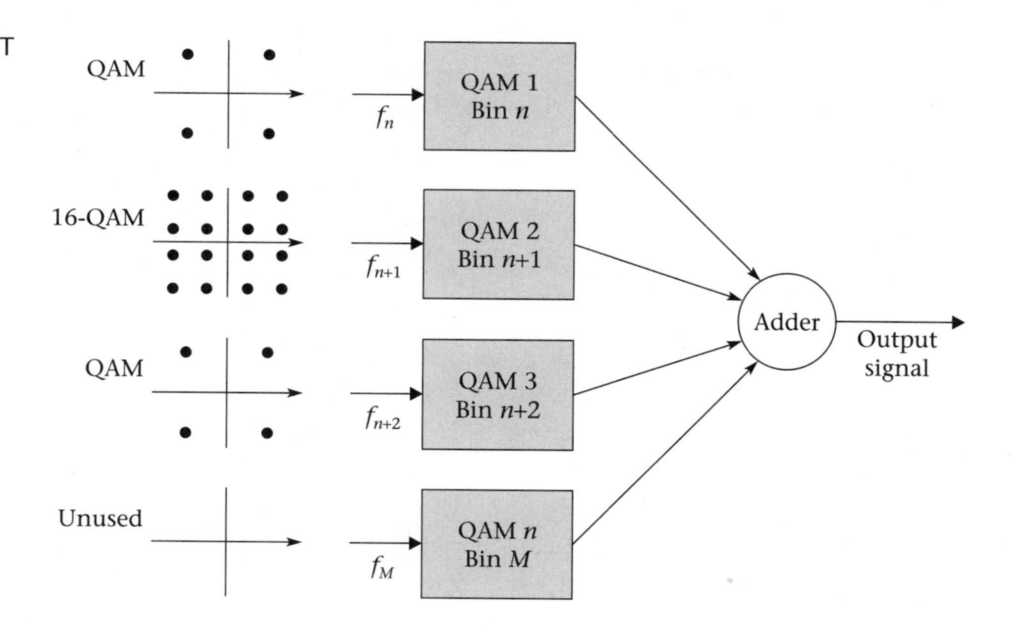

Each QAM transmitter is capable of up to 33.6 Kbps (V.34) or a reduced rate depending upon the quality of the particular channel bin. The ADSL system initially tests each channel bin. Based on the channel's frequency response or noise characteristics, the system will adjust the rate at which each QAM transmitter operates (if at all). Figure 7-21 shows the first and third QAM transmitters operating at 4-QAM and the second QAM transmitter operating at 16-QAM. The last transmitter has been disabled due to poor frequency response at its frequency bin. The top portion of Figure 7-22 shows system operation for a short run of a "clean" copper pair. The bottom portion shows the type of operation that might occur with various signal impairments. Notice how the system adapts itself for the best possible bit rate.

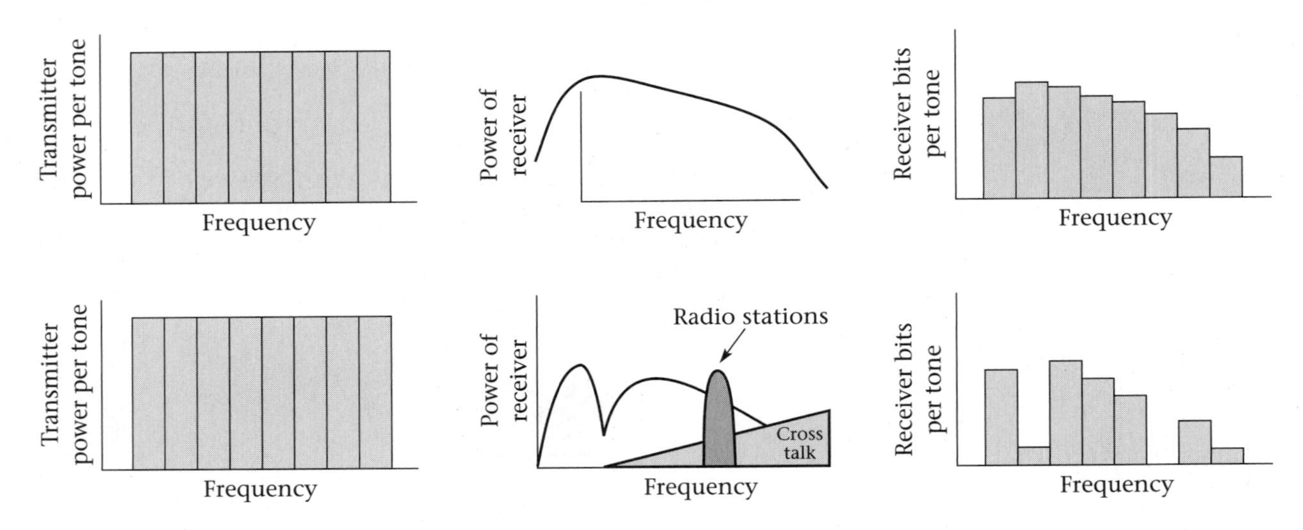

FIGURE 7-22 Typical ADSL operation. A "clean" copper pair and a copper pair with some signal impairments. (From *IEEE Spectrum Magazine*, vol. 36, no 9 © 1999 IEEE)

Summary

Over time, the telecommunications industry has attempted to use efficient modulation and multiplexing schemes to maximize the use of the channel or transmission media. The advent and evolution of digital modulation has continued to improve data rates over limited-bandwidth channels. Recently, clever access technologies like TDMA and CDMA have provided increased capacity for systems with limited radio-spectrum resources. WDM holds the promise of almost limitless bandwidth over new fiber-optic cables and already in-place infrastructure. Systems like ADSL are able to extend the useable bandwidth of what were once called "band-limited channels" by means of adaptive modulation schemes and complex coding algorithms to detect and correct errors in signal transmission. The "pico cells" of the new Bluetooth™ technology allow many users to share the same frequency band through geographic or spatial multiplexing. Telecommunications systems will continue to evolve and become more complex in an attempt to utilize the limited bandwidth channels provided by nature as efficiently as possible. Finally, governmental regulations will continue to evolve in an attempt to refarm frequencies already dedicated to other legacy services.

Questions and Problems

Section 7.1

1. Explain the purpose/function of multiplexing.
2. Why is it desirable to multiplex signals?
3. What is refarming? Do an Internet search of this concept. Start your search at http://www.FCC.org.
4. What is geographic or space multiplexing?

Section 7.2

5. Describe the process of frequency-division multiplexing (FDM).
6. Explain how cable TV uses FDM.

Section 7.3

7. Describe the process of phase-division multiplexing (PDM).
8. What are orthonormal signals?
9. What advantages do orthonormal signals have when used in the transmission of data?
10. Do an Internet search on "orthogonal signals." Report on other types of special signals that have this property.
11. Describe the process of time-division multiplexing (TDM).
12. Describe the T1-carrier system.
13. What is the difference between the T1 and E1 systems?
14. What is the purpose of the time-slot interchange switch?
15. How does the use of PCM at the central office (CO) restrict the overall bandwidth of the PSTN?
16. Describe higher order T1-carrier systems.
17. What is the function of an MXY multiplexer?
18. Describe the operation of an M13 multiplexer.
19. Describe the operation of an M35 multiplexer.
20. What type of transmission media is required for most higher order T1-carrier signals? Why?

Section 7.4

21. Explain the novel feature of spread-spectrum encoding.
22. Why is the bandwidth of a spread-spectrum signal larger that the original signal? What is the relationship between the pseudonoise code and the new signal bandwidth?
23. Explain the process of code-division multiple access (CDMA)
24. What advantages are there to the use of CDMA?
25. Why is CDMA referred to as an "access technology?"

Section 7.5

26. Explain the concept of a frequency-reuse pattern.
27. Think about the area where you live. Relate the demographics of your home location to the idea of cell splitting.

Section 7.6

28. What effect do WDM and DWDM have on the potential bandwidth of a fiber-optic cable?

Section 7.7

29. Explain the theory of operation behind digital multitone technology.
30. Explain the self-adapting features of ADSL.

Chapter Equation

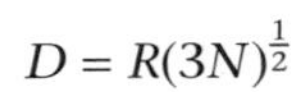

$$D = R(3N)^{\frac{1}{2}} \tag{7.1}$$

Transmission Media

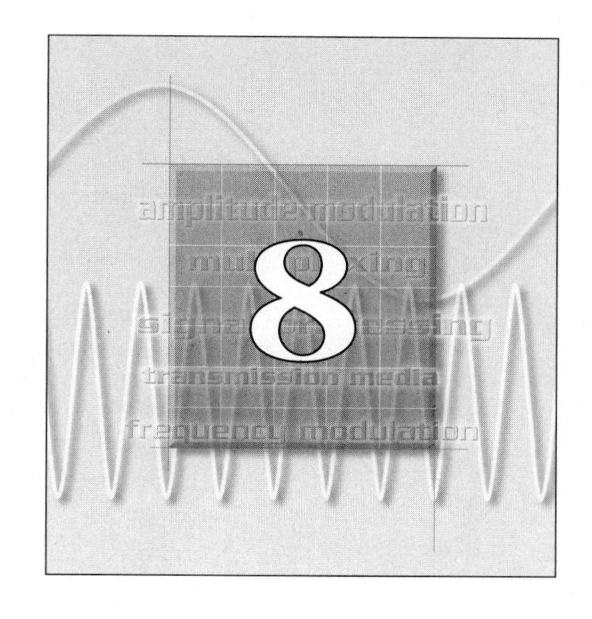

Objectives Upon completion of this chapter, the student should be able to:

- Describe the electromagnetic spectrum and a free-space model of an electromagnetic (EM) wave.
- Discuss and explain the concepts of polarization, plane and spherical waves, power density, electric field intensity, wavelength, and effective radiated power (ERP).
- Describe the mechanisms of EM wave propagation: reflection, refraction, diffraction, interference, absorption, and the Doppler effect.
- Describe the different types of transmission lines (TLs) and comment on their various uses and applications.
- Discuss the model of a TL and the propagation of EM waves on TLs .
- Understand the concepts of the reflection coefficient, the standing-wave ratio (SWR), and impedance matching.
- Discuss the uses of the Smith chart to effect TL/load matches.
- Discuss the characteristics of the PSTN local loop and modern cable TV systems.
- Discuss the characteristics of fiber-optic cables and their uses in the telecommunications infrastructure.
- Discuss the characteristics of fiber-optic systems, including repeaters and optical amplifiers, dispersion, and WDM and DWDM.
- Discuss TL testing with a time-domain reflectometer (TDR) and an optical time-domain reflectometer (OTDR).

Outline

Part I: Electromagnetic Wave Propagation

8.1 The Electromagnetic Spectrum
8.2 The Free-Space Model of EM Waves
8.3 Properties of EM Waves
8.4 The Propagation of EM Waves
8.5 Terrestrial Propagation and the Ionospheric Layers

Part II: Transmission Lines

8.6 Introduction to Conductor-Based TLs
8.7 Detailed TL Characteristics
8.8 Wireline TL Models
8.9 The Matching of a Source and Load
8.10 Propagation of Waves on TLs
8.11 TL Impedance Matching

Part III: The Local Loop, Cable-TV Plant, and Fiber-Optic Cables

8.12 The Local Loop of the PSTN
8.13 The Infrastructure of Cable-TV Systems
8.14 Fiber-Optic Cables and Systems
8.15 The Synchronous Optical Network (SONET)

Key Terms

absorption
antenna
balun
boundary conditions
cable modem
characteristic impedance
dielectric constant
diffraction
diffuse reflection
Doppler radar
electromagnetic spectrum
equipotential
free-space impedance
ground waves
guided waves
headend
interference
isotropic source
oblique incidence
permeability of free space
plane wave
polarization
quarter-wave transformer
Rayleigh fading
reflection
reflection coefficient
refraction
relative permittivity
skin depth
skin effect
sky waves
SONET
space waves
specular reflection
spherical wave
standing-wave ratio
standing waves
transmission coefficient
transmission lines
transverse
transverse EM
wavelength

Introduction

The channel of a telecommunications system is often defined as the means by which the signal propagates. Every physical medium has numerous different electrical properties that are important to understand because they limit the usefulness of the channel in terms of capacity and maximum distance of propagation. This chapter presents the student with information about the wireless, wireline, and wave-guiding channels. The chapter is broken up into three parts. Because there are many similarities between electromagnetic (EM) propagation in freespace and guided-wave propagation, the first part of the chapter will deal with an introductory treatment of free-space EM propagation. Included in this section will be coverage of the various different modes of EM propagation; different wavelength signals experience different modes of propagation due to interactions with the earth and the earth's ionospheric layers. The second part of the chapter will deal with transmission-line theory and include coverage of unshielded and shielded twisted pairs, parallel wire, local area network category-N twisted pairs, coaxial cable, waveguide, microstrip and stripline, and fiber-optic cables. The third part of the chapter will be an overview of the three popular wireline/guided-wave transmission-media infrastructure systems presently deployed. Transmission-line testing will be introduced as the final topic in the chapter.

This chapter is going to examine the transmission medium/physical channel in greater detail. We will start with a discussion of electromagnetic propagation and then look at wireline media and fiber-optic cables.

PART I

Electromagnetic Wave Propagation

8.1 The Electromagnetic Spectrum

The first topic of discussion is the **electromagnetic spectrum**. One might ask, What is the range of electromagnetic signals? Figure 8-1 on page 266 shows a graphical representation of the radio-frequency spectrum. As can be seen from the chart, radio waves are currently considered to range from extremely low frequencies to the terahertz (THz) frequencies. Radio waves are also described by their wavelength and are oftentimes also described as belonging to a certain wavelength regime (i.e., short wave, millimeter wave, and so forth).

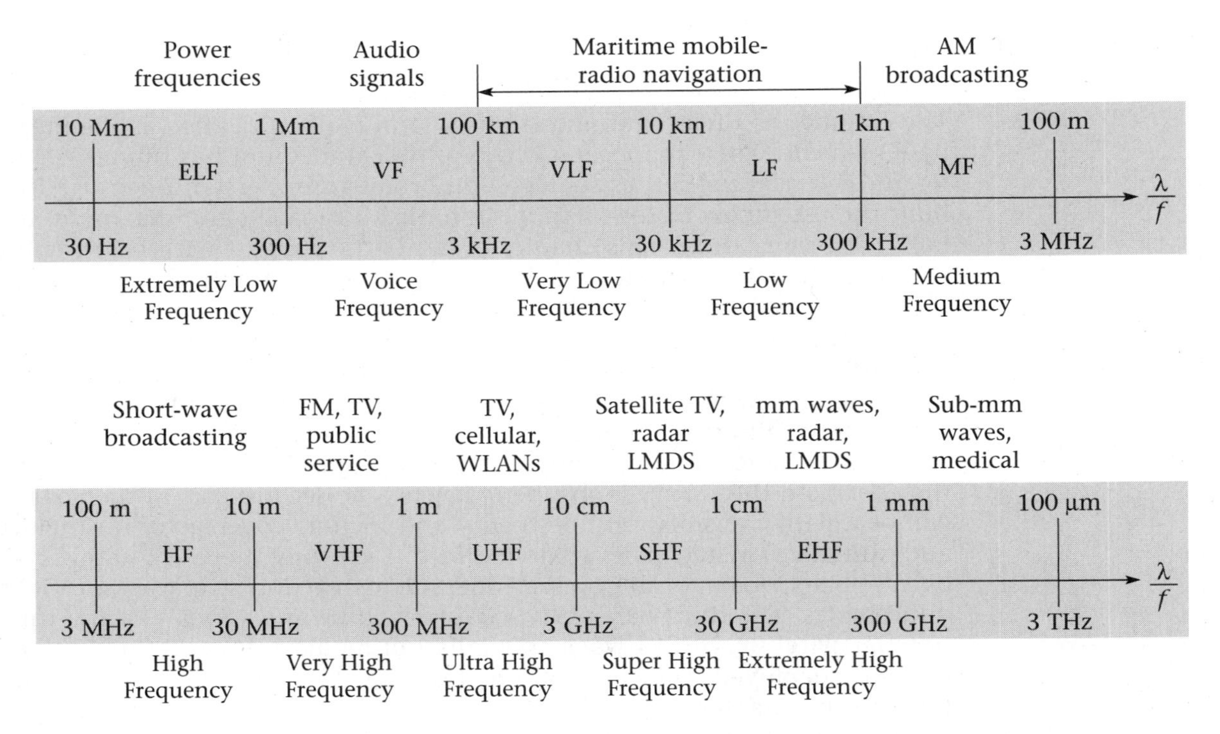

FIGURE 8-1 Radio-frequency spectrum shown with the standard acronyms for the various frequency bands

The Early Days of Wireless

Historically, Hertz verified Maxwell's theory of electromagnetic waves with very short waves at very high frequencies (VHF) and even showed that they were subject to the same laws of behavior as light waves (i.e., reflection, refraction, diffraction, interference, and so forth). Marconi started his wireless experiments in the VHF range also. However, Marconi and others quickly moved to lower frequencies and longer wavelengths for several reasons. The first reason had to do with their inability to generate sufficient power at higher frequencies; the second had to do with their lack of understanding of antenna and propagation theory; and the third had to do with their perceived success at lower frequencies. Marconi, in a classic paper from 1922 titled, "Radio Telegraphy," states:

> The progress made with the long-wave or antenna system was so rapid, so comparatively easy, and so spectacular, that it distracted practically all attention and research from the short waves. ...

It should be noted that the first reason they moved to lower frequencies and longer wavelengths remains a problem for us today. The progress achieved in the generation of RF energy at higher and higher radio frequencies has been slow and remains an active area of research in the field of semiconductor devices. The 1920s brought further understanding to the theory of antennas and EM propagation, and the transition to long-distance short-wave radio broadcasting was begun. At the present time a great deal is known about the propagation of EM waves. It is this knowledge that has guided the assignment of various frequency bands to particular types of radio services; these concepts will be discussed in this chapter. Presently, there is a great deal of active research in telecommunications systems designed to operate at 60 GHz and above. Why this interest? It is because of the large amount of bandwidth available (recall that data rate is directly proportional to bandwidth) and the very short propagation range (recall spatial multiplexing).

From ELF to EHF and Beyond

What about frequencies above 300 GHz? There is a great deal of active research at universities and laboratories around the world on the extremely high frequencies. With the advent of lasers and laser diodes, new applications of these devices in numerous fields are being developed at infrared through ultraviolet wavelengths and beyond. See Figure 8-2 on page 268 for a snapshot of the EM spectrum above radio waves.

8.2 The Free-Space Model of EM Waves

To start our discussion of EM waves we need to look at a model of such a wave in free space. Free space is chosen because otherwise the wave will interact with its surroundings. The free-space model of an EM wave is shown in Figure 8-3. There are currently many Web sites available with interactive java applets that one can use to examine many aspects of EM-wave behavior. A Web search of "physics applets" will lead the reader to many of these sites.

Figure 8-3 shows the universally accepted basic mathematical model of a **transverse EM** (TEM) plane wave as predicted by Maxwell's equations. This wave would have to be fairly distant from its source to have the **plane wave** characteristics displayed here. The term **transverse** refers to the perpendicular direction of the wave's electric field in relation to the direction of energy propagation. As shown in the figure, or in an appropriate Web applet about this topic, the electric field, E, and the magnetic field, H, are perpendicular to each other and to the direction of propagation. Another very common type of EM-wave configuration is the **spherical wave**, which emanates from a point source. The spherical EM wave also obeys Maxwell's equations and eventually turns into the plane wave at a sufficient distance

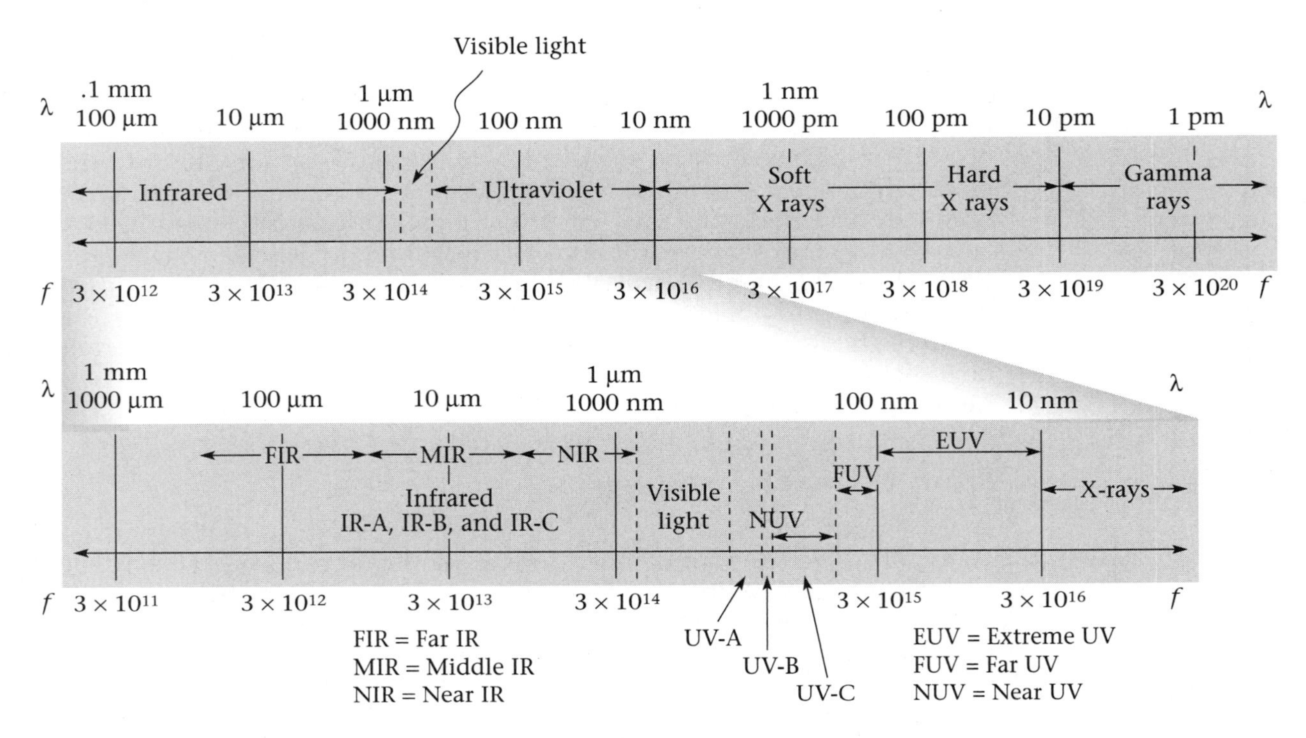

FIGURE 8-2 Electromagnetic spectrum above "radio waves"

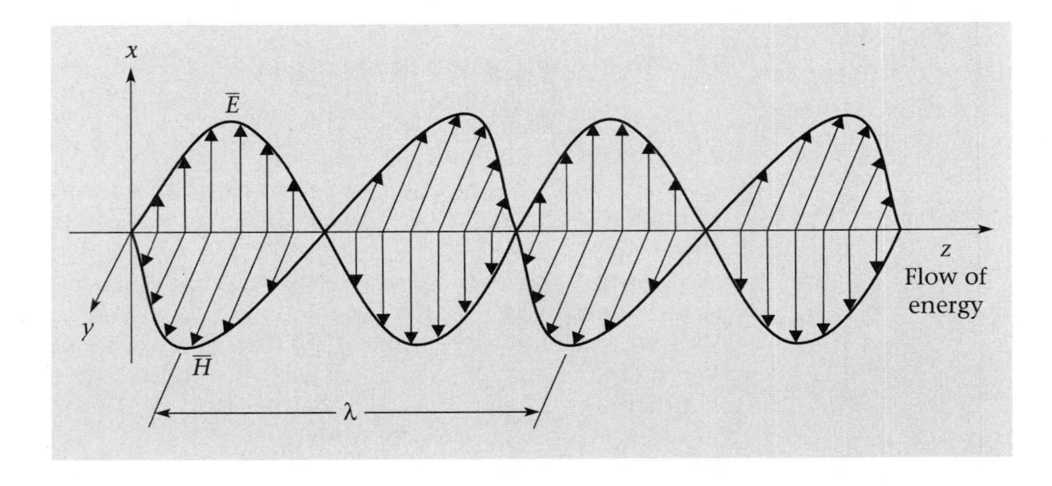

FIGURE 8-3 A transverse electromagnetic (TEM) plane wave

from its source. There are many other modes or configurations of EM waves, but their existence is associated more with guided EM waves and is closely linked to the geometry of the guiding structure.

8.3 Properties of EM Waves

At this time, we want to formally characterize EM waves. Recall how an EM wave is produced. Some source of electrical energy is applied to an antenna. The **antenna** is able to effectively convert a high-frequency voltage or current into a propagating EM wave (antennas and their characteristics are the subjects of chapter 9). Recall that the source of energy is usually a high-frequency sine or cosine wave or some type of pulsed waveform. In looking at Figure 8-3, one can find an obvious correlation between the EM wave and the sine wave applied to the originating antenna system. If we consider that the sine wave applied to the antenna is initially at 0 V, we will note that the intensity of the resulting EM wave will correspond exactly to the sine-wave amplitude as shown in the figure.

Wavelength

The resulting EM wave propagates away from the source (antenna) at a speed equal to the speed of light in free space, *c*, with each succeeding wave-front changing in amplitude as the applied sine wave changes its amplitude. This gives rise to the concept of wavelength. The **wavelength** of an EM wave is given by the distance between identical points on the wave. Another way to remember this concept is to consider the distance the initial part of the wave will travel as the applied sine-wave signal undergoes a complete cycle. Wavelength, or λ (lambda), is given in meters by:

$$\lambda = \frac{c}{f} \qquad \textbf{8.1}$$

where, $c = 3 \times 10^8$ m/sec, the velocity of light, and f is the frequency in Hz. If the EM wave is not in free space, then equation 8.1 is altered to include the effect of the transmission media on the velocity of propagation. A more general form that includes this effect is given by:

$$\lambda = \frac{c}{\sqrt{k}f} \text{ or } \lambda = \frac{c}{\sqrt{\varepsilon}f} \qquad \textbf{8.2}$$

where k is the **dielectric constant** of the medium *or* ε is the **relative permittivity** of the material, as appropriate.

EXAMPLE 8.1

Determine the frequency of the 1-m wavelength signals Marconi first experimented with.

Solution Rearranging equation 8.1 yields:

$$f = \frac{c}{\lambda}$$

and

$$f = \frac{3 \times 10^8 \text{ m}}{1 \text{ m}} = 300 \text{ MHz}$$

The Impedance of Free Space

Although not a direct characteristic of an EM wave, the value of the impedance of free space and the field intensities of an EM wave are directly related. **Free-space impedance**, Z_0, is given by:

$$Z_{0(\text{fs})} = \sqrt{\frac{\mu_0}{\varepsilon_0}} = 120\pi = 377\ \Omega \qquad \textbf{8.3}$$

where, μ_0 is the **permeability of free space** and ε_0 is the permittivity of free space.

The Meaning of Free-Space Impedance

Recall that the constants μ_0 and ε_0 are encountered in the study of inductors (magnetic fields) and capacitors (electric fields), respectively. It is not possible to simply measure the impedance of free space with a piece of test equipment. However, the value of Z_0 will allow us to calculate the values of the electric- and magnetic-field components of an EM wave. Z_0 is the ratio of the electric-field (E) component intensity (given in V/m) of the EM wave to the magnetic-field (H) component intensity (given in A/m).

Polarization of the EM Wave

The **polarization** of the EM wave is given by the direction of the electric-field component. The wave depicted in Figure 8-3 was vertically polarized. Usually, the construction and orientation of the antenna determines the electric-field polarization.

Types of EM Waves

There can be many types of EM waves in terms of geometric symmetries or field orientations relative to the direction of propagation. Guided waves can take on many configurations due to the **boundary conditions** imposed by the guiding structure. Figure 8-3 showed a TEM plane wave. Figure 8-4 depicts an EM wave with a spherical wave shape close to the source of energy.

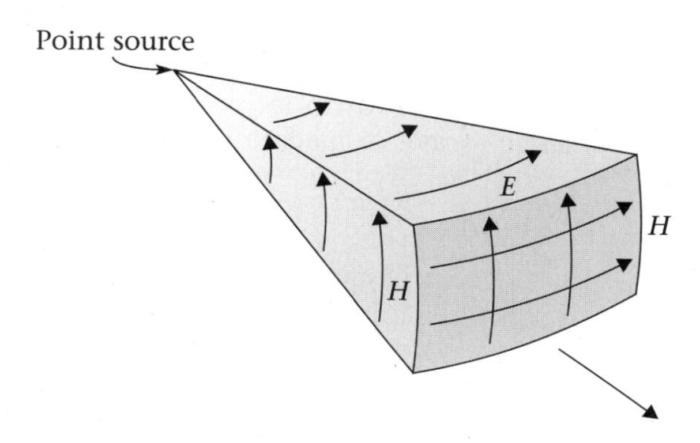

FIGURE 8-4 A spherical EM wave is emanated from a point source

As pointed out earlier, a spherical EM wave will transition into a plane wave at a distance far enough from the point source. A point source is also known as an **isotropic source**. An isotropic source radiates equally in all directions (see Figure 8-5).

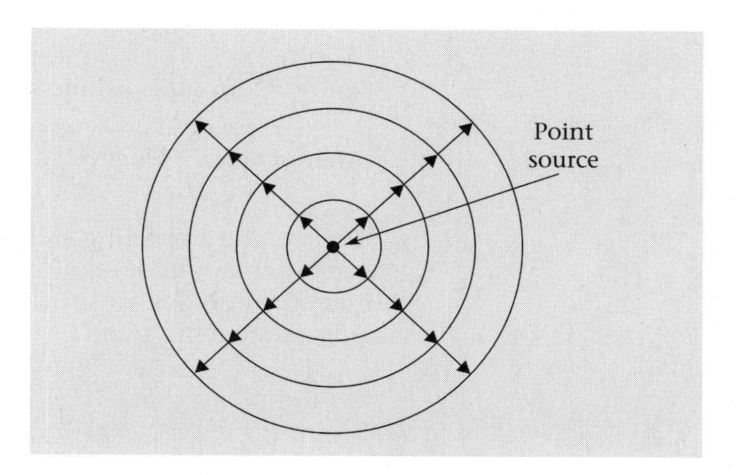

FIGURE 8-5 An isotropic source

EM-Wave Power Calculations

Let us perform some calculations to determine the power density and intensity of EM waves in free space. We can determine the power density of an EM wave in the following fashion: If the wave is emanated from a point source, it will propagate equally in all directions from the point source (see Figure 8-5), therefore, the power transmitted will spread out over the area of a sphere of radius r (area = $4\pi r^2$) and the power density, P_D, will be given by:

$$P_D = \frac{P_{\text{trans}}}{4\pi r^2}\ \text{W/m}^2 \qquad \textbf{8.4}$$

where r is the distance from the source in meters.

Additionally, the magnitude of the power flowing in the direction of wave propagation can be calculated by:

$$P = E \times H\ \text{W} \qquad \textbf{8.5}$$

The electric field intensity is given by:

$$E = \frac{\sqrt{30P_{\text{trans}}}}{r}\ \text{V/m} \qquad \textbf{8.6}$$

from equations 8.3 and 8.4.

Two additional common calculations involving EM waves are the effective radiated power and the path loss. Effective radiated power, ERP, or effective isotropic radiated power, EIRP, are both given by:

$$\text{EIRP} = \frac{\text{Antenna Gain} \times P_{\text{trans}}}{4\pi r^2} \qquad \textbf{8.7}$$

Path loss for an isotropic signal in free space is given by:

$$\text{Attenuation} = 20\log\left(\frac{r_2}{r_1}\right)\ \text{dB} \qquad \textbf{8.8}$$

Due to the spreading out of the signal energy, the EM wave undergoes an attenuation of –6 dB every time the distance the wave travels doubles. In equation 8.8, r_2 is the new location of the wave and r_1 is the wave's starting or reference point.

8.4 The Propagation of EM Waves

Now we will turn our attention to the propagation of terrestrial or earth-based EM waves. How these waves behave is of great interest to us because these types of EM waves are used for telecommunications and broadcasting services. As mentioned before, EM radio waves behave the same way that light waves do; therefore, their propagation is governed by the following properties (one should draw on his or her life experiences for examples of these properties):

- Reflection
- Refraction
- Diffraction
- Interference
- Absorption
- Doppler effect

Reflection of EM Waves

All EM waves (and any other type of wave for that matter) will undergo **reflection** if the medium that they are propagating in undergoes an abrupt change in its physical properties (i.e., impedance changes in the medium because either ε or μ or both have changed). Any reflection of EM energy is analogous to a sound echo. The more abrupt the "discontinuity," the more pronounced the echo. Let us investigate this property in greater detail. We start by looking at a special case, shown in Figure 8-6. This is known as normal incidence of an EM wave. In this case the transverse EM wave, E_{inc}, propagating in the z direction in medium 1 (lossless dielectric material number 1) encounters medium 2 (lossless dielectric material number 2) at the boundary or interface, located on the xy-plane between the two media at $z = 0$.

FIGURE 8-6 Normal incidence of an EM wave on a boundary

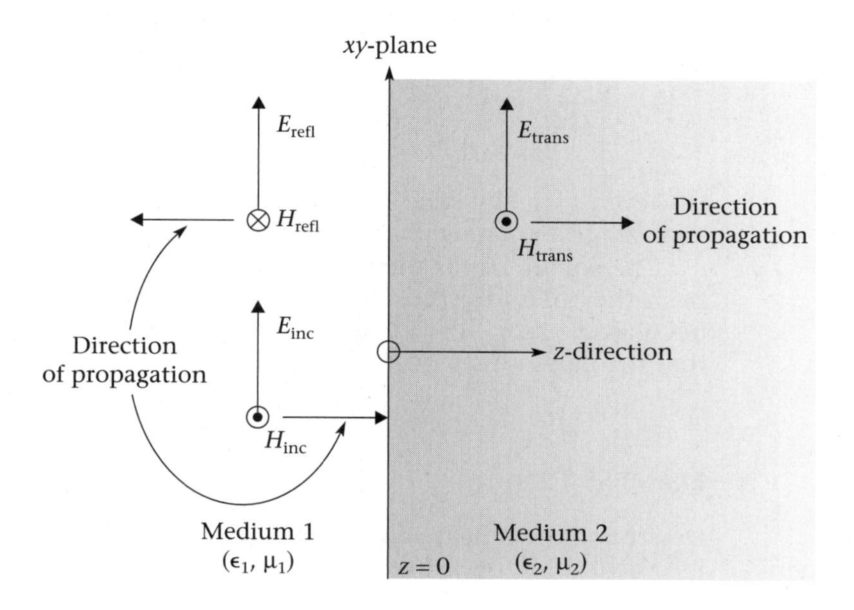

The figure further shows a partially reflected EM wave, E_{ref}, also in medium 1 and a transmitted EM wave, E_{trans}, in medium 2. The reflected wave or echo is caused by the change in media. Note that for lossless medium, usually only ε_n changes unless medium 2 has magnetic properties.

The Reflection Coefficient

It is possible to calculate the amount of reflected signal from the properties of the two media! The ratio of reflected electric-field intensity, E_{ref}, to incident electric-field intensity, E_{inc}, is given by Γ (Gamma), the **reflection coefficient**. The reflection coefficient can be calculated by:

$$\Gamma = \frac{E_{ref}}{E_{inc}} = \frac{\eta_2 - \eta_1}{\eta_2 + \eta_1} \qquad \textbf{8.9}$$

where η is the intrinsic impedance of the medium, given by:

$$\eta = \sqrt{\frac{\mu_R \mu_0}{\varepsilon_R \varepsilon_0}}$$

For free space, $\eta_0 = Z_0$. If we consider the usual case where $\mu_R = 1$, we get:

$$\eta_n = \frac{\eta_0}{\sqrt{\varepsilon_{R_n}}} .$$

The ratio of E_{trans} to E_{inc} is given by τ (tau), the **transmission coefficient**. The transmission coefficient can be calculated from the following equation:

$$\tau = \frac{2\eta_2}{\eta_1 + \eta_2} \qquad \textbf{8.10}$$

However, τ usually is not used as often as the reflection coefficient, Γ, when analyzing the propagation of an EM wave.

The equations for the reflection or transmission coefficients are derived from the "boundary conditions" of the physical situation. It is somewhat beyond the scope of this text to derive these equations, but we can note that the equations model how nature operates.

EXAMPLE 8.2

Determine the reflection coefficient for the case of normal incidence of an EM wave going from free space into a medium that has $\varepsilon_R = 4$.

Solution For this case, $\eta_1 = \eta_0$, and $\eta_2 = \dfrac{\eta_0}{2}$, therefore:

$$\Gamma = \frac{E_{ref}}{E_{inc}} = \frac{\eta_2 - \eta_1}{\eta_2 + \eta_1} = \frac{\dfrac{\eta_0}{2} - \eta_0}{\dfrac{\eta_0}{2} + \eta_0} = \frac{\dfrac{1}{2} - 1}{\dfrac{1}{2} + 1} = -\frac{1}{3}$$

What does this tell us? It says that the intensity of the reflected electric field is $\frac{1}{3}$ of E_{inc} and 180° out of phase!

Inspection of the equation for Γ reveals that it can take on values from −1 to +1. A minus value indicates a phase reversal of the reflected electric field and a positive value indicates no phase change. The phase of the reflected EM wave depends upon whether the EM wave is going from a less dense to a more dense medium or vice versa.

Standing Waves and the Standing-Wave Ratio

Returning to Figure 8-6, we note that if mediums 1 and 2 have different values of intrinsic impedance, there will be a partial reflection of the incident wave. We further note that in medium 1 there will now be two propagating waves, the incident wave and the reflected wave. These two waves will vectorially add together to give a resultant wave in medium 1. This resultant wave will have the characteristics of a **standing wave** (see Figure 8-7 on page 276). Standing waves occur quite often in nature due to incident- and reflected-wave energy from some physical process or scenario like the example we have been describing. The strings of a guitar or piano undergo vibrations when plucked or "keyed" due to the resulting standing-wave patterns. Actually, standing-wave vibrations are the basis for the operation of many musical instruments.

For our example, it is possible to calculate a measure of the intensity of the standing-wave pattern in medium 1. This measure of the standing-wave pattern is called the **standing-wave ratio** (SWR) and is given by the ratio of the maximum to minimum value of the electric-field intensity of a standing-wave pattern. The SWR can be calculated by the use of the following equation:

$$\text{SWR} = \frac{E_{max}}{E_{min}} = \frac{1 + |\Gamma|}{1 - |\Gamma|} \qquad \textbf{8.11}$$

Note a few special cases: For $\Gamma = 0$ (no reflection at the interface), SWR = 1 and all the incident power is transmitted into medium 2 ($\tau = 1$).

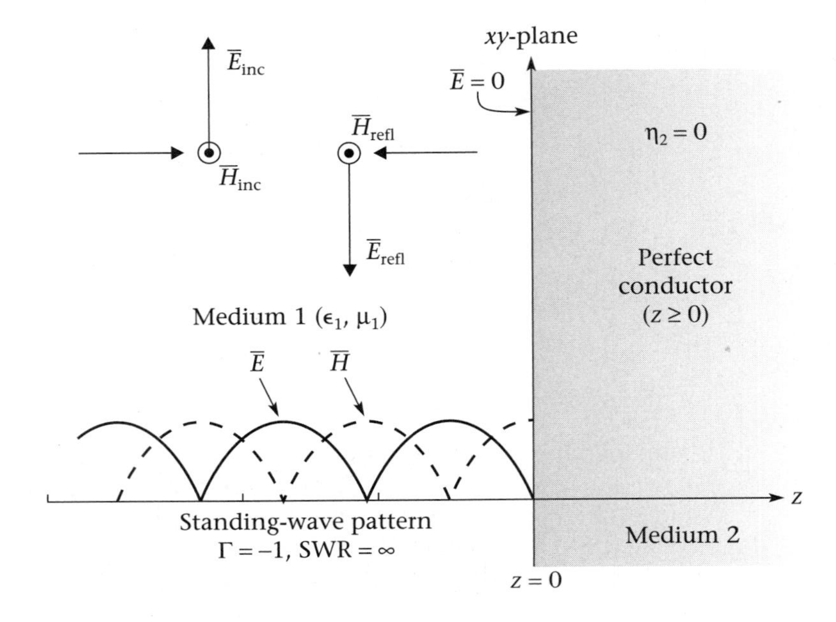

FIGURE 8-7 Normal incidence of an EM wave on a good conductor

For $\Gamma = \pm 1$, there is total reflection of the incident wave from the interface, $\tau = 0$, and the SWR $= \infty$.

To recap, Γ has a range of -1 to $+1$, τ can range from 0 to 2, and the SWR can take on values from 1 to ∞.

•—EXAMPLE 8.3

Determine the SWR from the value of Γ determined in example 8.2.

•—Solution From example 8.2 the value of $\Gamma = -\frac{1}{3}$, therefore:

$$\text{SWR} = \frac{1 + \left|-\frac{1}{3}\right|}{1 - \left|-\frac{1}{3}\right|} = \frac{\frac{4}{3}}{\frac{2}{3}} = 2.0$$

But what does this really tell us? An SWR of 2.0 does not relate much information unless we consult a table of "SWR to power-loss conversions." What we *can* tell is that there is some reflection of the incident wave, but not a total reflection. We will return to this point later on when discussing transmission lines.

A Special Case: Reflection from a Good Conductor

Let us examine what happens when an EM wave is normally incident upon a good conductor. In this case, $\eta_2 = 0$; as a consequence $\Gamma = -1$ and $\tau = 0$ or $E_{ref} = -E_{inc}$ and $E_{trans} = 0$. As was shown in Figure 8-7, an incident wave will undergo total reflection from a good conductor and also experience a phase reversal in the process.

Why does this phase reversal happen? A good conductor must have an **equipotential** surface. That means that charge does not gather or accumulate anywhere on the surface (any charge on the surface will spread out equally because like charges repel). Furthermore, the electric-field intensity must be equal to zero on the surface or at the interface, otherwise we could measure a voltage difference on the conductor. How is it possible that $E_{Z=0} = 0$ when we have an incident and reflected wave to the left of the conductor? Recall the phrase *boundary conditions* that was introduced earlier; for this example the boundary conditions mandate that the electric-field intensity at $z = 0$ must be zero. This is possible for the case $E_{ref} = -E_{inc}$, because:

$$E_{z=0} = E_{inc} + E_{ref} = 0 \quad \textbf{8.12}$$

Also notice the standing-wave pattern to the left of the conductor in Figure 8-7. For this case, because $\Gamma = -1$, the SWR $= \infty$. Due to the relative speed of the incident versus reflected wave there is a null of the electric field at the interface and the standing-wave pattern repeats itself every half wavelength back from the interface.

Reflection at an Angle

What happens in the general case of **oblique incidence**? If we have a good conductor, as shown in Figure 8-8, the wave is totally reflected and the angle of incidence equals the angle of reflection (Law of reflection). A smooth (mirror like) surface yields a **specular reflection**. If the surface is rough, a **diffuse reflection** results with the wave being scattered in many different directions.

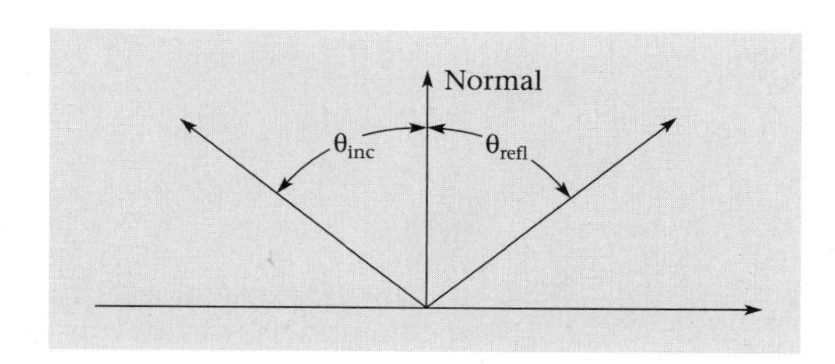

FIGURE 8-8 Specular reflection of an EM wave

Applications of Specular Reflections

A typical application of this type of reflection is in the construction of reflecting surface antennas. See Figure 8-9 for an example of this common type of antenna.

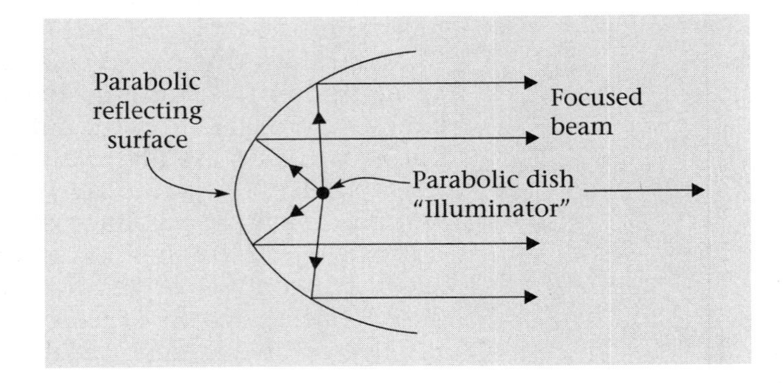

FIGURE 8-9 A reflecting-surface antenna (parabolic dish)

The Refraction of EM Waves

The next property of EM-wave propagation to be discussed is refraction. This occurs for the case of oblique incidence on a boundary between two different dielectrics. If we look at the simple case shown in Figure 8-10, we see that **refraction** occurs when an EM wave propagates into another medium at an angle.

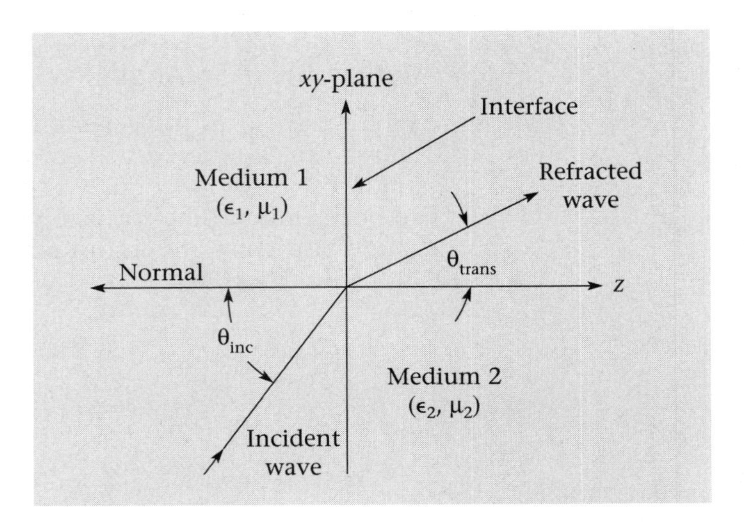

FIGURE 8-10 Refraction of an EM wave

For this case, Snell's Law predicts the new angle of the EM wave in medium 2. However, this is not always the case. Depending upon the polarization of the incoming wave relative to the interface, several other interesting possibilities exist. Although it is beyond the scope of what we want to discuss here, suffice it to say that these other special conditions can lead to occurrences of total internal reflection, partial polarization of the EM wave, and, at the Brewster Angle, total wave polarization. Lastly, it should be pointed out that if one considers lossy dielectrics for which the value of Γ can be complex (Γ has both a magnitude and a phase angle), the calculations involved can become very messy for both normal and oblique incidence.

The Diffraction of EM Waves

Diffraction is a subtle effect that causes EM waves to appear to bend around corners. The result of this effect is that the radio-wave shadow cast by an object is not necessarily a "sharp" shadow. Depending upon the circumstances, a diffracted radio wave can fill in the shadow. Figure 8-11 depicts this phenomenon. The incoming wave comes in contact with the "knife edge" of the obstruction. The effect of the sharp edge is to generate a weak point source of EM waves that can illuminate the shadow area.

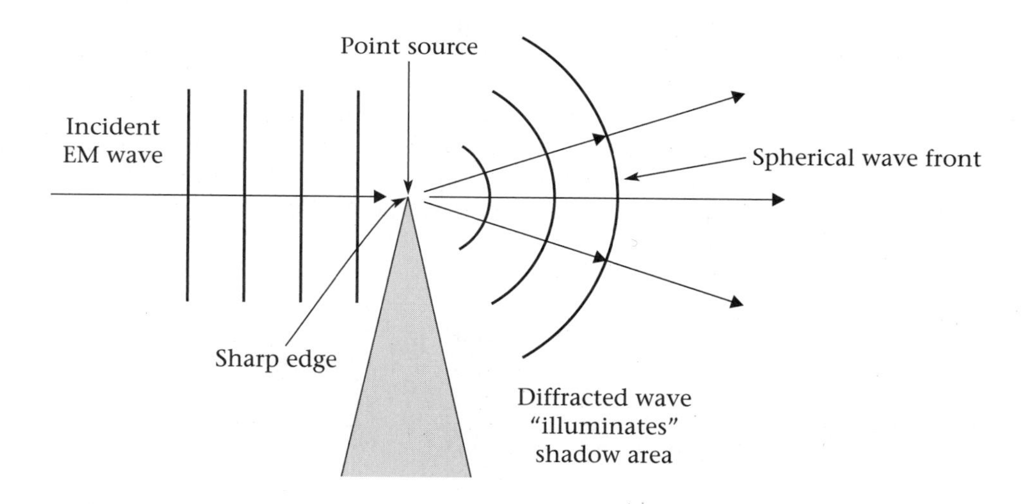

FIGURE 8-11 Diffraction of an EM wave

The Interference of EM Waves

An extremely common condition is that of EM-wave **interference**. This occurs when transmitted EM waves arrive at the same location by two or more different propagation paths. This can happen in numerous ways, one of which is shown in Figure 8-12 on page 280. This figure shows three waves arriving at the receiving location after traveling slightly different paths. Due to their phase difference, the waves can vectorially add either

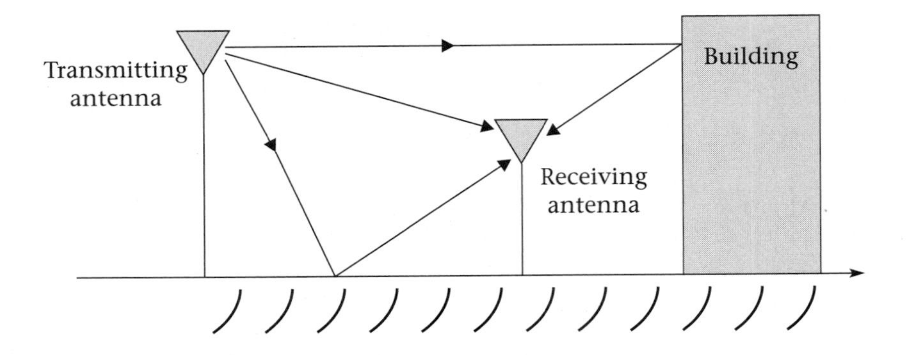

FIGURE 8-12
Interference of an EM wave

constructively or destructively at the receiver. Quite often the phase shift experienced by the propagating waves is time varying and causes a rapid fading of the signal. This is known as **Rayleigh fading**.

The Absorption of EM Waves

Another commonly occurring phenomenon in the propagation of EM waves is **absorption**. There are two important cases that should be considered—an EM wave incident upon a lossy media and the effect of the earth's atmosphere on propagating EM waves. Figure 8-13 depicts the first case. As shown, a normally incident EM wave propagates into lossy material (a building with a wall that has both wood and concrete sections). What occurs is an exponential decay of the wave energy as it travels into the materials. Eventually, the wave is either totally dissipated or it will reemerge from the material and continue its journey. The distance required for the wave intensity to become 37% of its incident value is called the

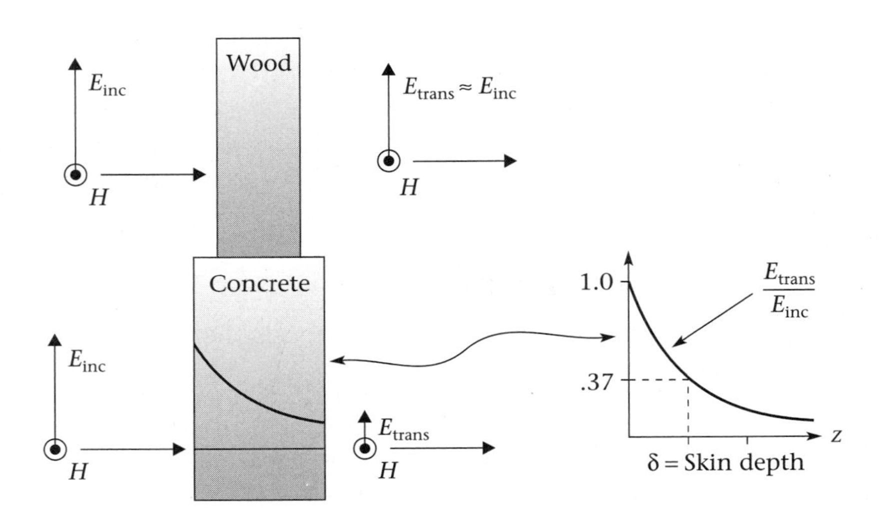

FIGURE 8-13
Absorption of EM waves by a structure. The concrete section of the wall absorbs quite a bit of the wave, while the thin, wood wall section causes only a slight attenuation of the wave.

skin depth, δ. Figure 8-13 shows how the concrete has absorbed quite a bit of the wave while an only slightly attenuated wave is able to propagate through the thin wood-wall section.

Figure 8-14 shows atmospheric absorption in dB/km at two different altitudes. Usually, atmospheric absorption is ignored below 10 GHz, however, new wireless applications for short ranges have opened up the spectrum above 10 GHz. Notice the windows of lower attenuation at certain frequencies. The attenuation peaks are due to resonances with water and oxygen molecules. Attenuation through the atmosphere is also dependent upon weather conditions (i.e., fair and dry conditions, drizzle, heavy rain, fog, snow, hail, and so forth). Our understanding of atmospheric attenuation and other propagation phenomena has given us **Doppler radar** for modern weather forecasting.

FIGURE 8-14
Average atmospheric attenuation versus frequency (horizontal polarization) at two different altitudes

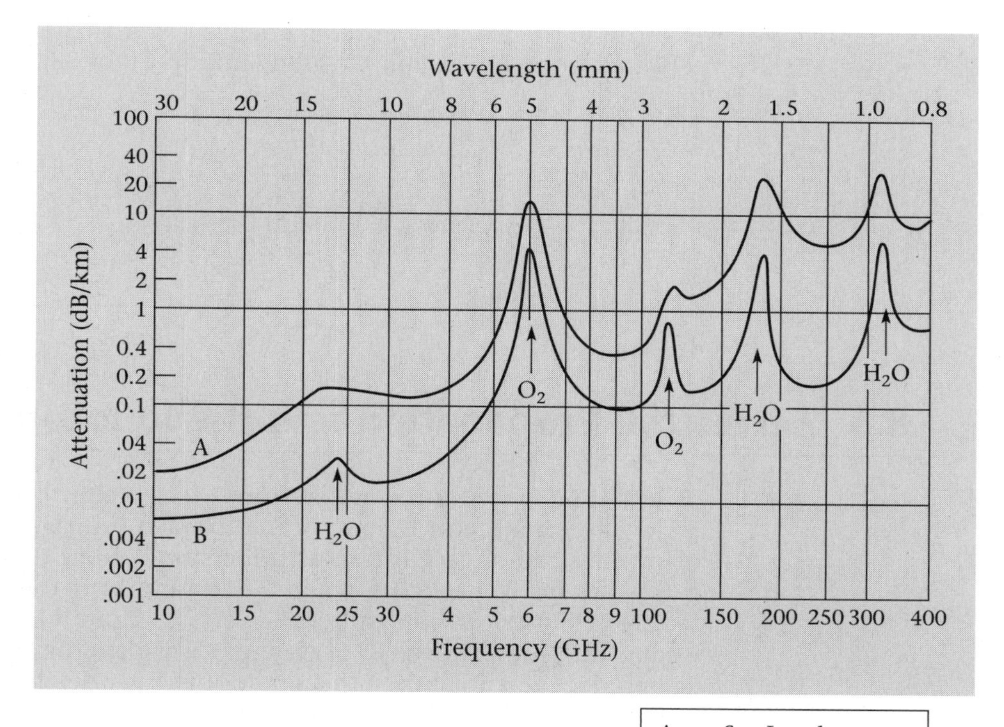

A:	Sea Level T = 20°C P = 760 mm $^{P}H_2O = 7.5$ gr/m^3
B:	4 km Elevation T = 0°C $^{P}H_2O = 1$ gr/m^3

The Doppler Effect

The Doppler effect is observed whenever there is relative motion between the source and the receiver. As one moves toward the source of an EM wave there is an apparent increase in frequency; as one moves away from the source there is a decrease in frequency. The relationship for the frequency observed from the source is given by:

$$f_{\text{doppler}} = f\left(1 \pm \frac{v_R}{c}\right) \quad \textbf{8.13}$$

where, v_R is the speed of the receiver/object relative to the source.

EXAMPLE 8.4

Determine the Doppler shift experienced by a 10.5-GHz radar signal reflected from an automobile going at a speed of 33.5 m/sec (approximately 75 mph):

Solution From equation 8.13,

$$\Delta f_{\text{doppler}} = 10.5 \times 10^{9}\left(\frac{33.5}{3 \times 10^{8}}\right) = 1172.5 \text{ Hz}$$

8.5 Terrestrial Propagation and the Ionospheric Layers

At this time we will turn our attention to the effects of the ionospheric layers of the earth's atmosphere on radio-wave propagation. In the early days of wireless radio, very little was understood about radio-wave propagation. It was not until the 1920s, when Sir Edward Appleton first theorized the effects of the various ionospheric layers on radio waves, that we started to understand how EM waves of different wavelengths and polarizations propagated. Figure 8-15 shows a diagram of the various ionospheric layers.

The ionospheric layers in the earth's atmosphere exist due to solar radiation. Our atmosphere receives sufficient energy from solar and cosmic radiation in the form of ultraviolet waves, x and γ (gamma) rays, and α (alpha) and β (beta) particles to dissociate the molecules in the upper atmosphere into ions. The different radiation energy levels and atmospheric properties give rise to four main ionospheric layers: the D, E, F_1, and F_2. The F_1 and F_2 layers combine into a single F layer at night and the ions in the D and E layers recombine so that the layers disappear. Note that the

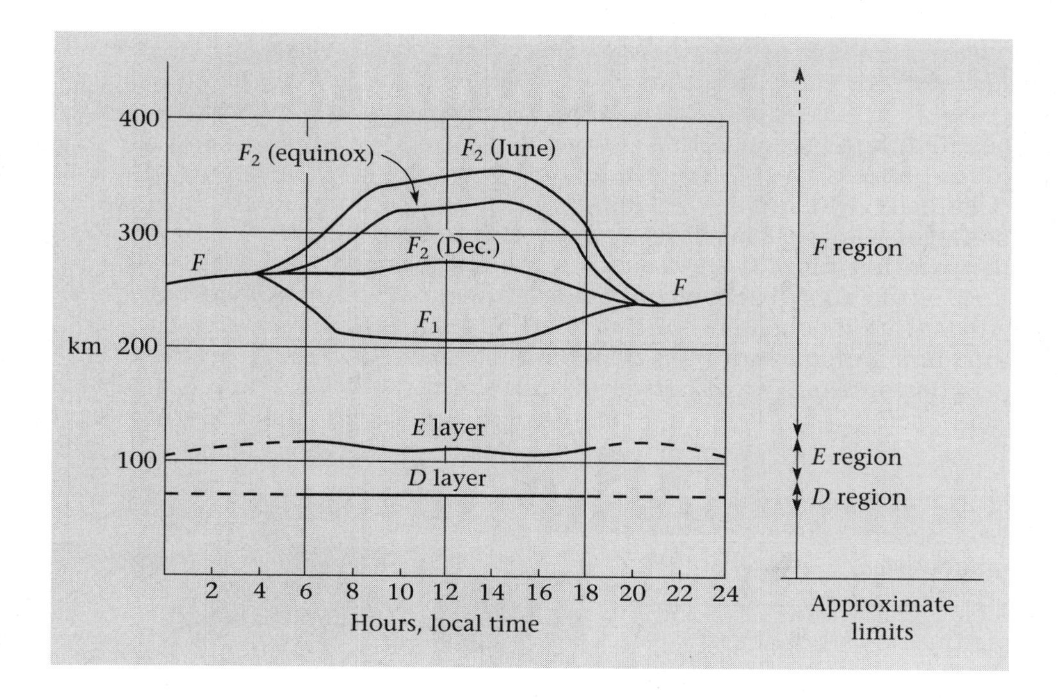

FIGURE 8-15 The ionospheric layers with daily and seasonal variations shown. (*Courtesy of Marconi plc*)

layers have seasonal and daily variations and can also experience sudden density changes as a result of variations in the sun's energy output due to sunspots or solar flares. Also, a sporadic E-layer of very high ionization density sometimes appears with the E layer. To propagating EM waves, the ionospheric layers act like either lossy dielectrics or good conductors, depending upon the density of ions in the layer and the angle of incidence between the propagating wave and the ionospheric layer. There are many interesting Web sites dedicated to this and other topics concerning the ionosphere and EM-propagation conditions.

The Effect of the Ionospheric Layers on EM Propagation

The radio spectrum can be divided into three frequency ranges that are characterized by different modes of propagation. These modes are known as **ground waves**, **sky waves**, and **space waves**. Historically, we have used these different frequency ranges for different services. As pointed out earlier, the first successful long-distance wireless telegraph systems used very low frequencies (VLF) for their operation. It was not until the 1920s that it was discovered that other modes of propagation allowed for long-distance EM wave propagation with much less transmitter power.

Ground Waves

Our first mode of propagation is called the ground wave. Frequencies below approximately 2 MHz must be transmitted using vertical polarization (hence, vertical antennas) or else the signal is quickly attenuated by absorption caused by the conductivity of the ground. These waves will propagate outward from the antenna but undergo refraction due to the variation in the density of the atmosphere. As the waves propagate further and further they eventually lay over and die (the electric-field component becomes parallel to the ground and is gradually absorbed). See Figure 8-16 for a depiction of this process. The figure insert shows how the tilted electric-field component of the wave consists of both a vertical and a horizontal element. The horizontal element is absorbed by ground conductivity.

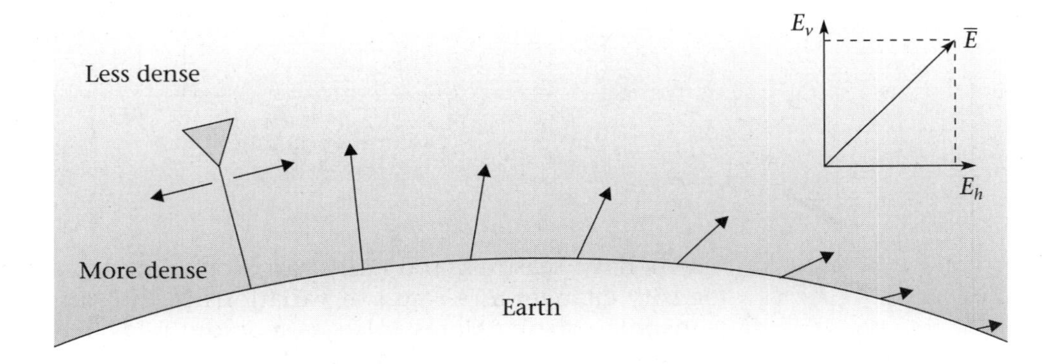

FIGURE 8-16 Propagation via the ground wave

For ground waves, an increase in transmitter power causes an increase in transmission distance. How do the ionospheric layers affect these waves? These waves tend to be confined by the lower ionospheric layers and the surface of the earth. At night, the D and E layers disappear and ground waves are able to propagate further. This enhanced propagation occurs because some of the energy launched by the antenna toward the sky can travel further before the effects of atmospheric refraction force it parallel to the ground. The legacy AM broadcast band uses ground-wave propagation. In the United States there are several "clear channel" AM broadcast stations (e.g., WBZ at 1030 kHz) that have output powers of 50 kW and can be received halfway across the country over ground-wave propagation during nighttime operation. Other AM stations, depending upon their class of operation, are required by the FCC to either maintain or reduce their nighttime output power or entirely cease operations at sunset. This is done to preclude interference to distant co-channel stations.

Sky Waves

Sky-wave propagation occurs for frequencies above approximately 2 MHz and below approximately 30 MHz (the upper frequency limit is extremely variable). EM waves in this range will propagate through the D and E layers, but will be reflected by the F layer(s). This mode of operation is called "skip" or "hop." Horizontal polarization is best for this type of propagation. See Figure 8-17 for a diagram of this effect.

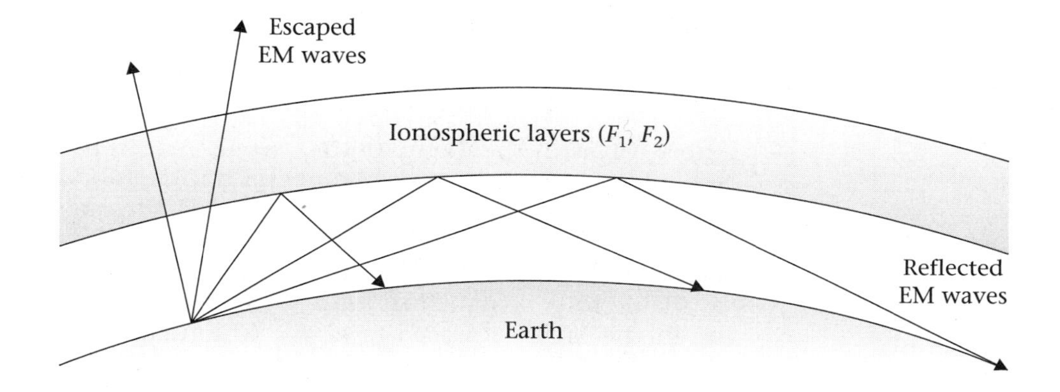

FIGURE 8-17 Sky-wave propagation

If at any given time a signal is transmitted vertically upward, there is a critical frequency, f_C, for which all frequencies lower than f_C will be reflected back to earth. There is also a maximum useable frequency (MUF) related to f_C by the angle of incidence for which reflection of an EM wave will occur (see Figure 8-17). The density of the ionospheric layers can be extremely variable; hence, the value of f_C and the MUF are constantly changing with time.

In this mode of propagation, the signal can bounce back and forth between the earth and the ionospheric layers, hopping around the entire world! See Figure 8-18 on page 286 for a diagram of this effect. In the early days of wireless, a lack of knowledge of the ionospheric layers and these propagation effects drove technology, leading to the development of extremely high-powered systems for low-frequency transmission in the 12,000- to 17,000-meter range. During World War I several members of the scientific community voiced concern regarding the lack of available bandwidth for more transmitting stations at these long wavelengths. Marconi had noticed the propagation effect of sky waves, but could not explain the theory behind it. In the early 1920s amateur radio operators successfully proved the feasibility of communicating over great distances with

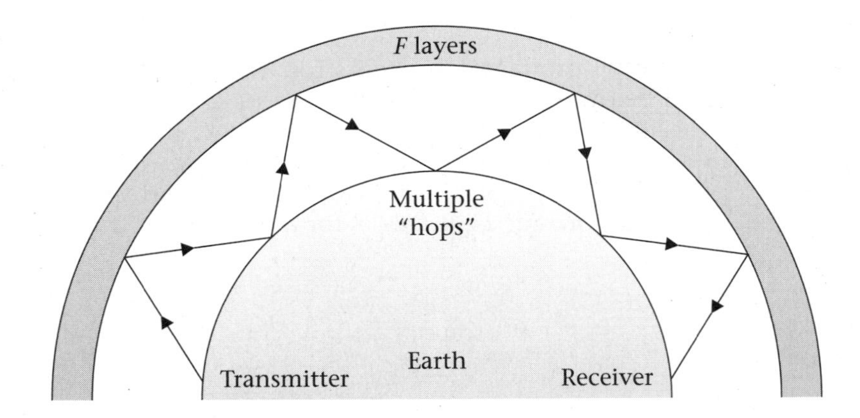

FIGURE 8-18 Multi-hop propagation

short waves in the 3- to 30-MHz range. A member of the radio club that was involved with these experimental transmissions was Edwin Armstrong.

As an aside, the variability of the critical frequency, f_C, and the appearance of the sporadic E layer is the reason why there is presently no Channel 1 in the VHF TV band. At one time, a Channel 1 at 48 to 54 MHz was available to be assigned to local TV stations. It was discovered that all too often the MUF increased to above Channel 1's frequency; therefore, the signal from Channel 1 in a distant city could propagate via a sky wave to another city where it would interfere with the local Channel 1 TV station. Because TV stations were licensed to serve the local community, the use of Channel 1 was discontinued by the FCC in the late 1940s. This discontinuation also occurred in Europe. The 48 to 54 MHz frequencies were assigned to other radio services.

At night, the attenuation effects of the D and E layers disappear, the two F layers combine, and propagation is enhanced due to a more definitive single F layer that acts like a good reflecting surface for propagating sky waves.

Space Waves

For frequencies above approximately 30 MHz or the MUF, line-of-sight propagation occurs. Frequencies above the MUF will propagate through the ionospheric layers (these signals are termed *trans-ionospheric*) and will not return to earth. Therefore, radio and TV broadcasting and other services that use frequencies above approximately 30 MHz are limited by line-of-sight conditions. The heights of the transmitting and receiving antennas determine the range of transmission. See Figure 8-19 for a diagram of this effect. Vertical polarization, horizontal polarization, or combinations of the two can be used for line-of-sight propagation. The radio horizon for a space wave is given by the following empirical equation:

$$d_{TA} = 4\sqrt{h_T} \qquad \textbf{8.14}$$

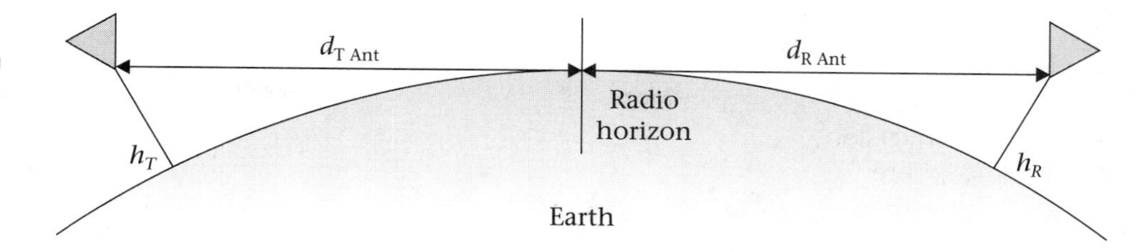

FIGURE 8-19 Line-of-sight propagation

where d_{TA} is the distance from the transmitting antenna to the radio horizon in kilometers and h_T is the height of the transmitting antenna in meters. A similar equation exists for the receiving antenna. Total propagation distance is determined by the sum of the two values, or

$$d_{\text{total}} = 4\sqrt{h_T} + 4\sqrt{h_R} \quad \mathbf{8.15}$$

For radio services that use space waves, antennas designed to radiate power outward and parallel to the earth's surface are typically used. Cellular systems employ antennas mounted with slight downtilting to ensure total coverage of their service area.

Trans-Ionospheric Propagation

Today, communication satellites are a large part of the telecommunications infrastructure. Although geosynchronous satellites have heretofore been the primary mode of operation for the delivery of various telecommunications services, there are plans for medium and low earth-orbiting satellite systems (MEOS and LEOS, respectively) for high data-rate delivery (the "Internet in the sky" idea). These systems all use frequencies in the GHz range that are able to pass through the ionospheric layers and that also provide the necessary bandwidth to deliver the services. Because the ionosphere imparts a rotation to the EM wave (known as the Faraday effect), circular polarization is employed by the transmitting and receiving antennas to minimize the negative aspects of this trans-ionospheric propagation.

Other Miscellaneous Propagation Modes

Several other propagation modes deserve some comment. Certain weather-related conditions can radically affect propagation conditions. Signals that normally only propagate in a line-of-sight manner might propagate hundreds of miles over the horizon due to super-refractive effects caused by temperature inversions or the approach of a weather front. See Figures 8-20a and 8-20b on page 288 for two examples.

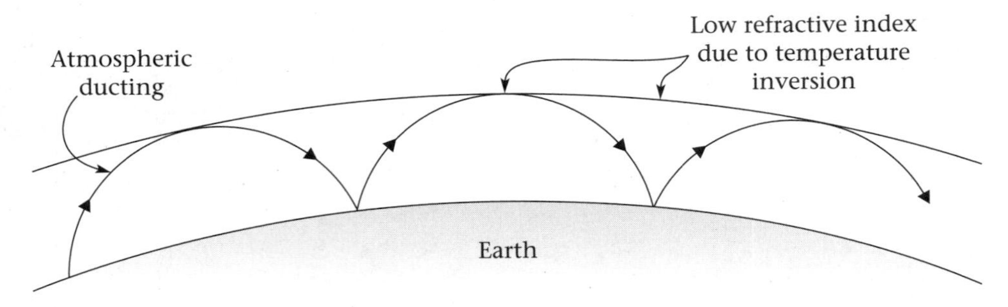

FIGURE 8-20a Super-refractive ducting caused by a temperature inversion in the atmosphere

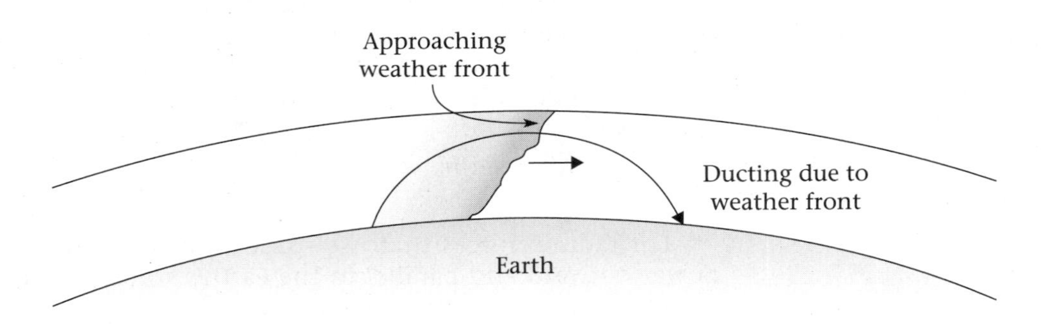

FIGURE 8-20b Ducting caused by the approach of a weather front between the transmitter and receiver

Summary of Part I

In part one, we covered most of the important aspects of electromagnetic propagation. It should be obvious that only in free space, which is devoid of interfering objects, do EM waves behave as simply as the models that have been presented. Terrestrial propagation is another story. The radio channel is extremely noisy and the ionosphere has certain fairly well understood but highly variable effects on long-distance communications. Most broadcasting utilizes the free-space channel and various radio services make do with the propagation modes that exist for the assigned frequencies of operation. Today, with the advent of fiber-optic cables, the vast majority of long-distance high-speed data transmission is being done over a telecommunications network of these cables that extends around the world. This network is extremely immune to external noise and is constantly being expanded and extended to reach every populated area of the globe. For other types of worldwide services, satellites offer good performance either on a global scale or when used in a broadcasting mode.

In fairly recent times, we have seen the use of short-range radio communications (the cellular-telephone system) become the method by which one achieves “mobility.” By this, we mean the ability to have access to the

PSTN without a wired connection. Even more recently, wireless LANs (WLANs) have become more commonplace and access to data over cell-phone systems has become more popular. The promise of high-speed data from third generation (3G) cellular systems, LEOS systems, and local multi-point-distribution systems (LMDS) that operate in the 10s of GHz range is coming closer to fruition.

Several companies have announced plans for nationwide networks of wireless LANs that will deliver high-speed data in designated "hot spots" across the United States to subscribers of this service. The initial hot spots will generally be located in high-traffic metropolitan areas.

The above being the case, the list of propagation effects studied earlier in this chapter becomes even more important in understanding how effectively these wireless systems will operate. Reflection, refraction, and diffraction all combine to scatter signals in the various environments in which use of these telecommunications services might be desired. The scattering and absorption of signals by these environments leads to greater path losses than those predicted by simple propagation models. The use of high-speed mobile data services in automobiles, trains, and planes is further restricted by fading and the Doppler effect.

An understanding of the basic principles of EM propagation will allow one to have a better comprehension of the operation and limitations of wireless systems.

As mentioned earlier, there are many interactive java applets available on the World Wide Web that demonstrate some of the principles of EM wave propagation presented here. A search for physics applets by any of the standard Web search engines (e.g., http:\\www.aj.com) will quickly bring one to several virtual physics labs with interesting java applets that simulate EM wave behavior.

PART II

Transmission Lines

8.6 Introduction to Conductor-Based TLs

At this time we will turn our attention to **guided-wave** transmission media, commonly known as **transmission lines** (TLs). Let us start this discussion by answering the question, Why are TLs needed? As electrical conductors become an appreciable fraction of a wavelength, they begin to radiate energy. This antenna effect is the basis for the need for transmission line structures that suppress this effect.

The purpose of a transmission line is to guide energy from point to point as efficiently as possible. At low frequencies (with extremely long wavelengths), current flows within the conductors and is not prone to radiate away from the TL. At higher frequencies the current flow takes place near the conductor surface (this is called the **skin effect**). At radio frequencies (RF) and higher (microwaves and millimeter waves) the transmission line acts as a structure that guides an EM wave.

Types of TLs

There are numerous types of wireline TLs available (note that wireline TLs have at least two conductors). Here is a list of some of the most common types:

- Unshielded twisted pair (UTP)
- Shielded twisted pair (STP)
- Parallel wire (twin lead)
- LAN category-*N* cable:
  - Category 1
  - Category 2
  - Category 3
  - Category 4
  - Category 5
  - Category 5e
  - Category 5e+
  - Category 6 (proposed), and so on
- Coaxial cable

Examples of Practical TLs

Figures 8-21 through 8-24 show examples of the common types of TLs.

There are other types of TLs for use at microwave/millimeter and optical wavelengths. Some of these TLs are:

- Waveguides
- Microstrip and stripline
- Fiber-optic cable
- Miscellaneous others including:
  - Dielectric
  - Co-planar
  - Slotline

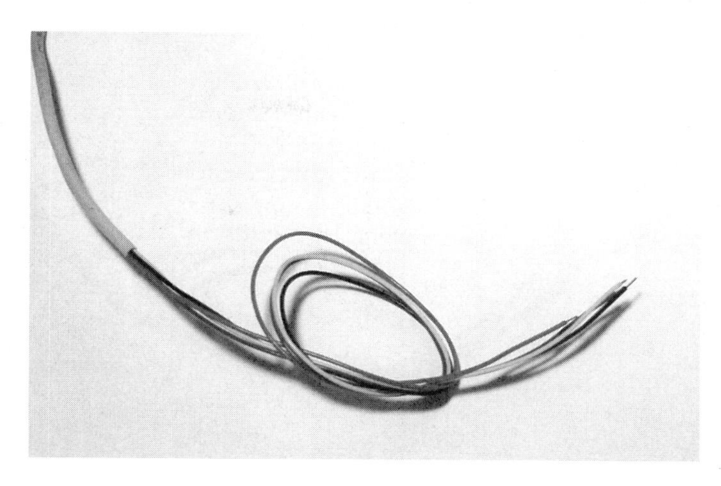

FIGURE 8-21 Simple unshielded copper pairs (Category 2)

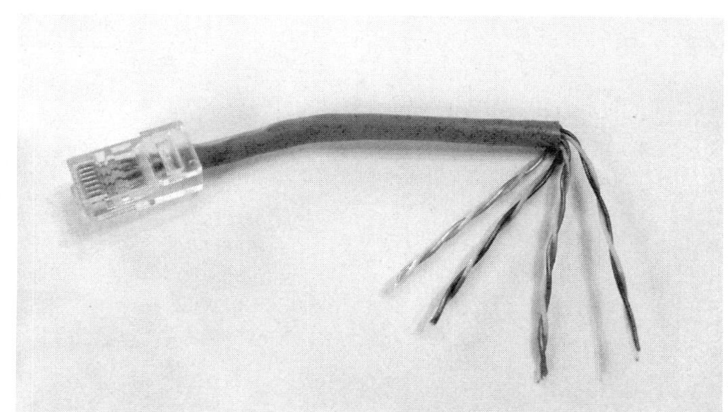

FIGURE 8-22 Category-5 networking cable for LANs. Four pairs of wire are shown with a standard RJ45 connector.

FIGURE 8-23 Typical cellular telephone tower coaxial feeder cables

FIGURE 8-24 Typical cable-TV coaxial cables and system components (top) and telephone cable (multiple copper pairs) below the coaxial cables

Some typical pictorial examples of these TLs are shown in Figures 8-25 through 8-29.

FIGURE 8-25 A typical millimeter-wave waveguide (hollow TL)

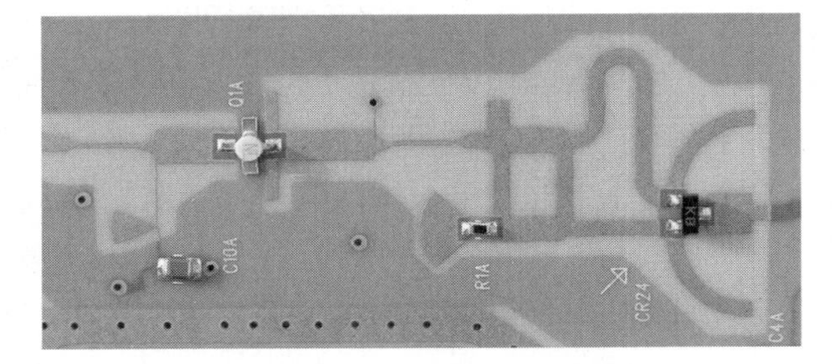

FIGURE 8-26 Typical stripline circuit assembly on a PCB

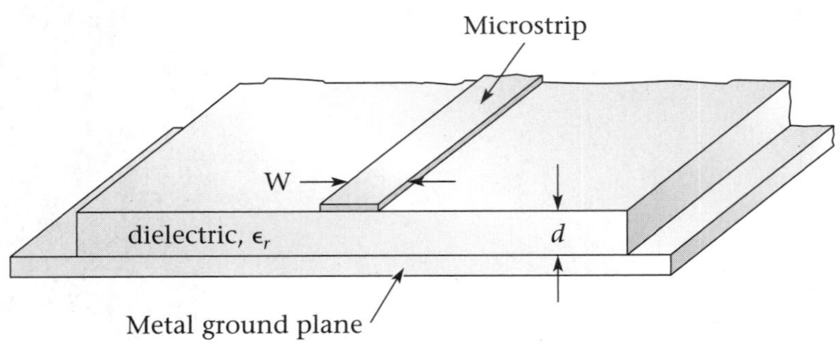

FIGURE 8-27 Typical microstrip TL

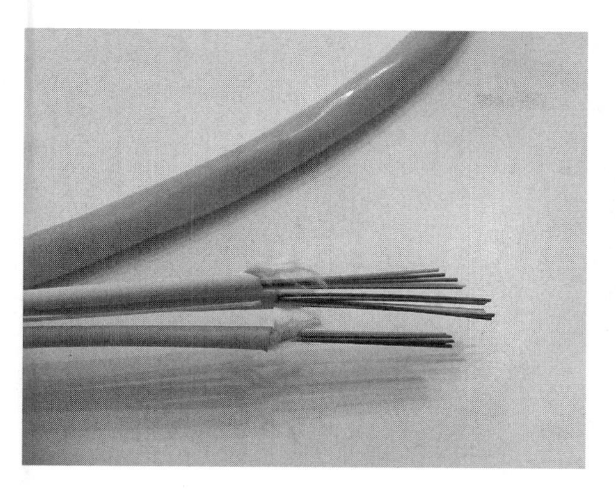

FIGURE 8-28a Typical fiber-optic cable or light guide

FIGURE 8-28b Typical ribbon fiber-optic cable

FIGURE 8-29a
Slotline TL

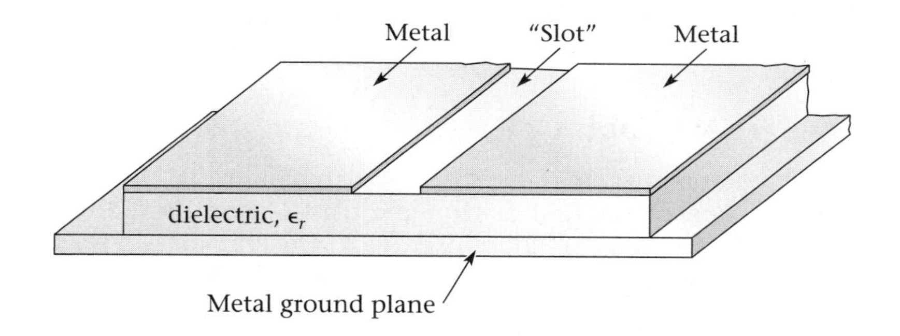

FIGURE 8-29b
Co-planar waveguide

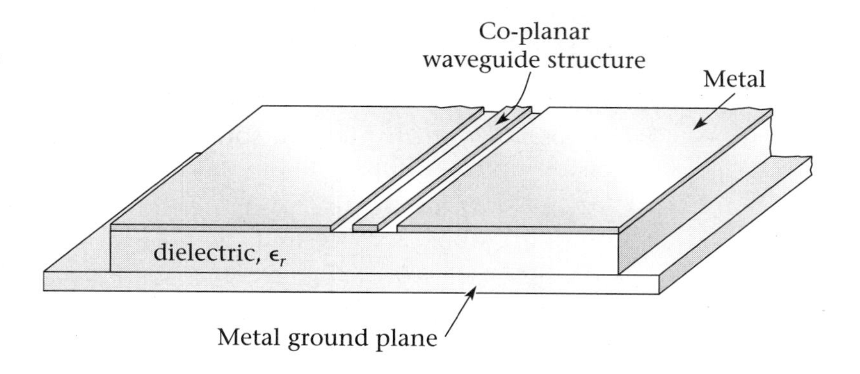

FIGURE 8-29c
Dielectric waveguide

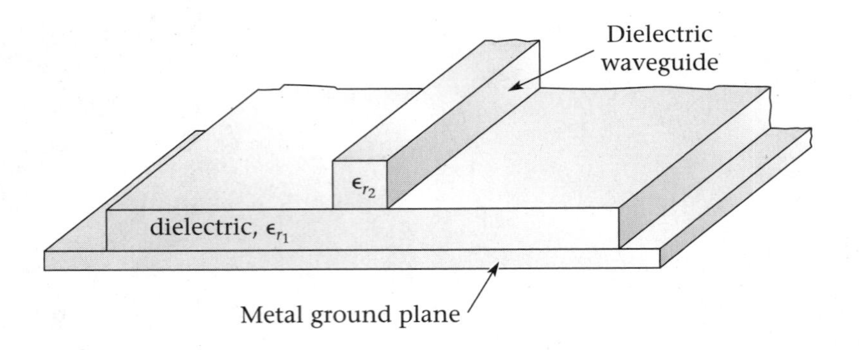

8.7 Detailed TL Characteristics

All TLs have common characteristics that are determined by their physical geometry, materials of construction, and frequency of operation. Depending upon their intended application, certain of these various characteristics may or may not be important. TLs can be used for such diverse functions as transmitting data in an LAN or supplying large amounts of RF power to an antenna.

LAN Cables

Returning to unshielded twisted pairs (UTP): We will consider LAN or data cables first as they are usually considered to be carrying data (baseband signals) as opposed to EM waves (passband signals). Keep in mind, however, that this point becomes blurred when one considers cabling for gigabit LANs.

The Electrical Industries Alliance (EIA) has defined five categories of UTP for telephone and data transmission and has proposed a sixth standard (250-MHz cable). Each of the first five categories' electrical performance is documented in a standard (TIA/EIA 568-A) developed by a joint industrial standards group. The present five categories are:

- *Category 1* Basic copper-wire twisted pair for telephone use.
- *Category 2* Four unshielded, solid twisted pairs. Good for data transmission up to 4 Mbps.
- *Category 3* At least three twists per foot. Certified for data transmission of 10 Mbps. If more than one pair, the pairs must have different twist rates.
- *Category 4* Similar to Category 3, but certified for data up to 16 Mbps. Also has more twists per foot.

- *Category 5* Three to four twists per inch. Data-grade cable certified for 100 Mbps. Used for most LAN installations. Note that there is a length limitation of 100 meters.
- *Category 5e* This is a modification to the Category 5 standard, and is covered by TIA/EIA 568-B.2.

Belden Corporation lists several products that surpass the 5e standard: *DataTwist* 5e+, *DataTwist* 350, with a bandwidth of 350 MHz, and most recently *DataTwist* 600e, with a bandwidth of 600 MHz.

What is the next step in this evolutionary improvement in LAN cables? Manufacturers will keep developing cables with higher speeds that will become the Category 6 (250 MHz) and the Category 7 cables of the future, to be used with gigabit LANs and other high-speed data applications.

General TL Characteristics

The three most important TL characteristics are:

- *Impedance* Each transmission line has a **characteristic impedance**, Z_0, expressed in ohms. This is the ratio of electric-field (E) strength in volts per meter to magnetic-field (H) strength in amperes per meter on the TL.
- *Loss with Distance* All transmission lines attenuate the signal with a loss proportional to the distance of transmission. This specification is usually given as the attenuation in dBs per meter (dB/m) or per foot (dB/ft). This attenuation factor is usually frequency dependent, so it is usually supplied as graphical data plotted as loss/distance versus frequency.
- *Power Capacity* TLs have power limitations determined by their size. Typical values are in watts for coaxial cables less than $\frac{1}{4}$ of an inch in diameter and kilowatts for cables greater than 1 inch in diameter. UTPs only carry low voltage levels and therefore do not specify maximum power levels.

8.8 Wireline TL Models

TLs can be either fractions of wavelengths or multiple wavelengths long; therefore, they are modeled as distributed-parameter networks with values of capacitance (C), conductance (G), inductance (L), and resistance (R) determined by physical shape and dimensions. Because these parameters are considered to be distributed over the length of the TL structure they are specified per unit length. A typical electrical circuit, on the other hand, consists of "lumped" components (*R*s, *L*s, and *C*s) interconnected by conductors. See Figure 8-30 for examples of TL equivalent circuit models.

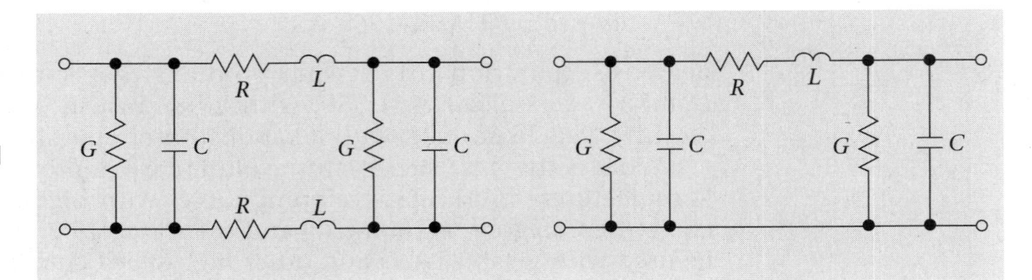

FIGURE 8-30
TL models (balanced and unbalanced lines). Note that the values of *C*, *G*, *R*, and *L* shown are per unit length (e.g., pf/m, Ω/m, and so forth).

Geometry Is Everything!

The values of capacitance, inductance, resistance, and conductance per unit length are functions of the physical dimensions of the TL and the materials used in its fabrication. The impedance, Z_0, of a TL is given by:

$$Z_0 = \left(\frac{R + j\omega L}{G + j\omega C}\right)^{\frac{1}{2}} \qquad \textbf{8.16}$$

In this equation, the value of Z_0 is the ratio of the electric- to magnetic-field intensities on the TL.

What happens if the TL is lossless? In this case *R* and *G* are both equal to zero and Z_0 is given by:

$$Z_0 = \sqrt{\frac{L}{C}} \qquad \textbf{8.17}$$

Again, the values of *L* and *C* are functions of the physical dimensions (the geometry) of the TL.

Parallel-Wire TLs

Parallel-wire TLs, depicted in Figure 8-31 have been used in the past, but are no longer very popular because they usually lack shielding. This lack of shielding is a definite disadvantage when installing the cable because it cannot come in close proximity of anything else that is conductive. We will see the reason for this shortly.

Figure 8-32 shows the end view of the cross-sectional geometry of a parallel-wire transmission line. Notice the "fringing" of the EM fields in the left portion of the figure.

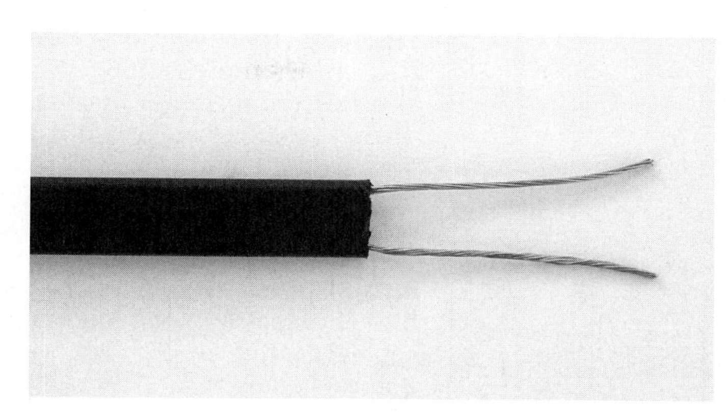

FIGURE 8-31 Typical parallel wire, 300-Ω TL

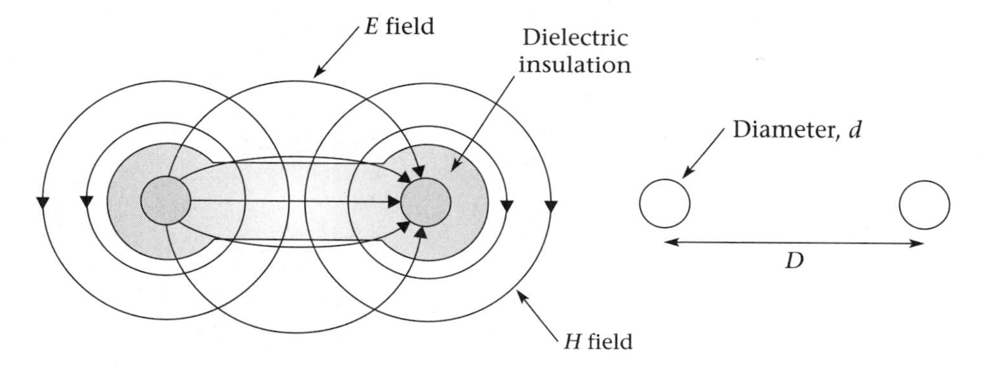

FIGURE 8-32 End view of parallel-wire TL geometry. Notice the "fringing" of the EM fields on the left figure.

The impedance of a parallel-wire TL is given by the following equation:

$$Z_0 = \frac{276}{\sqrt{\varepsilon}} \log\left(\frac{2D}{d}\right) \Omega \qquad \textbf{8.18}$$

where ε is the relative permitivity of the material between the conductors, d is the diameter of the conductors, and D is the spacing between the conductor centers.

Typical parallel-wire lines usually have values of Z_0 equal to 300 Ω. Twisted pairs (CAT-*N* type cables) are approximately 100 Ω. Recall the "fringing" of the EM fields for this type of geometry. When used for the transmission of RF energy, any close contact with a conductor will disrupt the flow of EM energy down the transmission line and cause a reflection of the propagating signal! In severe cases, all of the propagating energy might be reflected, resulting in no signal transmission at all.

Coaxial Cable TLs

Typical coaxial TLs are shown in Figures 8-33a and 8-33b on page 298 and 8-35 on page 300. Coaxial TLs usually consist of concentric conductors (see

Figure 8-34). The ratio of the diameters of the inner and outer conductors determines the characteristic impedance of the cable because the dimensions determine the values of C and L per unit length. Typical impedances of coaxial cables are 50 or 75 Ω. Note that for a coaxial TL all of the EM energy is contained within the coaxial structure; hence, with the EM energy shielded by the TL's outer conductor, there are few if any restrictions on the installation of coaxial cable.

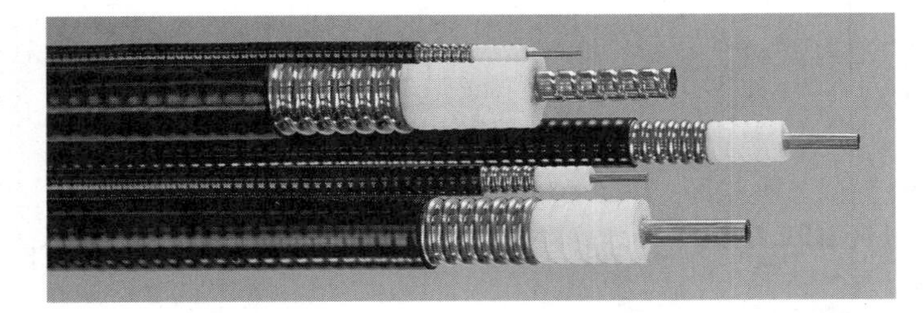

FIGURE 8-33a Solid dielectric coaxial cables (*Courtesy of Andrew Corp.*)

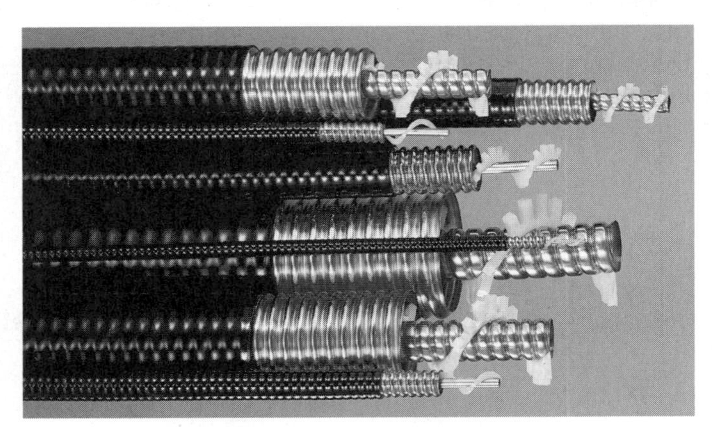

FIGURE 8-33b Air dielectric coaxial cables (*Courtesy of Andrew Corp.*)

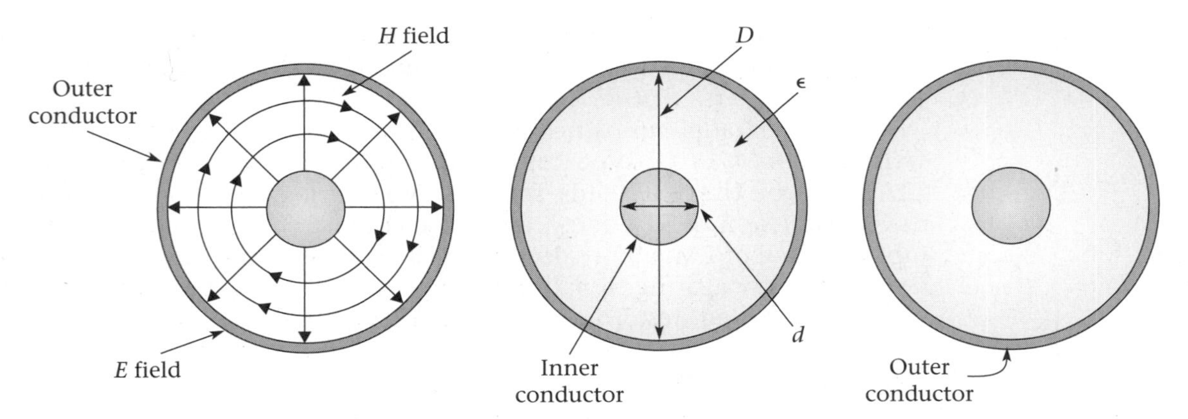

FIGURE 8-34 End view of the typical geometry of a coaxial cable and the electric- and magnetic-field patterns

For this geometry,

$$Z_0 = \frac{276}{\sqrt{\varepsilon}} \log\left(\frac{2D}{d}\right) \Omega \tag{8.19}$$

Typical Coaxial Cable Characteristics

Heliax coaxial cable, with foam dielectric, is a typical coaxial cable manufactured by the Andrew Corporation. Following are some of the electrical characteristics for two typical types, compliments of the Andrew Corporation. Other coaxial-cable characteristics can be seen in Figure 8-35 on page 300.

For $\frac{1}{4}''$ Superflexible, 50 Ω, Type FSJ1-50A:

- Attenuation at 100 MHz = 1.79 dB/100 ft
- Attenuation at 1 GHz = 5.96 dB/100 ft
- Attenuation at 10 GHz = 21.7 dB/100 ft
- Power rating at 1 GHz = 372 W

For $\frac{1}{2}''$ Superflexible, 50 Ω, Type FSJ1-50B:

- Attenuation at 100 MHz = 1.04 dB/100 ft
- Attenuation at 1 GHz = 3.60 dB/100 ft
- Attenuation at 10 GHz = 14.6 dB/100 ft
- Power rating at 1 GHz = 889 W

Notice the higher loss per unit length of the smaller cable and the lower power rating.

Other Characteristics

For a typical coaxial cable the velocity of propagation, v_p, will be approximately 67% to 95% of the velocity of light, c (v_p is dielectric dependent). Therefore, the wavelength of a signal on a coaxial cable will be different than its free-space wavelength. Other miscellaneous characteristics include jacket materials, fire retardant characteristics, direct burial properties, dielectric type, connector type, and flexibility.

8.9 The Matching of a Source and Load

At the beginning of our discussion of TLs, it was mentioned that the efficient transfer of energy was an underlying reason for the use of TLs. Total transmission efficiency is made up of three components. The first is the TL

Characteristics					
Maximum Operating Frequency, MHz	20400	13400	10200	13500	4900
Peak Power Rating, kW	6.4	13.2	15.6	15.6	90
Relative Propagation Velocity, %	84	83	81	85	88
Minimum Bend Radius, in (mm)	1 (25)	1 (25)	1.25 (32)	1.75 (45)	5 (125)
Attenuation, db/100 ft (db/100 m). Standard conditions: VSWR 1.0; ambient temperature 20°C (68°F).					
30 MHz	0.973 (3.19)	0.649 (2.13)	0.557 (1.83)	0.584 (1.92)	0.214 (0.702)
100 MHz	1.79 (5.89)	1.20 (3.94)	1.04 (3.41)	1.08 (3.56)	0.397 (1.3)
450 MHz	3.91 (12.8)	2.64 (8.66)	2.31 (7.59)	2.39 (7.83)	0.878 (2.88)
1000 MHz	5.96 (19.6)	4.06 (13.3)	3.60 (11.8)	3.68 (12.1)	1.36 (4.46)
2000 MHz	8.67 (28.5)	5.97 (19.6)	5.37 (17.6)	5.41 (17.8)	2.01 (6.59)
6000 MHz	16.1 (52.7)	11.3 (37.2)	10.5 (34.4)	10.3 (33.8)	—
10000 MHz	21.7 (71.2)	15.5 (50.8)	14.6 (47.9)	14.1 (46.3)	—
Average Power Rating, kW. Standard conditions: VSWR 1.0; ambient temperature 40°C (104°F); inner conducter temperature 100°C (212°F); no solar loading.					
30 MHz	2.28	3.97	5.76	3.99	12.3
100 MHz	1.23	2.14	3.09	2.15	6.62
450 MHz	0.567	0.975	1.38	0.978	2.99
1000 MHz	0.372	0.634	0.889	0.635	1.93
2000 MHz	0.256	0.431	0.598	0.431	1.31
6000 MHz	0.138	0.228	0.307	0.227	—
10000 MHz	0.102	0.166	0.220	0.165	—

FIGURE 8-35 Typical coaxial cables and their specifications (*Courtesy of Andrew Corp.*)

attenuation characteristic, the second is load matching, and the third is loss through radiation of EM energy (minimal for today's totally shielded coaxial cable). As the TL characteristics discussed previously indicate, TL attenuation is frequency dependent and proportional to length. This part of the efficiency equation is fairly well determined by the length of TL needed for the application. The second part of the equation is not so straightforward. Recall from basic circuit theory, the maximum power transfer theorem and load matching. For maximum power transfer we must have:

$$R_{source} = R_{load} \quad \text{or} \quad Z_{source} = Z^*_{load} \qquad \textbf{8.20}$$

where * indicates the complex conjugate. For example, if $Z_{source} = R \pm jX$, then $Z_{load} = R \mp jX$ for maximum power transfer.

With all TLs, we always try to achieve maximum power transfer. There are several reasons for this. The first is efficiency—it should always be as close to 100% as possible. A second reason that EM signal reflections are undesirable is that they can have an adverse effect on the signal-to-noise ratio (SNR), TL noise temperature, and the system bit error rate (BER). Additionally, if the signal reflection is severe enough it can possibly cause damage to or even destroy the active device used to produce the signal power.

TL Impedance Standards

Various de facto industry standards exist for TL impedance. The source impedance will be the same as the TL impedance. These nominal values are listed:

- RF signal delivery—50 Ω
- Cable TV—75 Ω
- Twisted pairs—100 Ω
- Audio applications—600 Ω

Note that these impedances are used in power calculations (i.e., 0 dBm, or 1 mW, for an RF system is referenced to 50 Ω).

8.10 Propagation of Waves on TLs

Let us now turn our attention to the propagation of EM waves on TLs. The theory of operation is the same as for the unguided EM-wave propagation covered earlier in this chapter. If the load (at $z = 0$) is not the same impedance as the TL's value of Z_0, a propagating EM wave will experience a reflection or echo. As with the unguided case, we can calculate a value of reflection coefficient, Γ, and a value of SWR.

The Reflection Coefficient and SWR

As with EM propagation, we define a reflection coefficient Γ (Gamma) that is the ratio of reflected electric-field intensity, E_{ref}, to incident electric-field intensity, E_{inc}. One can calculate Γ from the following equation (note the similarity to equation 8.9):

$$\Gamma = \frac{Z_L - Z_0}{Z_L + Z_0} \qquad \textbf{8.21}$$

where Z_L is the load impedance and Z_0 is the TL characteristic impedance.

The value of SWR is given by equation 8.10 from earlier in the chapter:

$$\text{SWR} = \frac{1+|\Gamma|}{1-|\Gamma|}$$

Two Special Cases of TL Reflection

There are two extreme cases of TL operation that must be discussed: an open load and a shorted load. Figure 8-36 shows the open TL case and Figure 8-37 shows the shorted TL case. For an open circuit, $Z_L = \infty$ and $\Gamma = +1$. This indicates a total in-phase reflection of the incident wave. The resulting standing-wave pattern is shown in Figure 8-36, in which one can see a maximum electric-field intensity at the interface and a current (magnetic field) null. The pattern repeats itself every $\lambda/2$. Note that a voltage maximum and current minimum at $z = 0$ yields an impedance (voltage-to-current ratio) of $Z = \infty$, which is what we would expect for an open circuit. Note that this is another example of a boundary condition being satisfied.

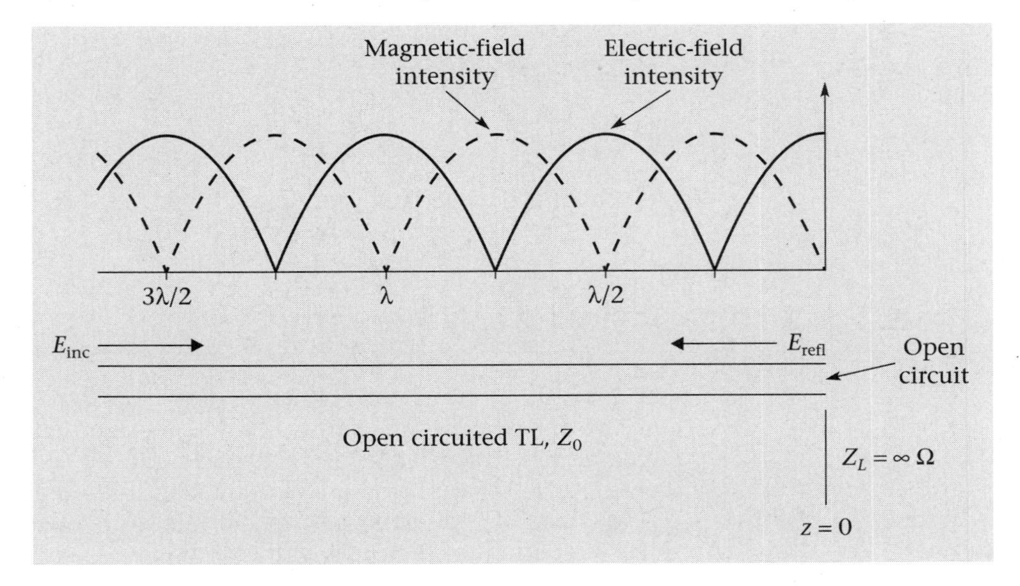

FIGURE 8-36 Standing-wave pattern on an open-circuited TL

For the short-circuit case, $Z_L = 0\ \Omega$ and $\Gamma = -1$. This also indicates a totally reflected wave with a 180-degree phase reversal. The resulting standing- wave pattern is shown in Figure 8-37. This figure shows a null at the interface and a repeating pattern every $\lambda/2$. Is this what should be expected for a short circuit (boundary condition) at $z = 0$? Yes. The voltage-to-current ratio would be equal to zero, which is the case for a short circuit, $Z = 0\ \Omega$.

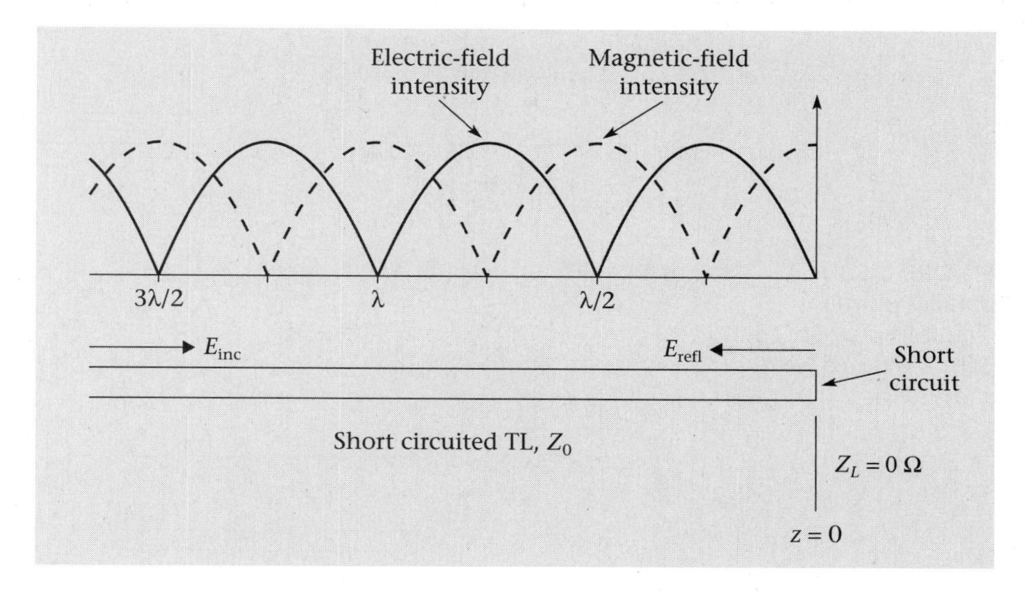

FIGURE 8-37 Standing-wave pattern on a shorted TL

Applications for TL Special Cases

There are applications for these two special TL cases that do not necessarily involve the transmission of power from one point to another. Using shorted or open-circuited TLs, it is possible to create equivalent circuits of inductance, capacitance, and even resonant circuits. See Figures 8-38a and 8-38b on page 304. As can be seen from the figures, different lengths of shorted or open TLs behave like inductors, capacitors, or resonant circuits constructed from *L*s and *C*s. At wavelengths in the centimeter and millimeter range, it must be pointed out that this can be one of the only ways to physically realize these circuits values.

How is it possible that shorted or open TLs can be equivalent to *L*s or *C*s or *LC*-resonant circuits? Consider a one-quarter wavelength ($\lambda/4$) shorted line. As shown in Figure 8-38a, a shorted quarter-wave TL looks like a parallel resonant circuit. The standing-wave pattern for this case, shown in Figure 8-38b, will indicate some value of current and zero voltage at the load (Z_L = short = $0\ \Omega$).

At the input, due to the standing-wave pattern generated, we will see just the opposite—zero current and some value of voltage. So at $z = \lambda/4$, the apparent value of line impedance is $Z_{line} = \infty$, or an open circuit. Since

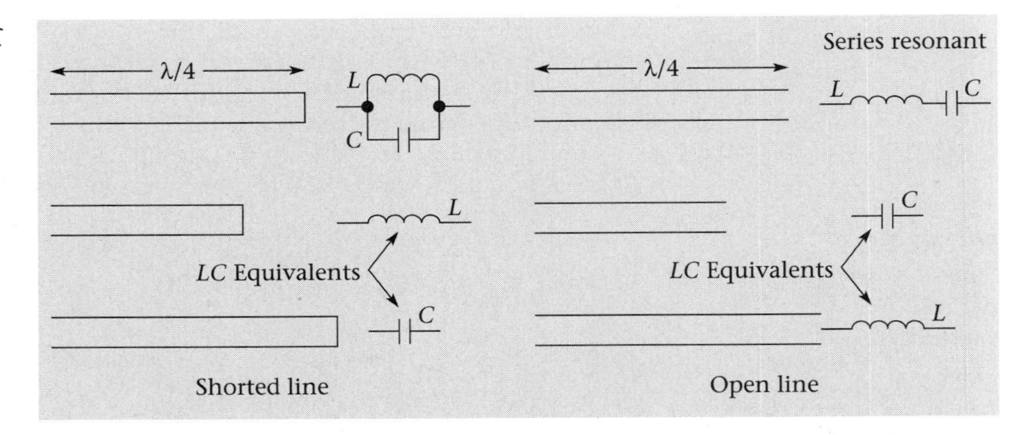

FIGURE 8-38a TL *LC* equivalents. Open or shorted TLs either shorter or longer than λ/4 act like either *L*s or *C*s.

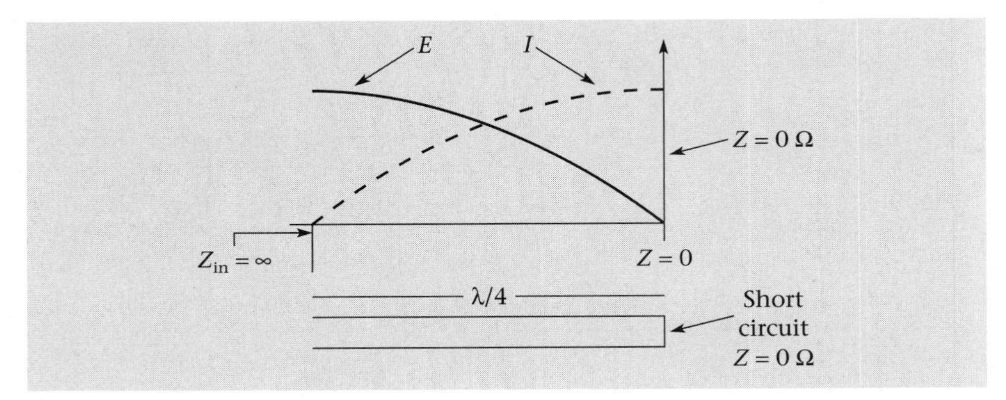

FIGURE 8-38b A parallel resonant circuit implemented with a quarter-wave shorted TL

this only happens at one input frequency or wavelength, it is therefore identical to the behavior of a parallel resonant circuit.

EXAMPLE 8.5

Determine the length of a λ/4 stub to act as a resonant circuit at 1200 MHz.

Solution Let $v_p = .8$. First, determine λ:

$$\lambda = \frac{.8 \times 300}{1200} = \frac{240}{1200} = 0.2 \text{ m or } 20 \text{ cm}$$

Therefore,

$$\frac{\lambda}{4} = 5 \text{ cm}$$

A 5-cm TL stub would be resonant at 1200 MHz.

TL Input Impedance

For a perfectly matched TL with no standing waves, the ratio of electric- to magnetic-field intensity is always equal to the value of Z_0. For any TL with a standing-wave pattern, there is an apparent line impedance, Z_{line}, given by the ratio of voltage to current intensity at that point ($z = -l$). This fact allows one to create various values of capacitance or inductance by using the correct-length "stub" (shorted TL). Practically speaking, the open-circuited TL is not used for these applications because it will radiate energy from its open end (Z_L is not really equal to ∞ for an open TL, for $Z = 377\ \Omega$ for free space). Some TL geometries are better suited than others for this type of application. Presently, microstrip TLs commonly use this technique to their advantage.

Power Relations on TLs

Let us turn to the use of TLs to provide power transmission from point to point. Consider the following two cases in which the transmission line does not see a load equal to Z_0. See Figure 8-39 for an example. For the case on the left, the EM propagating on the TL encounters a discontinuity due to a change in TL characteristics. For the case on the right, the load is not a short or an open; however, it is not equal to the TL impedance either, because $Z_{load} \neq Z_0$. In both cases, the mismatched conditions will cause a signal reflection to occur.

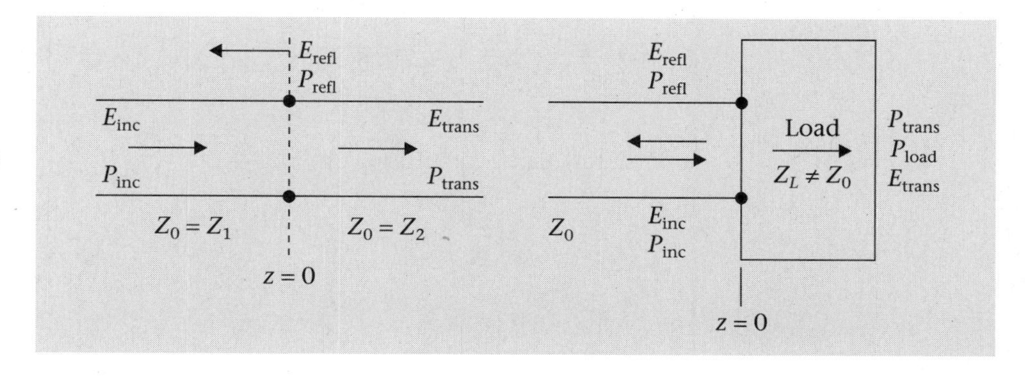

FIGURE 8-39 A TL feeding a TL with a different value of Z_0 (left figure) and a TL delivering power to a load with $Z_{load} \neq Z_0$

As can be seen, the same scenario exists for this situation as did for the normal incidence of a propagating EM wave at the interface to a new media (refer back to Figure 8-6). There will be an incident EM wave with some value of incident power, P_{inc}; a reflected EM wave with a value of reflected power, P_{refl}; and a transmitted EM wave with a value of P_{trans}. Also, we can calculate values of Γ, τ, and SWR from the parameters of the particular situation, as shown previously. It can be shown that one can use Γ to determine the amount of reflected or transmitted power for different scenarios. The relationships between Γ and the incident and reflected power are shown here:

$$P_{refl} = \Gamma^2 P_{inc} \quad \textbf{8.22}$$

$$P_{trans} = P_{inc}(1 - \Gamma^2) \quad \textbf{8.23}$$

Again note our special cases. For a short, $\Gamma = -1$ and SWR $= \infty$; for an open, $\Gamma = +1$ and SWR $= \infty$. In each case, $P_{refl} = P_{inc}$ and $P_{trans} = 0$. An example for $Z_{load} \neq Z_0$ might also be helpful in the understanding of these concepts.

EXAMPLE 8.6

A TL of $Z = 50\ \Omega$ is used to deliver RF power to a load of $100\ \Omega$ ($Z_{load} = 100\ \Omega$). Determine the value of Γ, SWR, the power delivered to the load, and the power reflected toward the source.

Solution First, calculate Γ and SWR:

$$\Gamma = \frac{100 - 50}{100 + 50} = \frac{50}{150} = \frac{1}{3}$$

$$\text{SWR} = \frac{1 + |\Gamma|}{1 - |\Gamma|} = \frac{1 + \frac{1}{3}}{1 - \frac{1}{3}} = \frac{\frac{4}{3}}{\frac{2}{3}} = 2.0$$

Then determine P_{refl} and P_{trans}:

$$P_{refl} = \Gamma^2 P_{inc} = \left(\frac{1}{3}\right)^2 P_{inc} = \frac{1}{9} P_{inc}$$

and

$$P_{trans} = P_{inc}(1 - \Gamma^2) = P_{inc}\left(1 - \left(\frac{1}{3}\right)^2\right) = \frac{8}{9} P_{inc}$$

From these calculations we can see that 11% of the incident power is reflected and only 89% reaches the load. The standing-wave pattern is shown in Figure 8-40 on page 307.

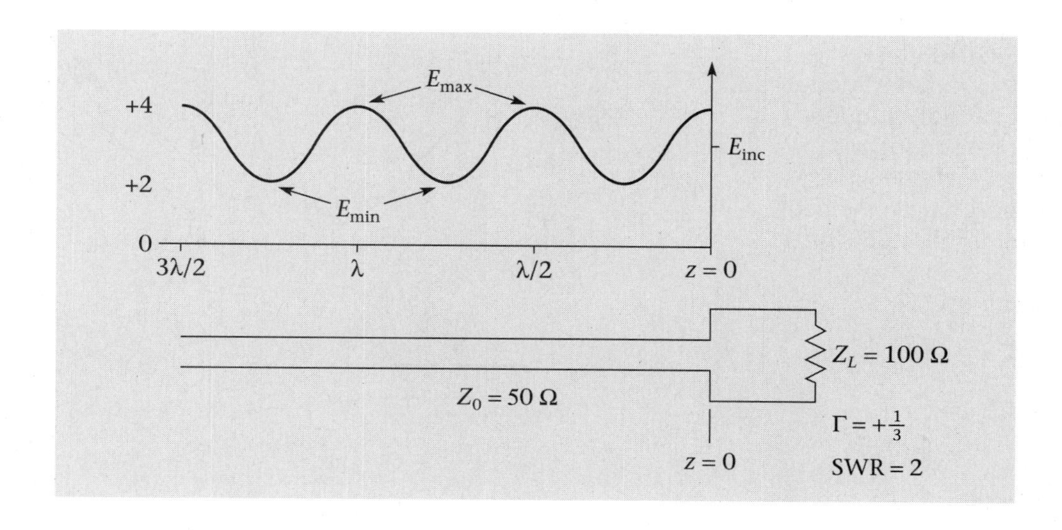

FIGURE 8-40 Standing-wave pattern for example 8.5

Let us make some final comments before leaving this topic. It is obvious that a TL "mismatch" affects the efficiency of a system, but that is not the only problem. If we consider the reflection of pulses or complex signals that represent symbols, we note that this problem will most likely impact the system bit error rate and limit the data rate due to the interaction of incident and reflected pulses. It could also cause serious damage to the source of power due to the reflected wave. So far we have shown simple examples in our treatment of this topic. It should be pointed out that the TL can be used to deliver power to a load with a reactive component, $Z_{load} = R \pm jX$ (actually, this is the case more often then not). This would certainly complicate the mathematics when calculating Γ, adding phase shifts into our solutions. Practically speaking, any phase angle associated with Γ would result in a shift of the peaks and nulls of the standing-wave patterns from $z = 0$.

S-Parameters and TLs

S-parameters are universally used to characterize any type of high-frequency component, including TLs and TL components, which will be introduced shortly. S-parameters treat a component as having input and output ports with incident, reflected, and transmitted signal intensity associated with each port. See Figure 8-41 on page 308.

For a TL, there is a one-to-one relationship between Γ and τ with two of the S-parameters: S_{11}, which is equal to the reflection coefficient, Γ; and S_{21}, which is equal to the transmission coefficient, τ. S_{11} is the ratio of reflected signal from port 1 due to a signal applied to port 1, and S_{21} is the ratio of the signal transmitted through the component from an applied

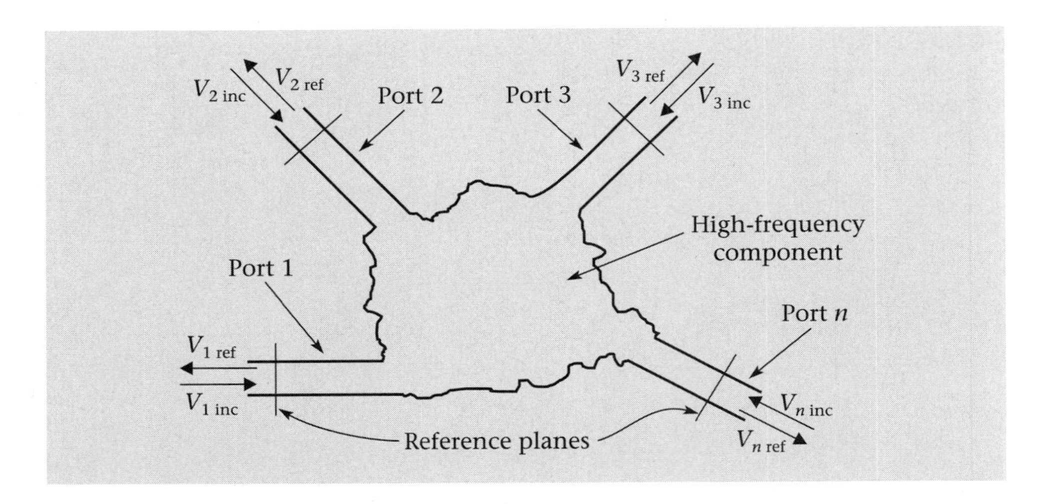

FIGURE 8-41 S-parameter model of a high-frequency component with several ports. Each port has an incident and reflected signal.

signal at port 1 to port 2. S-parameters are used by all manufacturers to characterize their components at high frequencies and by software-design packages to plot component characteristics from their design parameters.

8.11 TL Impedance Matching

Our coverage of TLs would be incomplete if we did not examine the solutions available for the problem of mismatches. We will first introduce the **balun** (*bal*anced-to-*un*balanced) transformer, which is used to match unbalanced to balanced lines. A popular commercial use of a balun was for older TV sets that needed to match the 75-Ω cable TV connection to a balanced 300-Ω input. See Figure 8-42 and Figure 8-43 for examples.

The Quarter-Wave Transformer

A quarter-wave ($\lambda/4$) transformer can be used to match certain types of loads. If the load is purely resistive, a **quarter-wave transformer** with the

FIGURE 8-42 A typical balun

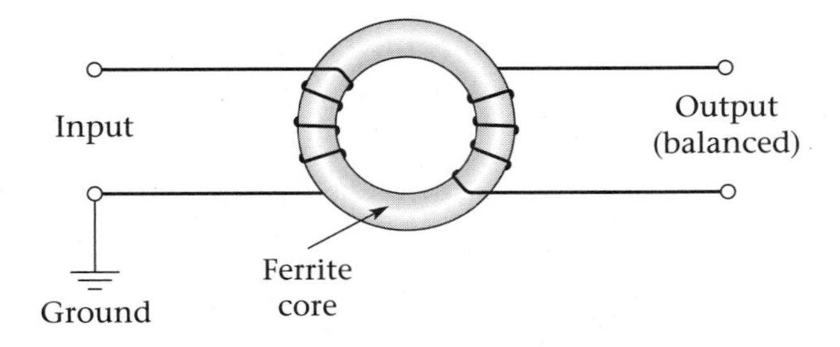

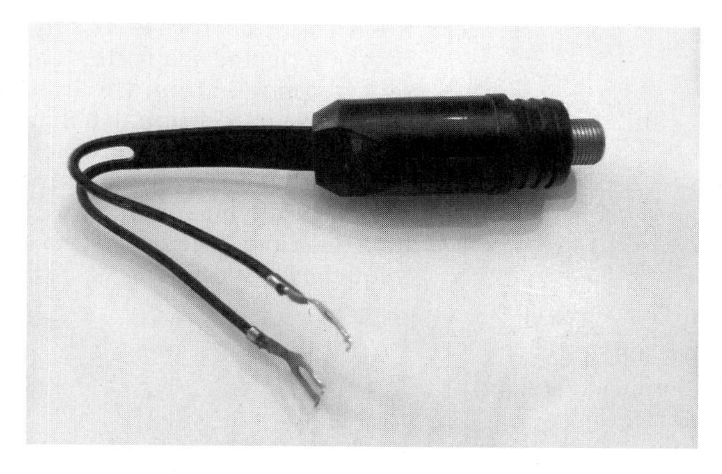

FIGURE 8-43 A typical 75-Ω coaxial cable to 300-Ω parallel wire balun

correct characteristic impedance can be used to provide a match, as shown in the following equation:

$$Z_{\lambda/4\,\text{trans}} = \sqrt{Z_0 \times Z_L} \qquad \textbf{8.24}$$

An example of a quarter-wave transformer is shown in Figure 8-44.

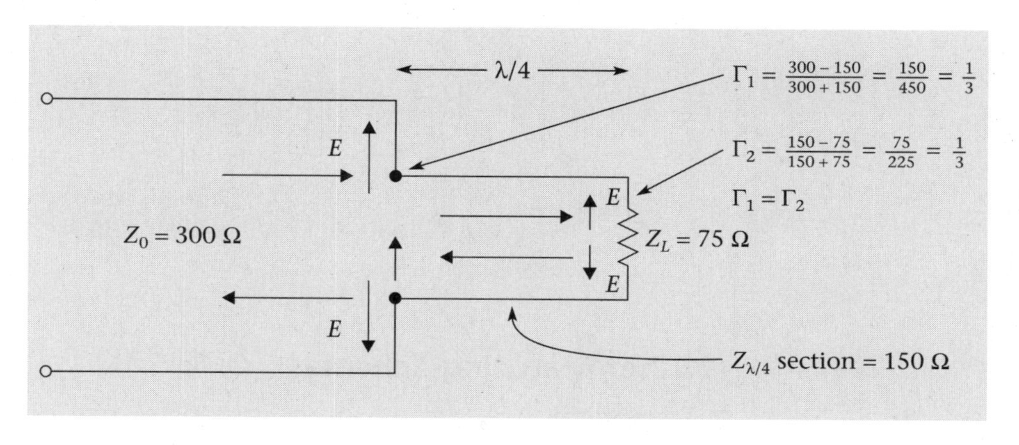

FIGURE 8-44 A typical quarter-wave transformer

As can be seen in the figure, the quarter-wave transformer, inserted between the load and the transmission line, only provides a match at one frequency or wavelength. The match is produced by the introduction of a mismatch from the transformer that cancels out the undesired mismatch from the load. Note that a signal propagating down the TL experiences a mismatch at the junction of the λ/4 transformer. The energy that is transmitted down the λ/4 transformer section also experiences a mismatch at the load. It can be shown that the magnitude of the mismatch, Γ, is identical in each case. It can further be seen that the phase of the signal power reflected from the load will be 180 degrees out of phase with the reflected power from the input to the $\lambda/4$ section, for this energy has had to travel an additional half wavelength. These two reflected signals will effectively cancel one another out and all of the power will reach the load.

Another form of matching can be accomplished through the use of a multisection binomial transformer. Note that through the use of several quarter-wave sections, seen in Figure 8-45, one can achieve a broader frequency response.

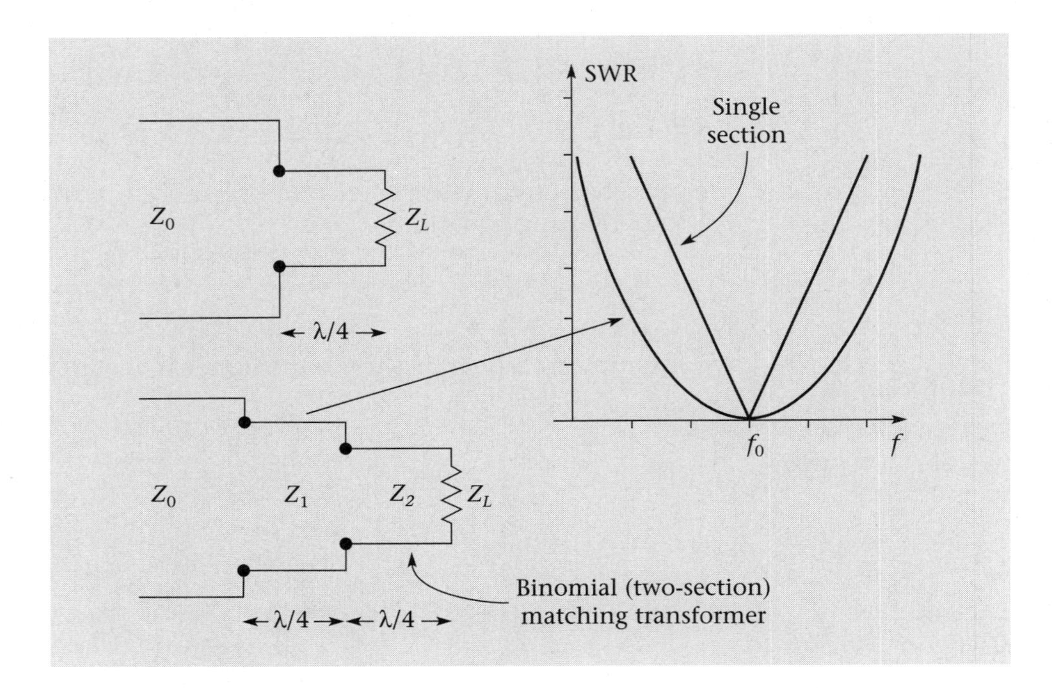

FIGURE 8-45 Binomial matching transformer

Other Matching Schemes

A graphical tool that has been used for many years in the analysis and correction of TL problems is the Smith chart, shown as Figure 8-46.

Using the Smith chart, for any type of load ($Z_L = R \pm jX$) we can determine the position and length of a stub (shorted TL) with the correct value of mismatch that needs to be inserted into the system to cancel the undesired mismatch. See Figure 8-47 on page 312 for two possible configurations. This solution to the problem of a mismatched load is similar to the solution afforded by the quarter-wave transformer. The signal propagating down the TL splits at the stub and the two signals continue onward. The signal incident upon the mismatched load will be reflected, as will the

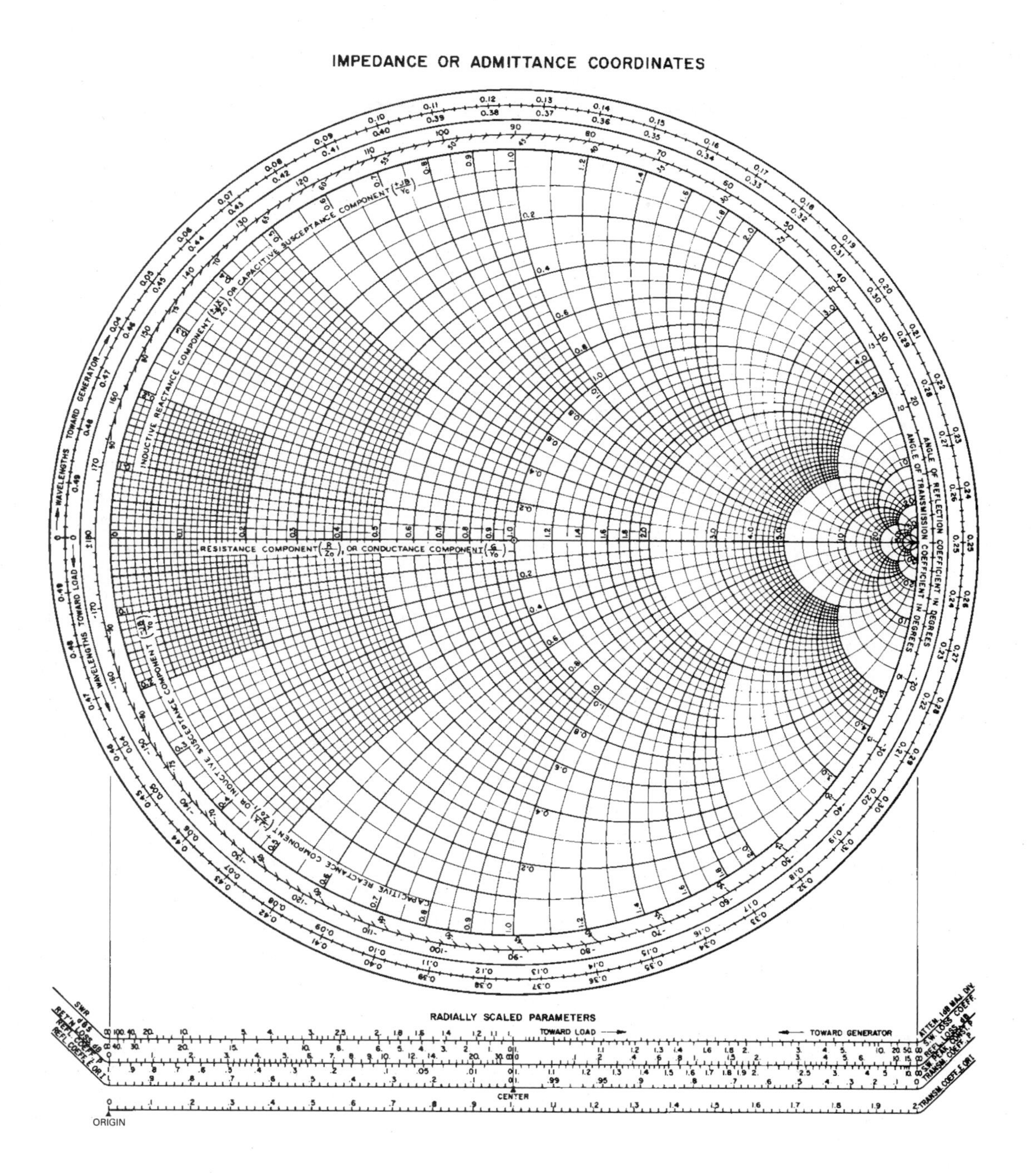

FIGURE 8-46 The Smith chart

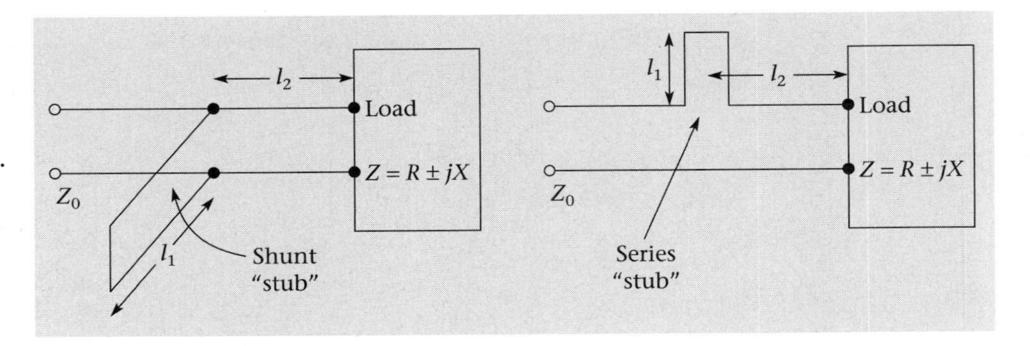

FIGURE 8-47 Stub matching. A shunt stub is shown on the left and a series stub is shown on the right.

signal from the short terminating the stub. The net result is that the two reflections arrive back at the junction of the stub and the TL in the phase and amplitude necessary to cancel one another out (see Figure 8-48).

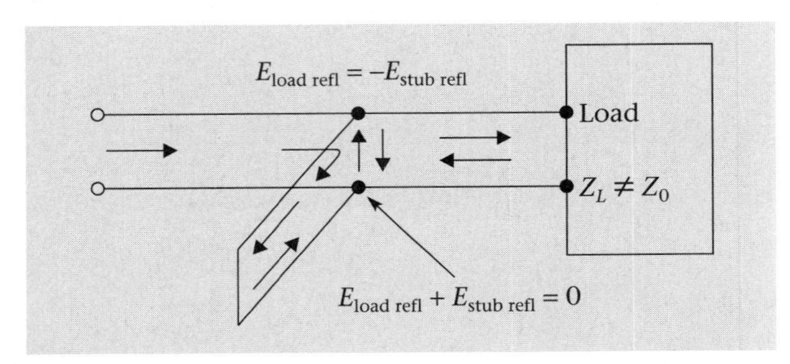

FIGURE 8-48 Single stub matching

Multiple-Stub Matching Schemes

For the case of multiple stubs, the position of the two stubs can be fairly arbitrary (see Figure 8-49) but their spacing is usually chosen to be $\frac{3}{8}\lambda$. This is a practical system usually employed to match antennas in the field. The first stub is adjusted for the lowest measured SWR and then the second stub is adjusted in the same manner.

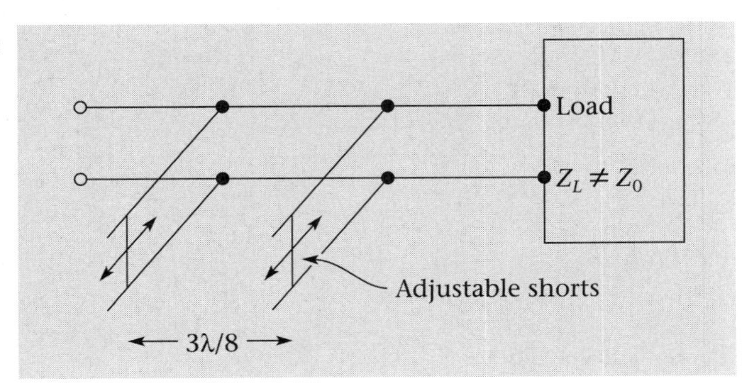

FIGURE 8-49 Multiple stub matching

In all cases of TL matching, there will exist standing-wave patterns to the right of the junction of the first stub or at the start of the quarter-wave transformer network; however, from the point of view of the source of power, the TL appears to be terminated by a matched load at this interface. Therefore, all the incident power from the source is absorbed by the load. The standing-wave patterns within the matching networks represent reactive power in the form of electric and magnetic fields.

TL Components and Connectors

Numerous standard TL components exist. Their function is usually denoted by their name:

- Power dividers (splitters)
- Attenuators (pads)
- Filters (blocks)
- Power combiners (adders)

There are also numerous standard TL connectors such as the BNC, Type-N, SMA, SSMA, APC-7, and so forth. The type of connector to use is almost always dependent upon the TL size and application. See Figures 8-50 and 8-51 for typical low-power coaxial-cable connectors.

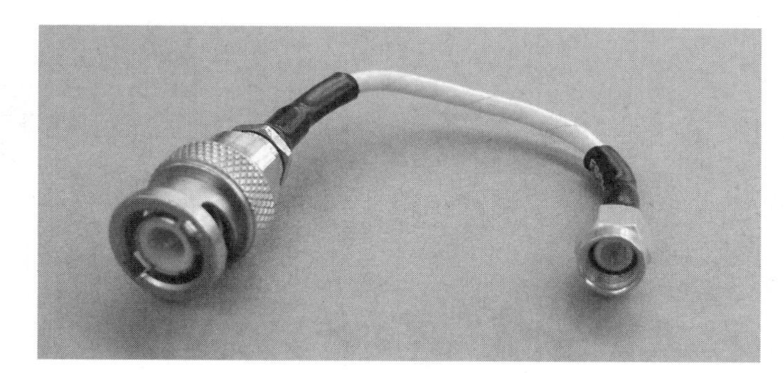

FIGURE 8-50 Male BNC (left) and SMA (right) connectors

FIGURE 8-51 Type-N male (right) to BNC female (left) connector

TL Testing

To determine TL faults a specialized type of "guided radar" has been developed (see Figure 8-52). A short pulse of RF or optical energy is launched down the TL to be tested, and the results are observed and measured. The length of time required for an echo to return to the time-domain reflectometer (TDR) can be used to determine the distance to a short or open in either a wireline or fiber-optic cable. However, this diagnostic technique does not solve the dilemma of the "buried loop of cable."

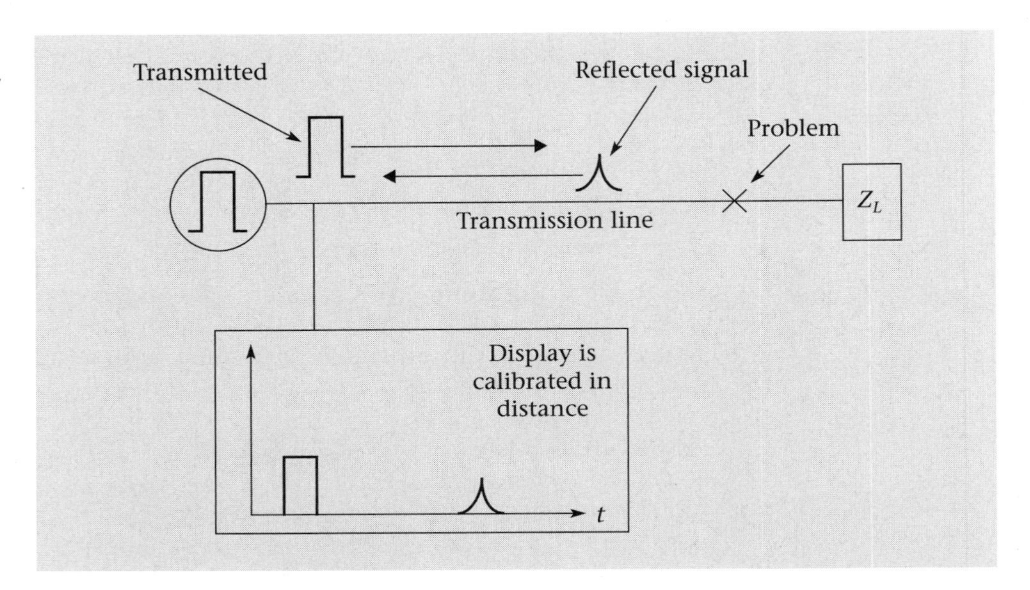

FIGURE 8-52 Time-domain reflectometry (TDR)

Software Resources for TL Theory

There are several software resources available for TL simulation on the World Wide Web. Two programs, TRLINE and BOUNCE, use computer simulation to explore TL phenomena and computer animation to show waves propagating on TLs. Both of these programs can be downloaded from the Web and used to explore TL behavior.

Summary of Part II

In part two of this chapter, we examined the need for TLs, the most common types, and their most important characteristics. So far we have seen that most telecommunications applications use conductor-based TLs for the delivery of either baseband data signals or passband RF signals. Category-*N* cables (multiple twisted copper pairs) are typically used with LANs and coaxial cables are used for the delivery of power in RF-based systems.

The efficient delivery of power and the matching of a source to a load were introduced, with examples taken primarily from the domain of coaxial TLs. It was shown that the behavior of EM wave propagation on TLs is analogous to the propagation of EM waves in free space. Similar equations can be used to calculate the reflection coefficient for the TL case of a mismatched condition as were previously used for the propagating EM wave encountering a change in media. The concept of standing-wave patterns was reintroduced for TLs and power relationships for mismatched TLs were examined through detailed examples.

After a short description of how electrically short TLs can be used to create equivalent circuit components, part two closed with a description of several methods that can be used to affect a TL-based source-to-load match where a mismatched condition had previously existed. Although one might mistakenly infer that TL matching is only important for the point-to-point transfer of power, it must be emphasized that any TL system is susceptible to the problems associated with a mismatched condition and that this situation should always be avoided.

PART III

The Local Loop, Cable-TV Plant, and Fiber-Optic Cables

The final section of this chapter consists of a short overview of the TL based point-to-point systems currently in operation. Introductions to the PSTN local loop, modern cable-TV systems, and fiber-optic cables will be given. Other texts devoted to these topics provide coverage in much greater depth.

8.12 The Local Loop of the PSTN

Recall the communications system model. Historically, the channel for the telephone system has been a copper-wire pair. The switching fabric embedded in the channel has evolved over time from manual switchboards to electronic switching systems (5ESSs or DMS-100s). The vast majority of telephones connected to the PSTN are analog. These analog telephones send electrical signals to the local exchange, at which point they are converted to digital signals and, depending upon their destination, may be further converted to light signals or EM waves. The local loop has the characteristic of being both narrowband and baseband in nature; the rest of the PSTN is wideband. Most of the PSTN carries high bit-rate digital signals over fiber-optic cables. Eventually, the ever-increasing demand for faster bit-rate transmissions to the home will cause the local loop to be transformed. Fiber to the home, wireless or broadband cable, or *x*DSL will change the present system. This evolution has been happening in a gradual fashion as fiber has been reaching out to increasingly greater distances from the local exchange. These fiber-optic cables serve as main feeder cables used to service a node or segment of the exchange territory. Their function is to replace copper pairs. Additionally, the other enabling technologies mentioned here are being rapidly deployed.

Local Loop History

Historically, open-wire systems were used with two wires (usually iron) needed per circuit. Copper wire was used between exchanges. Cross arms on telephone poles accommodated five pairs, so additional cross arms were added as additional circuits were needed. In rural areas up to ten telephones were serviced per line or pair of wires! As time went on, various types of insulation were employed to allow for copper pairs to be grouped into cables. These cables could then be more efficiently strung from poles or installed in underground ducting. Eventually, cables were introduced with plastic insulation and standard color coding to distinguish pairs.

Outside Plant Resistance Design

Due to the signaling and switching technology of the day, older telephones needed 23 mA to operate. Using 48 V of central office (CO) battery voltage lead to a goal of a maximum loop resistance of 1000 Ω and then 1200 Ω as improvements were made to the CO switching equipment; thus, the thickness of wire used depended upon these local-loop requirements. Newer equipment changed the loop-design criterion to 20 mA. Lower current requirements and an increase in battery voltage to 52 V allowed for the use of longer loops (1800 Ω). Historically, all of the technology-based switching-current requirements dictated the physical implementation of the local loop and the need for devices like range extenders and voice-frequency repeaters at the CO.

The CO, Exchanges, and Serving Area

There is usually a desire to limit the number of telephone lines, and hence the size of the main distributing frame (MDF), in the CO. The outside plant terminates on vertical terminal blocks on one side of the MDF while CO equipment is connected to the horizontal side of the MDF. Jumper wires connect a cable pair to any line circuit. To prevent cross talk between adjacent jumper wires the number of lines is usually limited, albeit to a large number. The number of telephone numbers assigned to each exchange (a particular NNX code) is limited to 10,000; therefore, a CO may house numerous exchanges.

In a metropolitan area with a high population density, numerous exchanges might be needed. As just pointed out, a CO may house numerous exchanges and several to many COs may exist. These COs (exchanges) are connected together by trunk lines that offer both high-capacity and high-speed interexchange connections. An exchange usually has a serving radius of 2 to 3 miles; therefore, all loops can be shorter than 18,000 feet and use less-expensive 26-gauge wire as compared to 19, 22, or 24 gauge. Exchanges in rural areas might need to use larger-gauge wire to meet the loop resistance design criteria for more distant customers.

Subscriber Carrier Systems

Using digital transmission techniques (T1s), the telephone companies have installed subscriber-line carrier systems with which 8 copper pairs can carry 96 (SLC-96) telephone calls. With the use of fiber-optic cables, it is not uncommon for these systems to be extended to DS3 rates of 672 calls (or higher) carried over a fiber cable to the remote node. The carrier serving area (CSA) is the area serviced by a subscriber carrier system that can support DS0 and ISDN service without special loop treatment. The CSA is limited to 12,000 feet using 24-gauge wire and 9,000 feet using 26-gauge wire. Additionally, all loops must be unloaded.

Frequency Response of the Local Loop

At this time, the frequency response of the local loop should be examined in greater detail. Recall our model of a TL. The local loop has low-pass filter characteristics. Historically, loading coils were used to improve voice communications. Placed at regular intervals on the cable, the inductance of the loading coils counteracted the capacitance of the line. There are various types of loading coils used for different applications. Figure 8-53 on page 318 shows a typical frequency response of a loaded and an unloaded pair. 88-mH loading coils are placed every 6,000 feet, yielding a f_{cutoff} of 3800 Hz, while 44-mH loading coils placed every 3,000 feet give a f_{cutoff} of 7400 Hz.

Effect of Loading Coils on Data Transmission

T1 signals require a much higher bandwidth than the 3800 Hz achieved with loading coils (recall from chapter 5 the frequency spectrum of typical

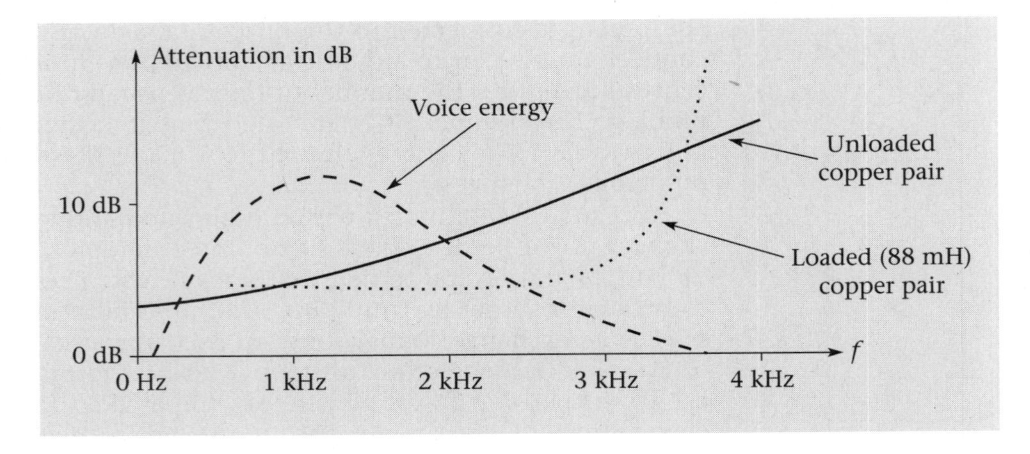

FIGURE 8-53 The typical unloaded and loaded frequency response of a copper pair

line codes). Pairs that will be used to carry data must have the loading coils removed and, depending upon the circumstances, a repeater or signal regenerator will replace the loading coil.

Power Loss and Noise in the Local Loop

The following nominal specifications are for the local loop: at 1000 Hz the power loss between the CO and the telephone should not exceed 8.5 dB; at 500 and 2700 Hz loss should be within 2.5 dB of the loss at 1000 Hz. Noise on the local loop should be less than –60 dBm or +30 dBrnC0 (1 nW), where 0 dBrnC0 is equal to –90 dBm. Noise on the local loop can come from many sources, such as cable-pair cross talk, induction, ground faults, and so forth.

Types of Cables Used in the Local Loop

Historically, main feeder cables consisted of 1800- to 3600-pair cables. These cables are usually installed underground and accessed through "subterranean vaults." Branch feeders are typically 900 pairs and distribution feeders (to the home) are 25- to 400-pair cables. No wonder fiber-optic cables are being installed by telecommunications providers to replace these huge copper cables!

Data on the Local Loop

Early modems for computer-data transmission were designed around the characteristics of the local loop. Modern modems use sophisticated digital-modulation techniques such as QAM that allow for faster bit rates over the local loop (V.90 is 56 Kbps). Digital-subscriber channel systems use modern technology in the form of digital-multiplexing techniques to achieve pair gain over existing copper pairs. Using standard 2B1Q encoding, commercial systems add two, three, or four POTS lines over existing nonloaded cable pairs. ADSL systems can provide even faster data transfer over copper-wire pairs.

In Conclusion: The Local Loop

The infrastructure of the telephone network is limited presently by the local loop, the low-bandwidth copper-wire pairs that connect the customer to the PSTN. The rest of the PSTN is a high-bandwidth high-speed system. ADSL holds some promise of overcoming this bandwidth limitation; however, over time the local loop will most likely be transformed into a small part of the overall network infrastructure in the United States by the continuing deployment of several different, innovative, and enabling technologies.

8.13 The Infrastructure of Cable-TV Systems

Historically, cable TV used all coaxial-cable systems. These systems used a network of cables and wideband amplifiers to deliver VHF TV signals (Channels 2 through 13) downstream from a "headend" of the system. The **headend** was the source of all television transmission from the cable operator. Trunk cables delivered the signals from the headend into the cable network, feeder cables connected neighborhoods to the trunk cables, line taps delivered signals to the subscribers through drop cables, and amplifiers were used to maintain the signal level throughout the system. This process is depicted in Figure 8-54 on page 320. The system needed to deliver a quality signal to the subscriber by overcoming the losses in the cable as the signal propagated from the headend to the subscriber. A variation to this architecture was the addition of a distribution hub between the headend and the cable plant. As time went on, the bandwidth of the cable system was more efficiently utilized by the use of midband frequencies and the channel capacity increased to 35 channels and then 55 channels with the introduction of a 400-MHz system. System bandwidth (not necessarily useable bandwidth) gradually increased to approximately 1000 MHz by the early 1990s.

Modern Hybrid Fiber-Coaxial Cable-TV Systems

Cable-TV systems have evolved from their early beginnings to become sophisticated and complex wideband systems designed to deliver analog

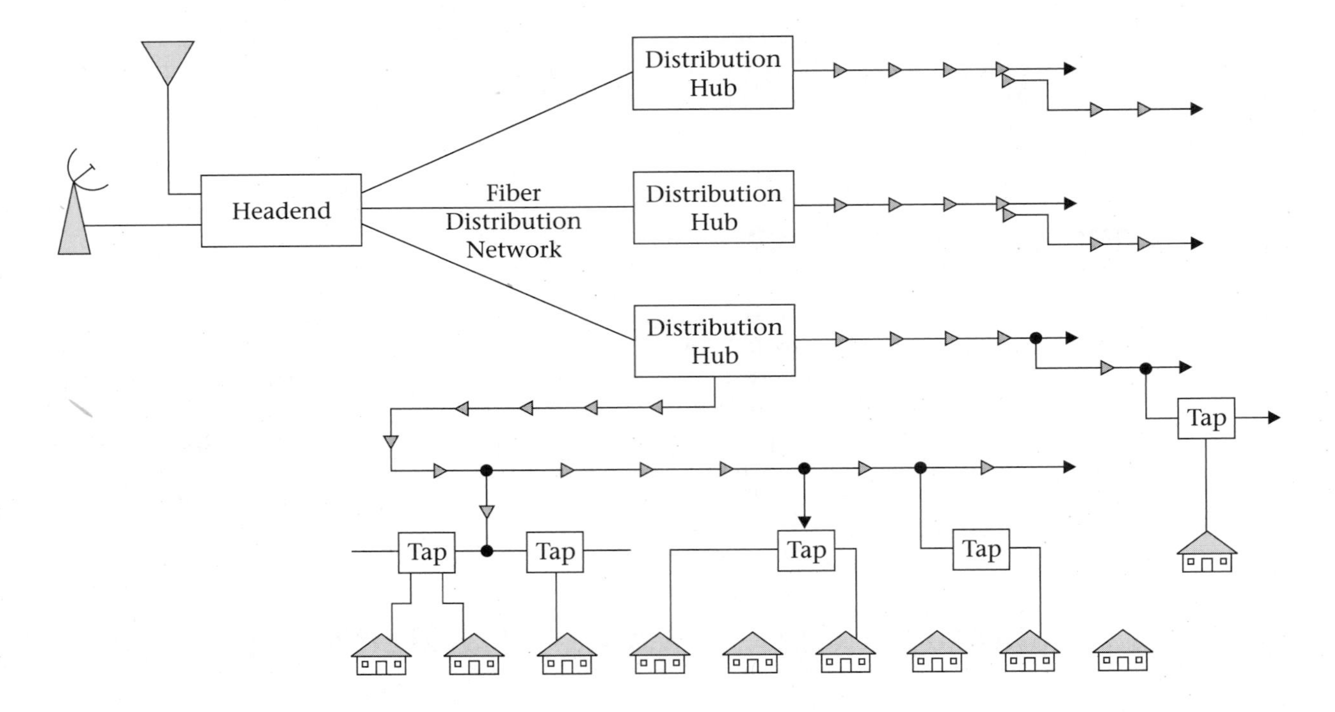

FIGURE 8-54 A typical early technology cable-TV system

and digital video signals, data, and plain-old telephone service (POTS) to the subscriber. The video content can come from local off-air TV stations, satellite feeds of network or distant-station program content, and local access locations. The data connection is normally to an Internet Service Provider (ISP) (a connection to the Internet) and telephone service is connected to the PSTN. The most important change in the cable-TV plant is the migration to the hybrid fiber-coaxial cable system shown in Figure 8-55.

The net effect of this upgrade to a hybrid fiber-coaxial cable plant has been the following:

- Signal quality has been improved with a reduction in maintenance costs.
- Frequency bandwidth has been expanded to 750 MHz.
- With a reduction in node size the number of subscribers per trunk has been reduced, resulting in more bandwidth per subscriber for upstream signals.

One aspect of the cable-TV system that has been slow to change is the upstream capability of the system. The use of frequency-division multiplexing

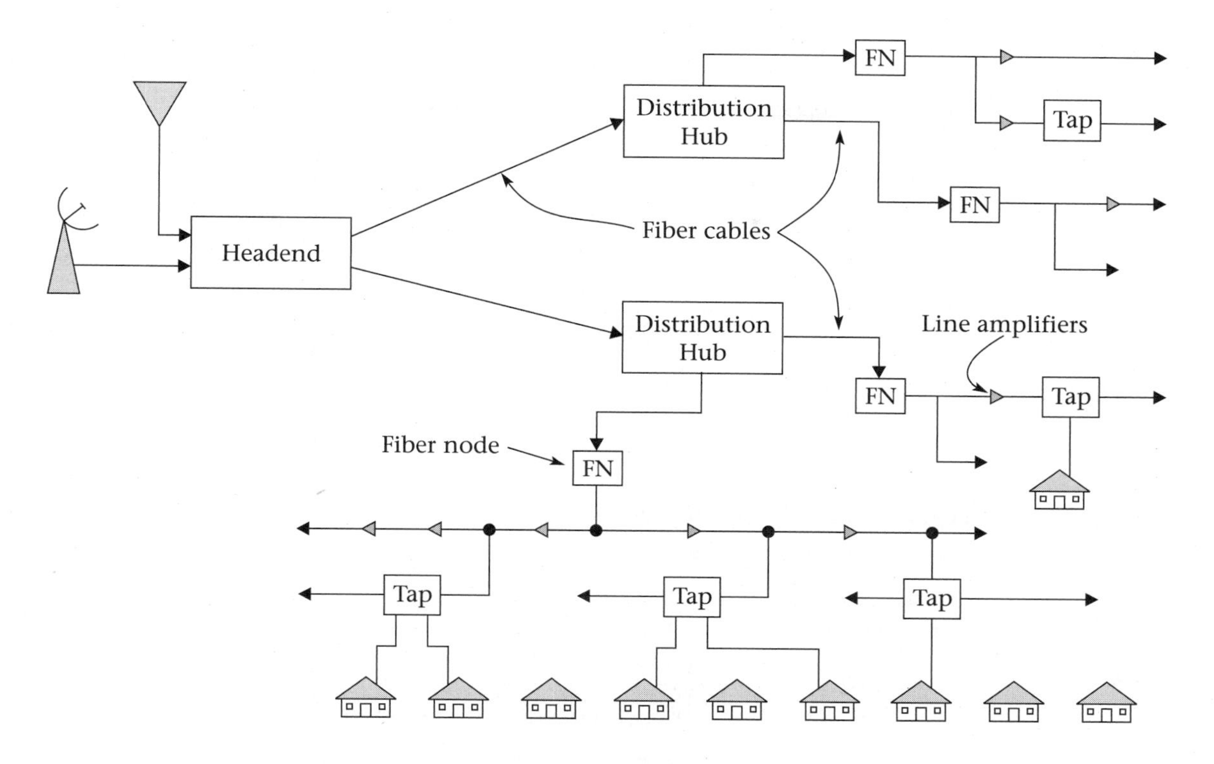

FIGURE 8-55 Modern hybrid fiber-coaxial cable-TV system with fiber nodes

allows for available upstream bandwidth in the 5- to 42-MHz range of the system's bandwidth.

The Cable Modem

Another important aspect to the evolution of the cable system is the evolution of the **cable modem**. The cable modem serves many purposes. In a modern cable-TV system the cable modem is used to perform the following:

- Interface IP packets to the cable system
- Handle complex digital-modulation formats
- Deal with pay-per-view services
- Allow the service provider to control access to the system, and so forth.

High-level modem operation as described here has been standardized by the data-over-cable-service interface specification (DOCSIS) project. This project, initiated in 1996, was intended to lead to multiple-vendor interoperability of cable modems. The DOCSIS specifications cover the RF and data-over-cable aspects of a cable-TV system. In the open-systems interconnect (OSI) model, the DOCSIS specifications cover the physical and media access-control (MAC) layers.

In Conclusion: Cable-TV System Infrastructure

The cable system brings many megahertz of presently available bandwidth to the network infrastructure. However, until the industry provides everyone with upstream capability and gains experience with data and telephony service, converting customers to a "single solution" over cable mindset will be a slow process.

8.14 Fiber-Optic Cables and Systems

The basic principles behind the use of fiber-optic cables have been known for many decades. In 1854, John Tyndall demonstrated before the British Royal Society the transmission of light through a dielectric. The 1950s brought the invention of the fiberscope, which was used for medical applications. In 1966 a theoretical paper predicted fiber-optic cables, but only if glass was pure enough. By 1967, fiber-optic cables with 1000 dB/km loss (1 dB/m) had been fabricated. By 1970, 20 dB/km loss (better than coaxial cable) was achieved. In 1976, 0.5 dB/km loss cable became available. By the 1980s 0.16 dB/km loss was achieved!

By 1988, Nippon Electric Company (NEC) demonstrated the transmission of a 10-Gbps signal over an 80-km fiber link. The SONET standards were published in the same year. During the 1990s and continuing today, more and more of the world has been connected by fiber-optic cables. DWDM has become practical and transmission rates of gigabits per second have become terabits per second over a single cable! At this point in time, fiber-optic cables offer potentially unlimited bandwidth for telecommunications purposes. A recently announced technological advancement dubbed hyperfine WDM (HfWDM), holds the promise of bringing fiber-optic cable into the "last-mile" equation. The goal of the fiber-optic industry is to achieve a totally optical network. Progress is being made on this front with the commercial availability of limited optical switches based upon MEMS technology.

Typical Fiber-Optic Systems

The typical fiber-optic system consists of a fiber-optic cable, essentially a waveguide for light frequencies, and an optical source and detector. Also needed are an electrical-to-optical (E/O) converter (modulator) at the transmitter end and an optical-to-electrical (O/E) converter (demodulator) at the receiver end (recall our model of a telecommunications system detailed in chapter 2). Presently, optical modulators and demodulators for 10 Gbps are commonplace.

The EM Spectrum and Light

Light has wavelengths of sub-millimeter (sub-mm) to nanometers (nm). Most of the telecommunications done over fiber-optic cable is at infrared (μm) wavelengths.

Typical Fiber-Optic Cable Construction

The typical fiber-optic cable consists of a core, a cladding, and a protective jacket, as can be seen in Figure 8-56. Fiber-optic cables have many advantages over coaxial cables or copper pairs. They are immune to electromagnetic interference (EMI) and cross talk from other cables. They can provide secure links and can operate in severe environments. Fiber-optic cables also provide size and weight advantages.

FIGURE 8-56 Typical fiber-optic cable construction

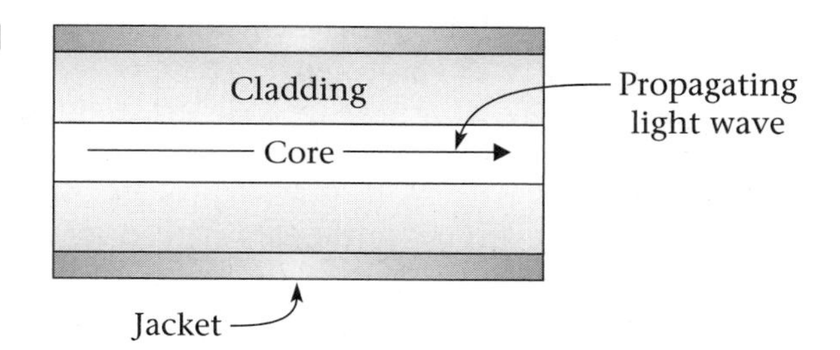

The actual fiber is a thin strand of glass or plastic. Because it has very little mechanical strength, it is usually encased in a protective jacket. Furthermore, numerous strands are generally included in a single round or flat (ribbon) cable.

Optical fibers work on the principle of "total internal reflection." Early fiber-optic cables allowed multimode operation. Modes of different orders will propagate down the fiber using paths of different lengths, resulting in different arrival times at the other end. This will cause dispersion or spreading of pulses, which will eventually limit the bandwidth or bit rate of a fiber. Single-mode fiber has less dispersion, and hence more bandwidth, than multimode fiber and is used for any high-performance installation. Single-mode fiber has the best characteristics, but is also the most costly. In some applications (e.g., low-speed LANs) low-cost plastic multimode cable can be used successfully.

Fiber-Optic Cable Limitations

As mentioned before, fiber-optic cable loss per kilometer is extremely low, but to enjoy this low loss means that certain, specific wavelengths must be used. Loss is least at 1550 nm (infrared) for glass fibers and at about 600 nm for plastic fibers. Also, in single-mode fibers "chromatic" dispersion occurs due to the bandwidth of the signal. Chromatic dispersion is caused by the fact that different light frequencies travel at different velocities. Dispersion will limit the fiber-optic link bit rate. A typical measurement of the effect of dispersion for a multimode cable is the "bandwidth-distance" product of a cable (given in MHz-km).

Fiber-optic cables can also experience losses due to both micro and macro bends. Macro-bending loss is due to overcurvature of the entire fiber-optic cable, while micro-bending loss is due to manufacturing imperfections. Losses in cable splices and connectors can be greater than the loss in the cable itself! Cable splicing and connector technology have matured to include fusion splicing and sophisticated optical connectors with less than one dB of loss per connector.

Optical Switches and Couplers

Optical-coupler and switching technology is currently in a rapid state of development. An optical coupler can be made by fusing together two fibers in a side-by-side arrangement. Optical switches have been commercialized through the use of MEMS technology. These systems use switchable micromirrors to redirect beams of light (a form of spatial switching) from fiber-optic cables to fiber arrays.

Optical Emitters and Detectors

The optical sources for the fiber-optic system can be LEDs, infrared LEDs, or laser diodes (LDs). LD performance has advanced greatly over the past decade, and our ability to control output wavelength and power have also improved. LDs are normally used in a pulse mode. Optical detectors are usually PIN diodes or avalanche photo diodes (APDs).

Optical Amplifiers

Fiber amplifiers have been commercialized that effectively eliminate fiber-optic cable loss and that use fibers doped with rare-earth elements like erbium. Erbium-doped fiber amplifiers consist of a doped fiber of several meters in length that receives a "pump" signal from a laser diode (typically a 980-nm pump source is used for a 1550-nm amplifier). The pump laser provides the power for the amplifier gain (see Figure 8-57).

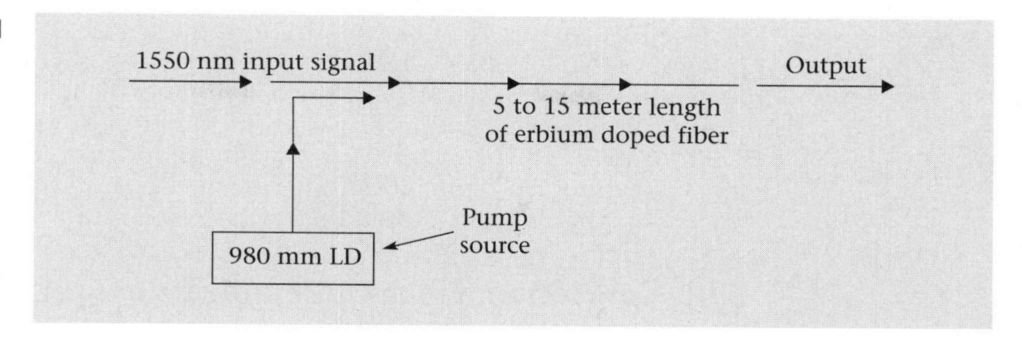

FIGURE 8-57 Typical erbium optical amplifier

WDM and Dense WDM

WDM is really the same as frequency-division multiplexing (FDM). Advances in laser-diode sources have allowed line spacing down to 50 GHz with the promise of even closer spacing in the near future. The ITU Grid is an international standard for wavelength usage in WDM systems and is based on the Krypton line at 193.1 THz. The frequencies run from 192,100 GHz to 196,100 GHz at 100-GHz spacing. Presently, there are numerous commercial systems of 16, 32, 64, and 128 wavelengths available.

Submarine Fiber-Optic Cables

Submarine fiber-optic cables have been in a state of rapid development and deployment over the last decade, starting with TAT-8 in December of 1988. TAT-8 is a first-generation system using 1.3-μm technology with 109 repeaters located every 70 km. The system uses two single-mode fibers and has a capacity of 295.6 Mbps. In 1992, a second-generation system, TAT-10, was put into service. It employed 1.55-μm technology with repeater spacing every 100 km. It has two times the data rate capacity of TAT-8. Third-generation technology is used in TAT-12/13. This system has 5-Gbps capability in each direction. Other systems that are either in operation or are planned include TPC-5 (Trans-Pacific Cable #5) that connects Japan and North America over four optical fiber circuits with a combined bidirectional capacity of 20 Gbps, China-USA and Atlantic Crossing #1, Fiber Link Around the Globe, Asia Pacific Cable Network, and the Africa Optical Network.

8.15 The Synchronous Optical Network (SONET)

A scalable digital transport system designed for fiber systems, the *s*ynchronous *o*ptical *net*work (**SONET**) is the North American standard and the *s*ynchronous *d*igital *h*ierarchy (SDH) is the European standard. The basic

SONET data rate is 51.84 Mbps. This is known as STS-1 or OC-1. Recall that the comparable DS3 wireline digital-signal rate is 44.736 Mbps. The reason for the difference is the amount of overhead used. SONET signals transport much more overhead information in the form of signal routing and setup information. This overhead is usually lumped together and designated as "Operation, Administration, Maintenance, and Protection" (OAM&P) and is much more sophisticated than what is used for DS*x* T-carrier signals.

In Conclusion: Fiber-Optic Infrastructure

Fiber is being deployed by every segment of the telecommunications industry. It is used in telephone toll circuits and fiber to the node. Cable-TV systems use fiber in their hybrid fiber systems. A fiber-optic cable system using coarse WDM (CWDM) is being proposed for 10-Gbps Ethernet. However, there is still very little installed fiber to the curb (FTTC). Eventually, for all tethered applications, fiber will become the ultimate solution.

Summary of Part III

The coverage of wireline-based infrastructure in part three only touched the surface of these topics. A student of telecommunications should be prepared to study each one of these technologies in much more detail. It should also be noted that this section did not contain any coverage of wireless-system infrastructure. This important and rapidly evolving technology needs its own text to be adequately covered.

Questions and Problems

PART I

Section 8.1

1. List the acronyms for the frequency bands of the electromagnetic spectrum. What do they stand for?
2. What wavelength signals are used for short-wave broadcasting?
3. Do an Internet search to learn which wavelength signals are used for fiber-optic communications.

Sections 8.2 and 3

4. What is the relationship of the electric field to the magnetic field in a propagating transverse EM wave?
5. What is the relationship between frequency and wavelength?
6. What does the impedance of free space represent?
7. What determines the polarization of an EM wave?

8. What is the relationship between a spherical EM wave and a plane EM wave? Draw a diagram.
9. Calculate the power density of a 100-W transmitter at a distance of 1000 m. Repeat the calculation for 10 m. What is the dB difference in your two calculations?
10. Determine the electric-field intensity for a 5,000-W AM transmitter at a distance of 100 m. Repeat the calculation for 10 m. Does the second calculation give you any reason for concern?

Section 8.4

11. What do the reflection coefficient, Γ, and the transmission coefficient, τ, represent?
12. What does the standing-wave ratio (SWR) represent? What is its possible range of values?
13. What is meant by the phrase *boundary conditions* when discussing EM-wave propagation?
14. Describe specular and diffuse reflections.
15. Describe the diffraction of EM waves.
16. Describe the process of interference of EM waves.
17. Can you determine what is special about the frequency of 60 GHz? Explain what it is.
18. Describe the Doppler effect.
19. Do an Internet search to find the best "virtual physics lab" site devoted to java applets that display the properties of EM waves.

Section 8.5

20. Why do some of the ionospheric layers undergo daily variations?
21. Why do sky waves have the ability to travel further distances than ground waves?
22. What is the MUF?
23. Do an Internet search and find a site that displays daily propagation forecasts.
24. Explain the idea of the radio horizon as it pertains to the propagation of EM waves.
25. Describe the potential effect of the weather on propagation conditions.

PART II

Sections 8.6 and 7

26. Explain the purpose/function of a transmission line.
27. What are the most important characteristics of transmission lines?
28. What effect does the twisting of wire pairs have on the maximum transmission rate for data?
29. Do an Internet search on LAN category-*N* cable. Report on the results of your search.

Section 8.8

30. What effect do the distributed values of *R* and *G* have on wireline transmission lines?
31. What advantage does coaxial cable have over parallel-wire transmission lines?
32. Try to design a high-impedance coaxial cable ($Z > 100\ \Omega$). Report on your results.
33. What is the reason for the construction of "air-dielectric" coaxial cables?

Sections 8.9 to 8.11

34. Determine the value of Γ and the SWR for a 75-Ω transmission line terminated by a 40-Ω load.
35. What is the range of values that Γ can take on?
36. For a certain transmission line, Γ is equal to –0.4. Find the SWR. Also determine the amount of power that is reflected from the load and the amount of power that is absorbed by the load.

37. Describe the principle of matching used by a quarter-wave transformer. What is the major disadvantage of using this type of matching?
38. Match a source of 250 Ω to a load of 100 Ω at 600 MHz using a quarter-wave transformer. What is the length of the transmission line section? Assume a velocity of propagation on the line of 0.8 c.
39. Compare a binomial matching transformer to a quarter-wave transformer. Which type of match would work the best with CDMA modulation?
40. The Smith chart can be used to determine transmission line matches for a myriad of mismatched conditions. Do an Internet search on the Smith chart and report on your findings. Are there Smith-chart simulation programs?

PART III

41. For your local area, try to determine how many central offices exist. Talk to someone who works for the local telephone company.
42. Determine approximately how far you are located from your CO. Call your local telephone company for information. Try to determine whether or not you are on a subscriber carrier system.
43. Determine some of the details about your local cable-TV system. What type of system is it? Is cable modem (Internet) service available? Where is the system headend located? What type of split-frequency plan does it use? Call your local company to learn the details.
44. Do an Internet search on wavelength-division multiplexing. Report on your findings.
45. Do an Internet search on plastic optical fibers (POF). What frequencies do they transmit best?

Chapter Equations

$$\lambda = \frac{c}{f} \tag{8.1}$$

$$\lambda = \frac{c}{\sqrt{k}f} \text{ or } \lambda = \frac{c}{\sqrt{\varepsilon}f} \tag{8.2}$$

$$Z_{0(\text{fs})} = \sqrt{\frac{\mu_0}{\varepsilon_0}} = 120\pi = 377\ \Omega \tag{8.3}$$

$$P_D = \frac{P_{\text{trans}}}{4\pi r^2}\ \text{W/m}^2 \tag{8.4}$$

$$P = E \times H \text{ W} \tag{8.5}$$

$$E = \frac{\sqrt{30P_{\text{trans}}}}{r} \text{ V/m} \tag{8.6}$$

$$\text{EIRP} = \frac{\text{Antenna Gain} \times P_{\text{trans}}}{4\pi r^2} \tag{8.7}$$

$$\text{Attenuation} = 20\log\left(\frac{r_2}{r_1}\right) \text{ dB} \tag{8.8}$$

$$\Gamma = \frac{E_{\text{ref}}}{E_{\text{inc}}} = \frac{\eta_2 - \eta_1}{\eta_2 + \eta_1} \tag{8.9}$$

$$\tau = \frac{2\eta_2}{\eta_1 + \eta_2} \tag{8.10}$$

$$\text{SWR} = \frac{E_{\text{max}}}{E_{\text{min}}} = \frac{1+|\Gamma|}{1-|\Gamma|} \tag{8.11}$$

$$E_{Z=0} = E_{\text{inc}} + E_{\text{ref}} = 0 \tag{8.12}$$

$$f_{\text{doppler}} = f\left(1 \pm \frac{v_R}{c}\right) \tag{8.13}$$

$$d_{TA} = 4\sqrt{h_T} \tag{8.14}$$

$$d_{\text{total}} = 4\sqrt{h_T} + 4\sqrt{h_R} \tag{8.15}$$

$$Z_0 = \left(\frac{R + j\omega L}{G + j\omega C}\right)^{\frac{1}{2}} \tag{8.16}$$

$$Z_0 = \sqrt{\frac{L}{C}} \tag{8.17}$$

$$Z_0 = \frac{276}{\sqrt{\varepsilon}} \log\left(\frac{2D}{d}\right) \Omega \tag{8.18}$$

$$Z_0 = \frac{138}{\sqrt{\varepsilon}} \log\left(\frac{D}{d}\right) \Omega \tag{8.19}$$

$$R_{\text{source}} = R_{\text{load}} \quad \text{or} \quad Z_{\text{source}} = Z^*_{\text{load}} \tag{8.20}$$

$$\Gamma = \frac{Z_L - Z_0}{Z_L + Z_0} \tag{8.21}$$

$$P_{\text{refl}} = \Gamma^2 P_{\text{inc}} \tag{8.22}$$

$$P_{\text{trans}} = P_{\text{inc}}(1 - \Gamma^2) \tag{8.23}$$

$$Z_{\lambda/4\,\text{trans}} = \sqrt{Z_0 \times Z_L} \tag{8.24}$$

Antennas

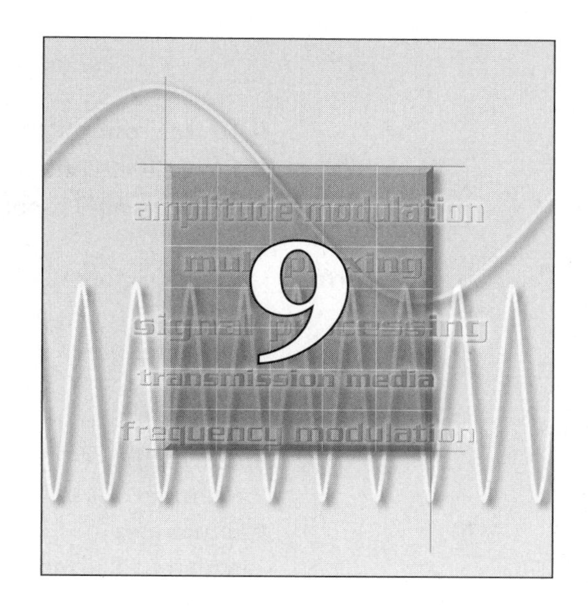

Objectives Upon completion of this chapter, the student should be able to:

- Explain the basic principles of antenna operation.
- Define antenna characteristics such as gain, directivity, efficiency, beamwidth, polarization, bandwidth, and impedance.
- Calculate effective isotropic radiated power from antenna characteristics.
- Identify common antenna types and be able to calculate the dimensions of these antennas for a given frequency.
- Discuss the operation of different types of array antennas.
- Make gain and beamwidth calculations for parabolic-type antennas.
- Understand radiation-pattern plots for antennas and antenna arrays.
- Discuss the common types of microwave horn antennas.
- Discuss the construction and characteristics of microstrip patch antennas and arrays.

Outline

9.1 Antenna Definitions
9.2 Antenna Basics
9.3 The Half-Wave Dipole
9.4 Universal Antenna Characteristics
9.5 Types of Antennas
9.6 Practical Antenna Examples
9.7 Antenna Directivity Patterns
9.8 Antenna Arrays
9.9 Reflecting Surface Antennas

9.10 Microstrip Patch Antennas and Arrays
9.11 Cell-Site Antenna Technology
9.12 Software Resources for Antenna Theory

Key Terms

antenna element
aperture
array antennas
arrays
azimuthal
beamwidth
broadside
capture area
circularly polarized
cylindrical symmetry
directivity
driven element
effective area
electrically small dipole
elevation
end-fire
fan-beam
feed point
folded dipole
ground plane
half-wave dipole
helical antenna
input impedance
isotropic radiator
log-periodic array
loopstick antenna
microstrip antenna
microstrip patch antenna
monopole array
patch antennas
parasitic element
pencil-beam
point source
polar plot
power density
quasi-optical antennas
radiation pattern
vertical monopole
Yagi-Uda array

Introduction

The antenna is the interface between free space and the TL. This chapter will present the basic operating principles of antennas and examine the parameters that describe their performance. Antennas are reciprocal devices and have essentially the same characteristics for receiving as for transmitting. For receiving, the antenna's task is to convert incident EM energy into antenna current; for transmitting, a current applied to the antenna is converted into EM radiation that propagates away from the antenna. Depending upon the design of the antenna, it might radiate the EM energy in everything from an omni-directional pattern to a narrow, pencil-shaped beam. The material in this chapter is developed by first presenting basic antennas—the isotropic radiator and the half-wave dipole—and then building from their characteristics. Topics covered include:

- The half-wave dipole antenna
- Radiation patterns
- Gain and directivity
- Effective isotropic radiated power (EIRP)

- Impedance
- Polarization
- Ground effects
- Antenna matching
- Typical antenna calculations

Next, the student is presented with examples of practical single-element antennas and their operational characteristics. Included in this section are:

- The folded dipole
- The Marconi antenna
- Ground-plane antennas
- Loops
- Helical antennas

How basic antenna elements are formed into arrays is discussed by way of example with the study of Yagi-Uda, log-periodic, and monopole arrays. Lastly, the student learns about microwave and millimeter-wave antennas that are used for broadband wireless applications. Antennas discussed for these frequency ranges include:

- Reflecting surface antennas such as the parabolic dish
- Aperture antennas such as horns
- Microstrip patch antennas and arrays

9.1 Antenna Definitions

This chapter will present an overview of antenna theory. We start with a definition of an antenna:

> An *antenna* is a device that converts energy propagating on a TL into energy propagating in free space or vice versa. The first process is called the transmission or launching of the electromagnetic (EM) wave while the second process is called reception.

In the case of the antenna used for transmission, RF power from a transmitter is usually applied to a TL connected to the "**feed point**" of the antenna. The power applied to the antenna causes a current to flow in the antenna structure, which results in the launching of EM waves into free space.

In the case of the antenna used for reception, the antenna will intercept energy from a propagating EM wave. This intercepted energy will excite a small current in the antenna structure, which now acts as a source of

RF power to feed a TL, which in turn delivers this typically small received signal to the receiver.

9.2 Antenna Basics

Concepts of antenna operation are usually introduced through simple physical models. The rationale for proceeding in this fashion is based on the inherent difficulty in determining an equivalent circuit for the antenna. Extremely advanced and sometimes intractable mathematics, a lack of adequate knowledge of the current distributions existing over the antenna structure, and complex interactions between **antenna elements** are but a few of the reasons why the operation of all but the simplest antennas is extremely difficult to model.

The Isotropic Source

The first antenna model to be presented is the ideal *isotropic source*. This is the easiest model to understand because it draws on one's life experiences. The isotropic source, depicted in Figure 9-1, is one that emits or radiates energy away from itself equally in all directions. An isotropic source is also known as a **point source** or an **isotropic radiator**. This model fits our own impressions of the operation of light sources that we have observed daily in our own lives. Lightbulbs, candles, and the sun, to name but a few examples, all appear to behave as isotropic sources. Certainly, stars and nearby planets appear as point sources of light as we observe them in a night sky.

If one wants to calculate the **power density** received from an isotropic source, one can write an equation to calculate this value fairly easily. The **power density** of the EM wave received at some distance from the isotropic

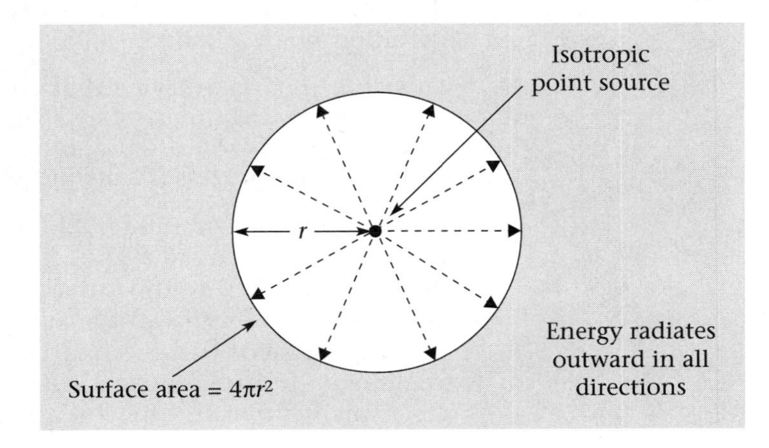

FIGURE 9-1 An isotropic (point) source

source will be equal to the source power divided by the area it illuminates (a sphere of area $4\pi r^2$). This is written as follows:

$$P_D = \frac{P_T}{4\pi r^2} \text{ W/m}^2 \tag{9.1}$$

where r is the distance from the point source

Example 9.1

What is the power density in W/m^2 from a 100-W isotropic source at a distance of 1000 m?

Solution From Equation 9.1:

$$P_D = \frac{P_T}{4\pi r^2} = \frac{100}{4\pi(1000)^2} = 7.96\ \mu\text{W/m}^2$$

The Electrically Small Dipole

The second antenna model is that of an **electrically small dipole** (ESD) antenna. This is a very thin and electrically short wire of length l, where $l << \lambda/2$, with a source feeding it at its midpoint. If the length of this electrically small dipole is very short, we can say that the current that flows in it as a result of the source voltage is maximum at the feed point and goes to zero at each end. See Figure 9-2 for an example.

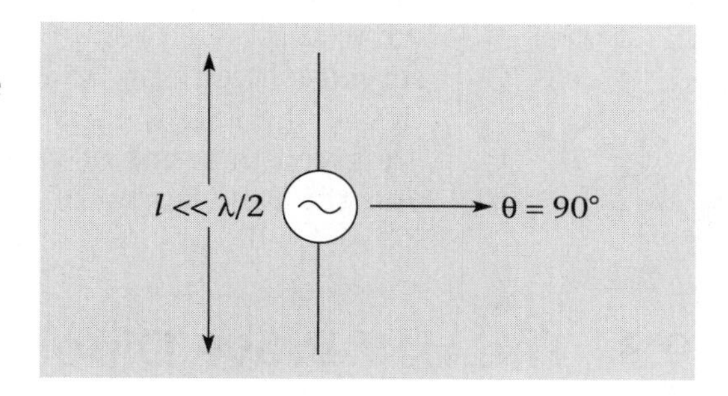

FIGURE 9-2 An electrically small dipole

This model will yield a simplified mathematical analysis of the radiation properties of the ESD. However, even this simplified analysis is beyond the scope of this book. The goal of this chapter is to present material about antenna properties and characteristics and how antennas can be used in telecommunications systems. Therefore, only the end result of the mathematical analysis is of importance to us. For the case of the ESD, a fairly uncomplicated radiation pattern with a Sin θ dependence results. If we examine a polar plot of the pattern predicted by the model, we see that it varies as Sin θ with maximum radiation intensity perpendicular to the ESD ($\theta = 90°$) and radiation nulls off of the dipole ends ($\theta = 0°$). See Figure 9-3.

FIGURE 9-3 An electrically small dipole-antenna radiation pattern looking from the side of the dipole

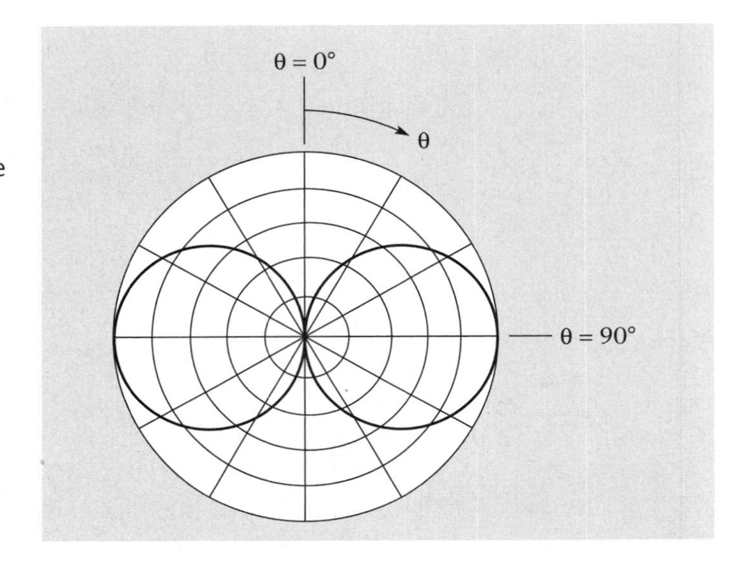

As we can see from the figure, the radiation pattern of the ESD is plotted on a **polar plot** and indicates that the dipole radiates with maximum power perpendicular to its length and has "nulls" or zero radiation off of its ends. Because the ESD has **cylindrical symmetry** (it does not matter what angle we look at it from), we observe that the radiation pattern observed is the same in all directions around the ESD. The other characteristics of the ESD (i.e., radiation resistance, radiation efficiency, and so forth) are given in terms of very complex equations with the length of the dipole and the wavelength of operation as important determining variables.

9.3 The Half-Wave Dipole

The third antenna model used is that of a practical antenna: the **half-wave dipole**. The $\lambda/2$ dipole can be physically realized for most frequencies and is used extensively as a basic building block for array antennas. This antenna can be developed by considering an open-circuited TL that has a flared end. Theoretically, there would be total reflection of the propagating energy and a resulting standing-wave pattern (SWR = ∞) for an open line. As a quarter-wave section of the end becomes flared, as shown in Figure 9-4, energy will tend to radiate away from the structure due to the impedance of the flared section approaching that of free space. Eventually, if one flares the ends further, they become perpendicular to the TL. Now the current in both sections of the ends of the TLs are in the same direction and appear to form a structure similar to the ESD. However, this structure (called a dipole) is now one-half of a wavelength long, which gives it special properties. The $\lambda/2$ dipole has a similar radiation pattern to the electrically short dipole.

FIGURE 9-4 A half-wave dipole antenna

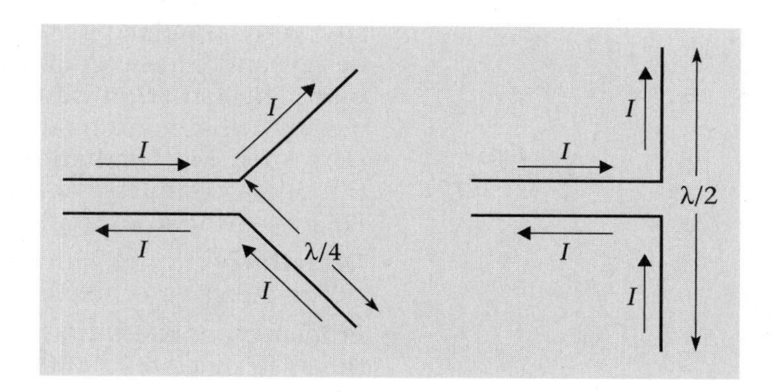

Example 9.2

Determine the length of a half-wave dipole for 100 MHz. Neglect the width of the conductors used to fabricate the structure.

Solution Since $\lambda = \frac{c}{f}$, plugging in the frequency yields:

$$\lambda = \frac{300}{100} = 3 \text{ m}$$

and

$$\frac{\lambda}{2} = 1.5 \text{ m}$$

The half-wave dipole can also be developed by simply lengthening the ESD. As the length approaches $\lambda/2$, the input impedance to the dipole becomes approximately 75 Ω.

9.4 Universal Antenna Characteristics

All antennas have the following characteristics of operation:

- An **input impedance** given by the ratio of feed voltage to feed current. This input impedance is usually frequency dependent, affected by the antenna's proximity to other objects (other metal conductors, distance from the earth, and so forth) and can have a reactive component.
- A **radiation pattern**, which is the variation in radiated power density with angular position. It is determined by antenna type and design and can be graphically represented as a radiation pattern or plot. The radiation pattern can be three-dimensional (3-D) but is usually two-dimensional (2-D) in either the electrical (*E*) or magnetic (*H*) plane. Most radiation patterns are plotted as polar graphs that provide a complete view of the antenna's operation. Refer back to Figure 9-3 for an example.
- **Directivity**, or the gain of the antenna over some standard antenna in the direction of maximum radiation.
- Efficiency, or the ratio of power applied to power radiated. Usually, this is fairly close to 100%.
- Gain, which is the product of directivity and efficiency.
- Bandwidth, which is the useable frequency range (very similar to the resonant curve of an *LC* circuit).
- **Beamwidth**, which is an angle defined by the half-power points of the radiation pattern. The half-power points are where the radiation intensity has fallen to .707 of maximum or where the power density has fallen by 3 dB from maximum.
- Polarization, the direction of the launched electric field, *E*. This is determined by the antenna orientation. Vertical conductors will yield vertical polarization and horizontal conductors yield horizontal polarization.

9.5 Types of Antennas

Many varieties of antennas with numerous geometries have been developed over the years; however, there is one universal constant for all antennas that are excited by an electrical current. For efficient radiation, an antenna's size must be approximately one-half of an electrical wavelength or of the same order of magnitude as a wavelength.

Recall wavelength, $\lambda = \frac{c}{f}$. For practical antennas this equation is altered slightly by a correction factor derived from the physical width of the antenna element. The actual antenna length is therefore slightly shorter than that calculated from the wavelength equation.

A listing of the various antenna types or geometries is given here:

- *Wire* These were the earliest antennas. They were usually for low frequencies. Some of the different types of wire antennas are the dipole, the loop, and the helix.
- **Aperture** These antennas are usually for microwave and millimeter-wave frequencies. Some of the more common types are the horn, the open waveguide, and the slot.
- *Reflector* These are dish-type antennas with feed elements and reflector surfaces.
- **Microstrip/Patch Antennas** This is the newest type of antenna; it has many applications. Conformal (shaped) designs are possible!
- *Lens Antennas* EM waves are refracted by lenses to form radiation patterns. For this reason they are commonly referred to as **quasi-optical antennas**.
- **Array Antennas** Several to many basic antenna elements are arranged together to enhance some particular antenna characteristic.

9.6 Practical Antenna Examples

Over the next several pages, diagrams and pictures of commonplace practical antenna structures will be shown. Included with the antenna sketches are the approximate directivity patterns. The reader will recognize some of these antennas, but because of the specialized nature of their applications, others will not be familiar.

Wire Antennas

Figure 9-5 on page 340 shows several examples of wire antennas and their respective directivity patterns. Figures 9-6a and 9-6b on page 340 show vertical wire antennas that radiate in an omnidirectional pattern. In all cases, the antenna's length is proportional to the wavelength of the signal being transmitted.

Aperture Antennas

Horns, open waveguides, and slot geometries are typical aperture antenna configurations. Figure 9-7 on page 341 shows several of these antennas, which are usually used at microwave and millimeter-wave frequencies. The

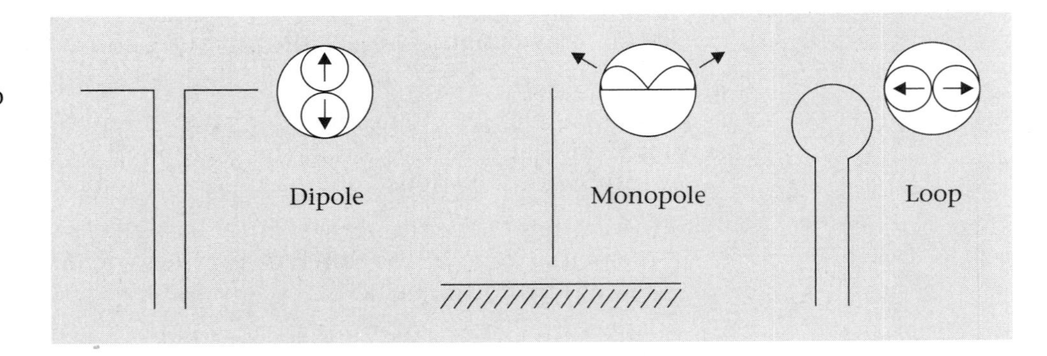

FIGURE 9-5 Wire antennas: dipole, monopole, and loop

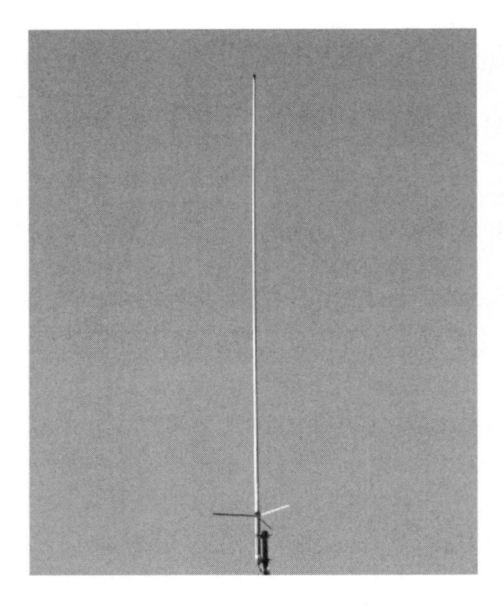

FIGURE 9-6a A typical monopole wire-type antenna

FIGURE 9-6b A monopole antenna used by an AM broadcast station

aperture antenna is sometimes best suited for applications involving high-speed aircraft. These types of antennas can be integrated into the structure of the airplane and will present minimal wind resistance.

FIGURE 9-7 Aperture antennas: horn, open-ended waveguide, and ground-plane aperture

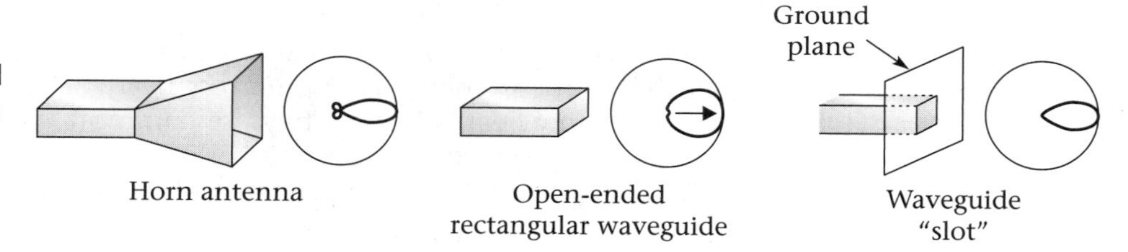

Microstrip and Patch Antennas

Microstrip and patch antennas are shown in Figures 9-8a and 9-8b. Microstrip and patch antennas are fast becoming popular for applications in the microwave and millimeter-wave range. They can be fashioned into arrays and are compatible with planar microwave technology, which relies heavily on microstrip TLs. These types of antennas can be integrated into cellular telephone products.

FIGURE 9-8a Microstrip patch and printed dipole antennas

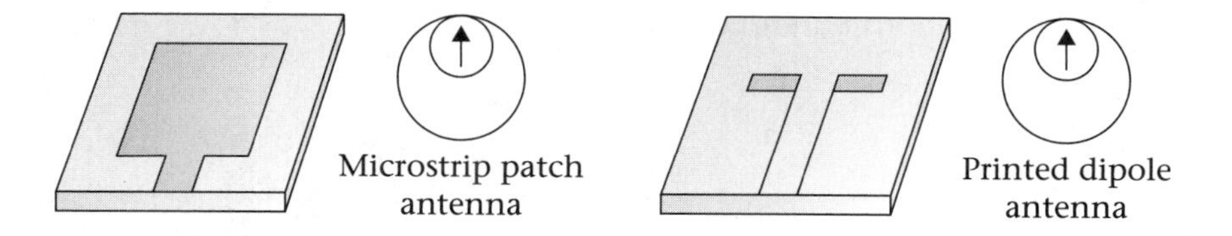

FIGURE 9-8b Microstrip patch antenna array (*Courtesy of L-3 Communications Randtron Antenna Systems*)

Reflector Antennas

Reflector antennas are used extensively at microwave and millimeter-wave frequencies to focus energy into a narrow beam. These antennas have high gains and narrow beamwidths. They are common today due to their popular use for the reception of direct satellite broadcasts. Figures 9-9a, 9-9b, and 9-9c show examples of hogs-horn and parabolic-reflector antennas.

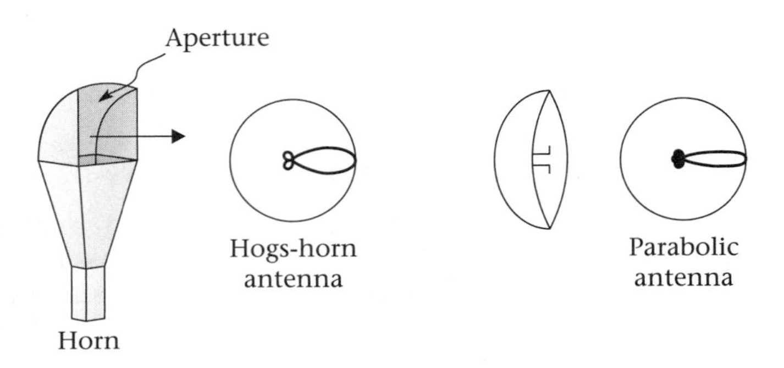

FIGURE 9-9a Reflector antennas: hogs-horn and parabolic-reflector

FIGURE 9-9b Hogs-horn antenna

Optical/Lens Antennas

Lens antennas use the property of refraction to focus an EM wave. These antennas tend to be used at microwave and millimeter-wave frequencies for special applications. The source is usually a horn antenna. See Figure 9-10.

Array Antennas

Array antennas are constructed from numerous basic antenna elements formed into an array of elements. Figure 9-11a shows a wire antenna-type array and Figure 9-11b shows an aperture-array antenna.

FIGURE 9-9c Parabolic-reflector (satellite up-link antenna)

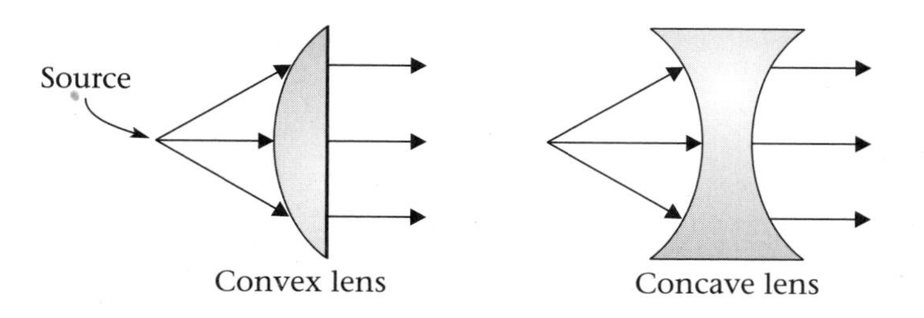

FIGURE 9-10 Lens antennas: convex and concave lenses

FIGURE 9-11a Wire-array antennas for use at VHF wavelengths

FIGURE 9-11b Aperture/slot array antenna (*Courtesy of L-3 Communications Randtron Antenna Systems*)

9.7 Antenna Directivity Patterns

Let us return to the directivity patterns of antennas. See Figure 9-12, which shows a three-dimensional (3-D) plot of the radiation pattern of a half-wave dipole superimposed upon the radiation pattern of an isotropic source. The isotropic source is omnidirectional and favors no particular direction, while the dipole pattern is "squinting" the energy toward the *xy* plane. The dipole antenna is vertically polarized and lies along the *z* axis.

FIGURE 9-12
Dipole directivity pattern

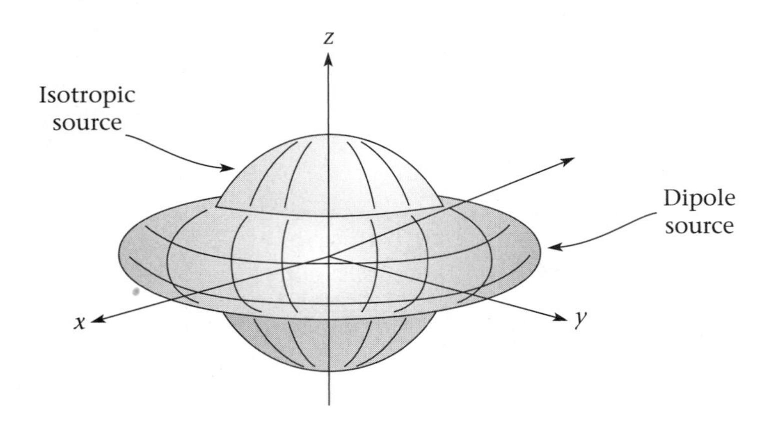

Review of Coordinate Systems

First a word about the rectangular (x,y,z) coordinate system and the spherical (r_S,θ,ϕ) coordinate system. Most antennas demonstrate some form of symmetry; therefore, the radiation pattern is commonly a function of cylindrical or spherical coordinates. See Figure 9-13 for a comparison of the rectangular and the spherical coordinate systems. The value of r_S represents the distance from the origin; θ represents the **elevation** and is referenced to the *z* axis; and ϕ represents the **azimuthal** direction or orientation on the *xy* plane and is referenced to the *x* axis.

Antenna Radiation Patterns

One of the most important characteristics of an antenna is its radiation pattern. The radiation pattern of an antenna is a plot of the magnitude of the field intensity at a fixed distance from the antenna. Depending upon the antenna polarization, the pattern is usually plotted as a function of elevation, $F_\theta = (\theta,\phi)$, or of azimuth, $F_\theta(\theta,\phi)$. The radiation pattern is therefore usually shown as a polar plot showing 360 degrees of variation in

FIGURE9-13 A comparison of the rectangular (x,y,z) and spherical-coordinate (r_s,θ,ϕ) systems

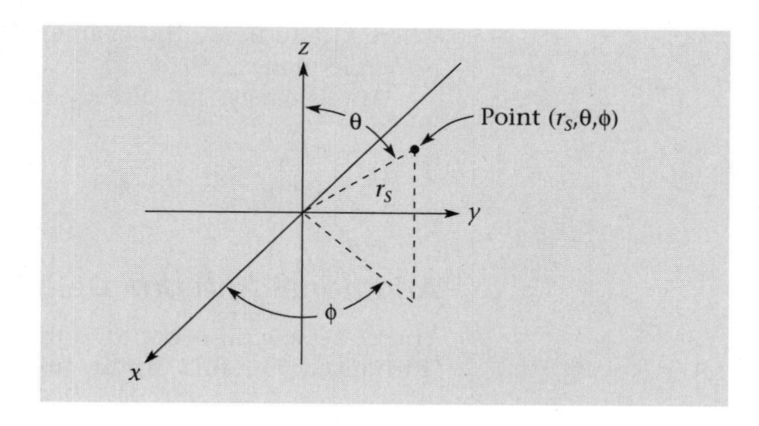

either θ or ϕ. The radial scale of the plot is usually in dB and calculated from the radiation pattern.

The antenna beamwidth may be different in the two principle planes (defined as electric-field direction and magnetic-field direction). A narrow beamwidth in both principle planes results in a **pencil-beam** antenna pattern. A beamwidth that is narrow in one plane and wider in the other results in a **fan-beam** antenna pattern.

The radiation pattern of an antenna can also be plotted linearly (usually for a particular principle plane) with radiation intensity plotted in dB versus angle θ or angle ϕ. A typical linear plot of an antenna radiation pattern is shown in Figure 9-14.

FIGURE 9-14 Antenna radiation-pattern linear plot

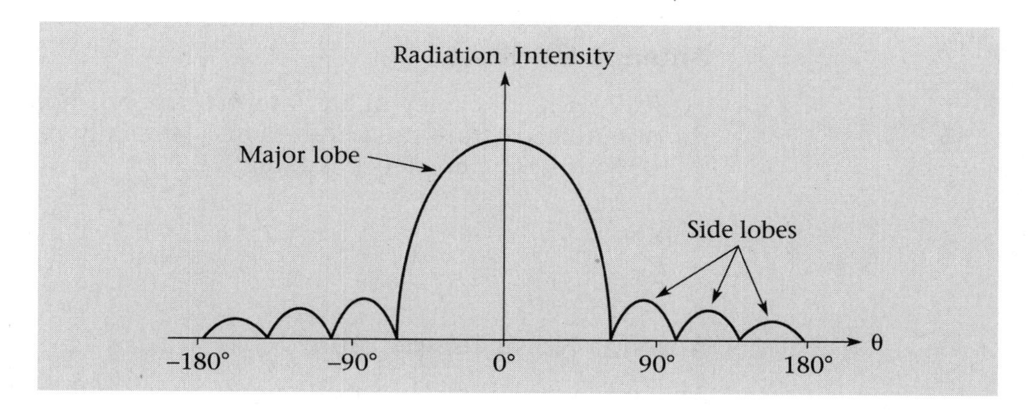

Directivity Calculations

The directivity (D) of an antenna can be approximated by the following calculation:

$$\text{Directivity } (D) = \frac{32{,}400}{\theta_1 \times \theta_2} \tag{9.2}$$

where θ_1 and θ_2 are the beamwidths of the antenna (in degrees) in the two principle planes.

Power density has already been shown in chapter 8 to be given by:

$$P_D = \frac{P_{\text{trans}}}{4\pi r^2} \text{ W/m}^2$$

Additional Antenna Definitions and Concepts

There are several other antenna definitions and concepts that need to be introduced: antenna capture area, efficiency, and noise temperature.

Antenna Capture Area

The **effective area** of an antenna is the same as the **capture area** of an antenna. This measurement relates to how well the antenna either focuses the energy for transmitting or gathers the energy for receiving. Typical values for this parameter range from 0.6 to 1.0, or 60% to 100% of the actual physical area of the antenna.

There is a relationship between an antenna's effective area, A_{eff}, and the directivity of an antenna. It is given by equation 9.3:

$$D = \frac{4\pi A_{\text{eff}}}{\lambda^2} \qquad \textbf{9.3}$$

Antenna Efficiency

Typically, the efficiency of an antenna depends upon the conductivity of the antenna structure itself. Values of η generally range from 60% to close to 100%. Antenna efficiency is given by:

$$\eta = \frac{P_{\text{rad}}}{P_{\text{in}}} \qquad \textbf{9.4}$$

Antenna Noise Temperature

Any antenna will pick up noise power from both natural and man-made sources, and unless the antenna is 100% efficient it will also generate some thermal noise itself. The noise picked up by an antenna is also a function of the antenna angle with respect to the radio horizon. An antenna looking straight up will have an approximate noise temperature of 5° K, essentially due to cosmic noise, while an antenna pointed at the Earth's horizon will have a noise temperature of 100 to 150° K. Antenna noise temperature is also a function of the antenna's wavelength of operation.

Antenna Polar Plots

Let us return to the λ/2 dipole radiation pattern. See Figure 9-15 for a 3-D sketch.

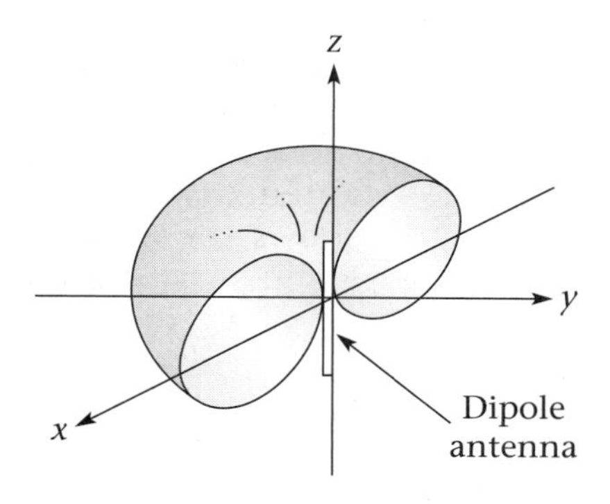

FIGURE 9-15 λ/2 dipole radiation pattern. Note the doughnut shape of the pattern.

As indicated earlier and as shown in several examples, most antenna radiation patterns are plotted in polar form. Again, let us compare the λ/2 dipole with the isotropic source. See Figure 9-16.

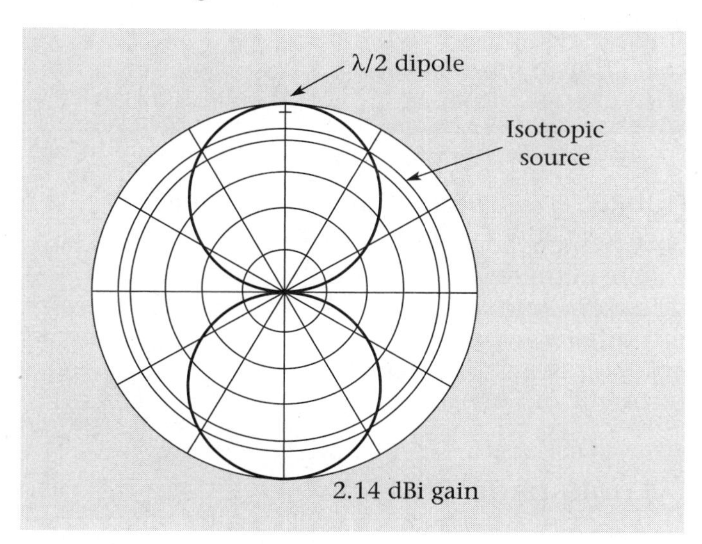

FIGURE 9-16 λ/2 dipole antenna pattern versus isotropic source

This 2-D figure shows that in the direction of maximum radiation the dipole has a gain of 2.14 dB relative to the isotropic source. For this plot, the dipole is oriented in a horizontal direction.

Effect of the Ground

What is the effect of the earth or ground on vertical antennas? The earth usually acts like a "good conductor," therefore reflecting energy propagating toward the ground from a vertical antenna. The reflected energy and the nonreflected energy from the antenna form interference patterns. These patterns result in the typical overall radiation pattern for a vertical antenna located close to the ground that is shown in Figure 9-17a. The exact pattern produced depends upon the height of the antenna above the ground and its length in wavelengths. Vertical antennas are usually constructed so that the maximum radiation is in the horizontal direction, parallel to the earth's surface. Figure 9-17b shows an approximate radiation pattern for a vertical antenna with its base half a wavelength above the ground; Figure 9-17c shows the approximate radiation patterns of three antennas of differing lengths that have their bases located near the ground. Many AM broadcast stations opt for the radiation pattern shown for an antenna of length $\frac{5\lambda}{8}$.

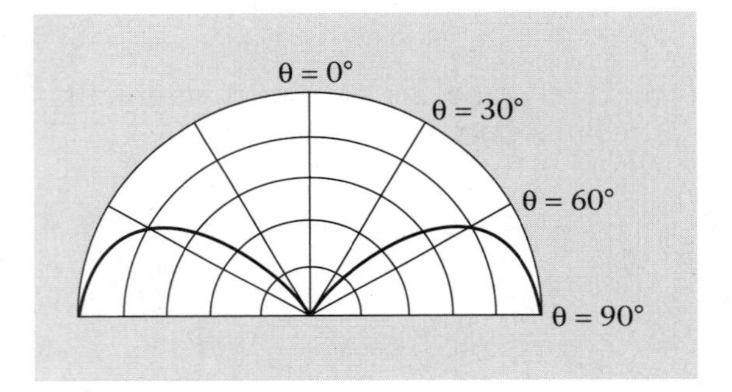

FIGURE 9-17a Effect of the ground on the radiation pattern of a vertical antenna

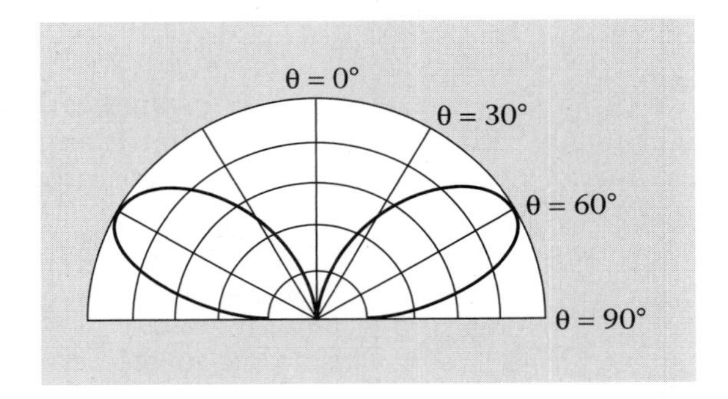

FIGURE 9-17b Approximate radiation pattern of a half-wave dipole located half a wavelength above the ground

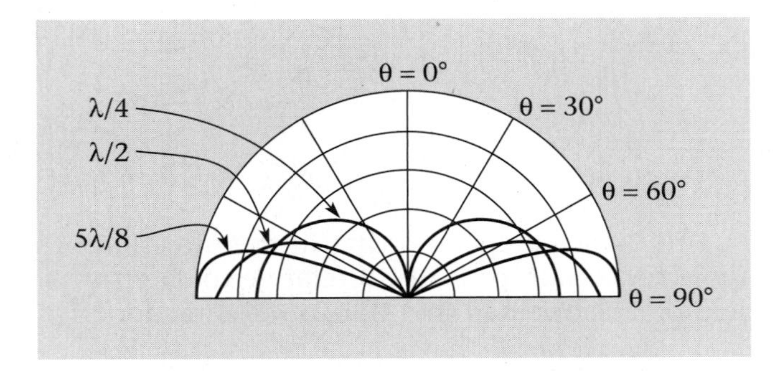

FIGURE 9-17c Approximate radiation patterns for vertical antennas located near the ground with lengths as shown

Other Simple Dipole Antennas

There are several other types of simple dipole antennas. Some of the more common ones will be described here.

Folded Dipole

The **folded dipole** antenna is one of the more common ones; an example is depicted in Figure 9-18. While the λ/2 dipole antenna has an impedance of approximately 75 Ω and is fairly narrowband in its operation, the folded dipole has an approximate input impedance of 300 Ω and has a broader bandwidth. The common FM "T" antenna is an example of a folded dipole.

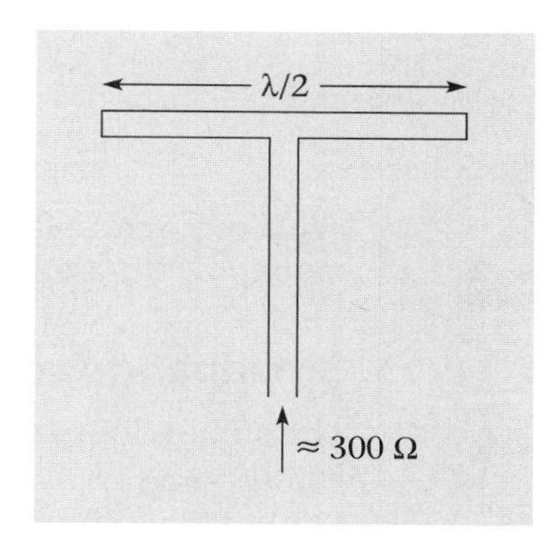

FIGURE 9-18 The λ/2 folded dipole

The Vertical Monopole Antenna

The **vertical monopole** (sometimes known as a Marconi antenna) is a variation of the vertical λ/2 dipole. Using a conductive structure λ/4 in length and isolated from the ground, this antenna will have a radiation pattern similar to that shown previously in Figure 9-17. A typical monopole antenna is shown in Figure 9-19 on page 350.

This type of antenna is usually employed by legacy AM broadcast-band radio stations and will typically be situated in wet or moist areas so that the earth's conductivity will be high. If the conductivity of the earth is poor (sandy soil), a mat of radial conductors will need to be put in place to form a **ground plane**. This type of vertical monopole appears to be twice its physical length because of the reflection of energy from the ground plane. Imagine that the ground appears to be a mirror. Anything placed on the mirror's surface will appear twice as long as a result of the reflection of the original object's image. A variation on this antenna is the "sleeve" monopole antenna. These types of antennas are used extensively over a wide range of frequencies and can be mounted on towers or other supporting structures to extend their transmission range. Often, one will observe a

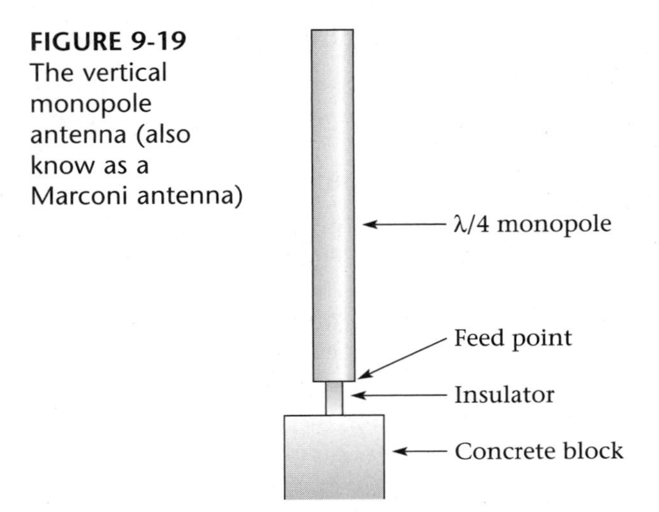

FIGURE 9-19 The vertical monopole antenna (also know as a Marconi antenna)

peusdo ground plane in the form of several radial wires at the base of the monopole for a tower-mounted antenna.

Broadband Antennas

The simple loop antenna was one of the first types of antennas used (see Figure 9-20). A **loopstick antenna** utilizes a ferrite core to increase the loop's effective size. Maximum radiation is in the plane of the loop and broadside to the loopstick dipole. In both cases the loop antenna has a very broad bandwidth of operation.

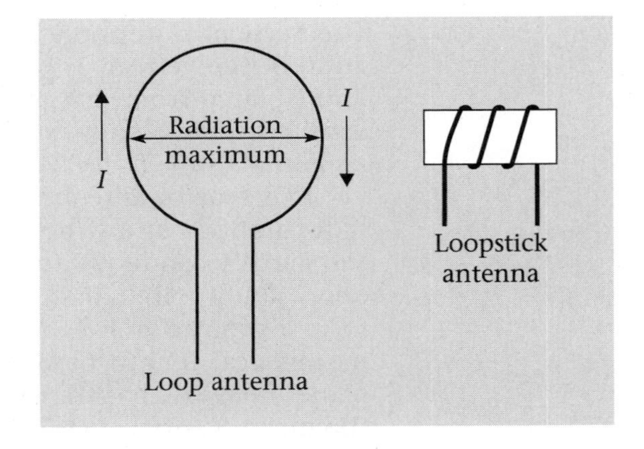

FIGURE 9-20 The loop and loopstick dipole antennas

The Helical Antenna

The **helical antenna** is another broadband antenna and is usually used for VHF and UHF applications (see Figure 9-21). Interestingly, the helix antenna transmits a **circularly polarized** wave (a combination of vertical and horizontal polarization). The direction of maximum radiation is usually off the end of the helix. The direction of the helix (clockwise or counterclockwise) determines whether the EM wave is launched with right-hand circular polarization (RHCP) or left-hand circular polarization (LHCP). For reception, the antenna must have the same orientation as the incoming wave.

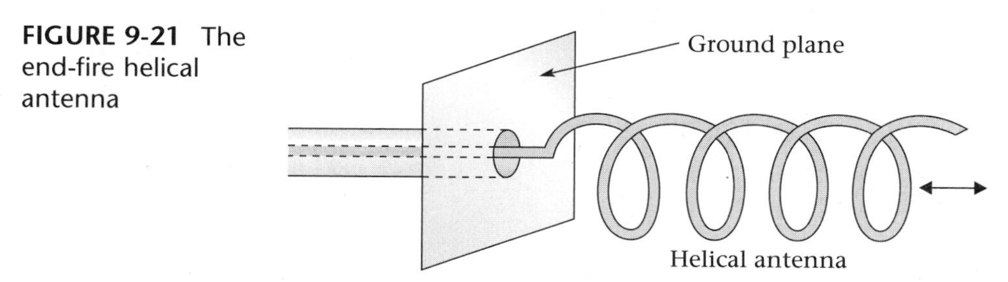

FIGURE 9-21 The end-fire helical antenna

Other Antennas

There are many other types of antennas that have been used for wireless applications over the years. However, their importance to present-day applications is minimal; so, at this point we will end our discussion of wire antennas.

9.8 Antenna Arrays

One can take and arrange patterns of basic antenna elements to create antenna **arrays**. These wire arrays can be used to maximize various characteristics of the resulting antenna structures. Typically, the parameters that are most commonly maximized are the bandwidth and the directivity of the antenna. For radiation patterns, arrays can be organized to maximize radiation either **broadside** to the array or along the length of the array (known as **end-fire**). See Figure 9-22.

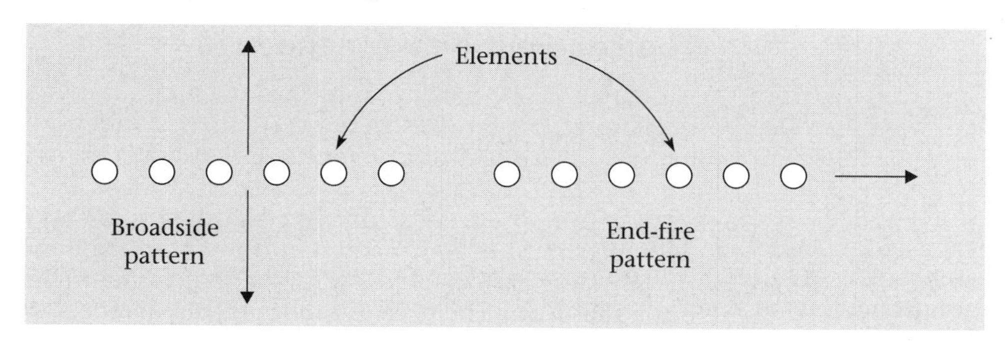

FIGURE 9-22 Broadside or end-fire radiation from an array of elements

Two popular arrays used from HF to UHF wavelengths are the **Yagi-Uda** and the **log-periodic arrays**. The basic Yagi-Uda array is shown in Figure 9-23; the basic log-periodic array is shown in Figure 9-25 on page 353.

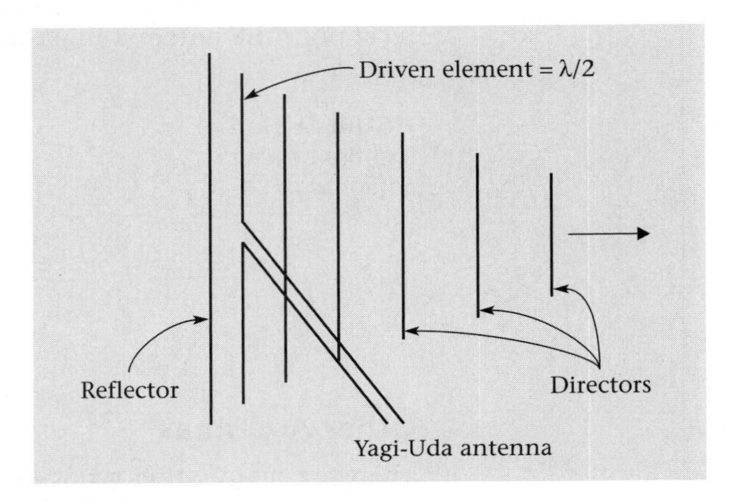

FIGURE 9-23 The Yagi-Uda antenna array

The Yagi-Uda Antenna Array

The Yagi-Uda antenna consists of a **driven element** (DE) and several to many **parasitic elements**. Any element longer than the driven element acts like a reflector and elements shorter than the driven element act like directors of EM energy. Usually, the director elements will decrease in size further from the DE and their spacing will increase for maximum gain. Some of the radiation characteristics of interest for a Yagi array are forward and backward gain, bandwidth, front-to-back ratio, input impedance, and *minor-lobe* or *side-lobe* magnitude. A typical radiation pattern for a Yagi is shown in Figure 9-24. Yagi antennas are capable of gains well in excess of 10 dB depending upon the number of elements used. Note the side and back lobes in the radiation pattern.

The Log-Periodic Antenna Array

Unlike the fairly narrowband Yagi-Uda array, the log-periodic antenna array displays characteristics that are frequency independent. It consists of numerous $\lambda/2$ dipoles of different size that are cross connected, as is shown in Figure 9-25.

The antenna feed is connected to the small end of the array and the radiation pattern is also maximum off of the small end. Unlike the Yagi-Uda, the physical dimensions of a log-periodic antenna are related in a predetermined fashion. The element lengths, diameters, spacing, and gap spacings

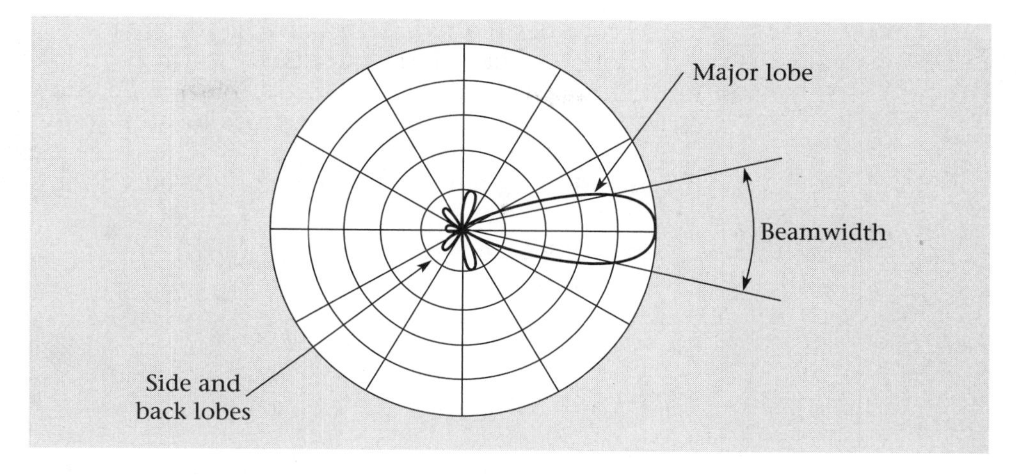

FIGURE 9-24 Typical Yagi-Uda radiation pattern

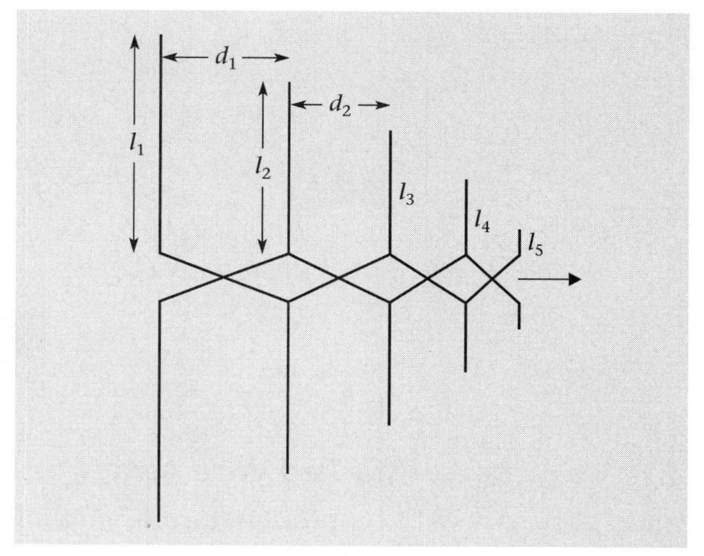

FIGURE 9-25 The log-periodic antenna

are all related to a design variable, τ (tau). Figure 9-25 shows the relationships for element length and spacing. Similar relationships can be written for the element diameters and gap spacings.

$$\frac{1}{\tau} = \frac{l_1}{l_2} = \frac{l_2}{l_3} = \frac{l_n}{l_{n+1}} \quad and \quad \frac{1}{\tau} = \frac{d_1}{d_2} = \frac{d_2}{d_3} = \frac{d_n}{d_{n+1}} \qquad \textbf{9.5}$$

Other Types of Arrays

There are many other examples of array antennas. Our final example will be that of a **monopole array**. This type of array is used at low frequencies to produce desired radiation patterns for broadcast stations. In the case of a monopole array, the elements are vertical and usually the array size is limited to two or three elements. Figure 9-26 shows a picture of a three-element monopole array of antennas located some distance from the city that the AM broadcast station serves. The three elements serve to squint the radiation pattern in the direction of the station's service area.

FIGURE 9-26 AM radio station monopole array

Stacked Wire Arrays

It is possible to take and effectively "stack" array antennas to achieve increased gain. This idea is shown in Figure 9-27 and previously in Figure 9-11a for several wire arrays. This is a practical approach to increasing antenna gain since both the Yagi and log-periodic antenna have practical limits to their total possible number of elements.

For the configuration shown in Figure 9-27 the total gain would be approximately 3 dB greater than the gain of a single array. To increase the gain by 3 dB, one must double the effective receiving elements (area); therefore, to increase the gain by 6 dB requires 4 stacked antennas (refer back to Figure 9-11a). The gain of a stacked system can be calculated by:

$$\text{Total Gain in dB} = \text{Gain of Basic Antenna} + 3\text{ dB} \times \log_2 N \qquad \textbf{9.6}$$

where N is the number of stacked arrays and must be a power of 2 (i.e., $N = 2^n$, and $n = 1, 2, 3, \ldots$).

FIGURE 9-27
Stacked Yagi-Uda arrays

EXAMPLE 9.3

A basic Yagi antenna has a gain of +12 dB. Determine the gain for 8 stacked Yagis.

Solution

$$\text{Total Gain in dB} = 12\text{ dB} + 3\text{ dB} \times \log_2 8$$
$$= 12\text{ dB} + 3\text{ dB} \times 3 = 12\text{ dB} + 9\text{ dB} = 21\text{ dB}$$

Example 9.3 shows how a fairly substantial gain can be achieved by stacking arrays. If one attempted to construct a single Yagi antenna to provide this much gain, it would most likely result in a physical structure with unwieldy dimensions.

Review of Wire Arrays

Antenna arrays can be constructed by the physical arrangement of basic antenna elements such as half-wave dipoles or monopoles that are in close proximity to each other. These arrays can be designed to maximize certain antenna characteristics such as directivity or bandwidth and to control the direction of maximum radiation. Arrays can be stacked to effectively increase the antenna capture or effective area, thus increasing total gain in a practical fashion. There are numerous other types of antenna arrays including phased arrays, but they will be left to a more advanced treatment of antennas.

9.9 Reflecting Surface Antennas

The operation of reflecting surface antennas is fairly straightforward. For transmitting, a basic antenna element (usually either a half-wave dipole or an illuminating element like a horn antenna) is located at the focal point of the antenna. This basic antenna element radiates energy that is focused by the reflecting surface into a narrow beam of energy. For receiving, the antenna surface focuses EM energy onto the basic antenna element located at the feed point.

The Parabolic Reflector

Figure 9-28 shows a typical parabolic dish antenna. Several variations exist for the transmitting feed or the receiving element of the parabolic antenna. Figure 9-29 shows an active receiving element while Figure 9-30 shows a Gregorian feed. For the Gregorian feed, a horn antenna illuminates a small reflector, which in turn illuminates the rest of the antenna surface.

The gain of a parabolic antenna can be determined by the following equation:

$$\text{Gain} = \frac{\pi^2 D^2}{\lambda^2} \qquad \textbf{9.7}$$

where D is the antenna diameter in meters and λ is the wavelength of operation in meters.

The antenna beamwidth, θ, of a parabolic antenna is given in degrees by:

$$\theta = \frac{70\lambda}{D} \qquad \textbf{9.8}$$

EXAMPLE 9.4

Determine the gain and the beamwidth for a $\frac{1}{2}$-meter dish antenna used at 18 GHz.

Solution The gain is given by:

$$\text{Gain} = \frac{\pi^2 D^2}{\lambda^2} = \frac{\pi^2 \left(\frac{1}{2}\right)^2}{\left(\frac{300}{18{,}000}\right)^2} = \frac{\pi^2(.25)}{\left(\frac{1}{60}\right)^2} = \pi^2(3600)(.25) = 8882.6$$

FIGURE 9-28 A typical parabolic dish antenna

FIGURE 9-29 An active parabolic dish-antenna receiving element

FIGURE 9-30 A Gregorian feed

Or the gain in dB is:

Gain in dB = 10log(8882.6) = 39.5 dB

The beamwidth is given by:

$$\theta = \frac{70\lambda}{D} = \frac{70 \times \left(\frac{1}{60}\right)}{\frac{1}{2}} = \frac{70}{30} = 2.33 \text{ degrees}$$

These values are representative of the current receiving equipment used for direct satellite broadcasting systems.

Figure 9-31 shows a typical radiation pattern for a parabolic dish antenna. Note the pencil-beam pattern and the numerous side and back lobes caused by spill over.

FIGURE 9-31 Typical parabolic dish-antenna radiation pattern

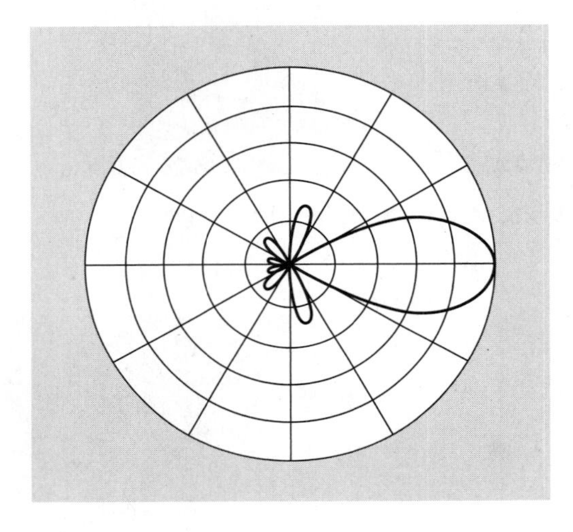

Microwave Horn Antennas

Horn antennas are the oldest type of microwave antenna. They are basically flared waveguides or TLs. The gradual changes in dimension match the impedance of the waveguide to that of free space. Horn antennas are very broadband and can provide fairly large gain over their range of operation. Applications include serving as feeds for large reflector or lens antennas and serving as a universal gain standard for antenna calibration. Horn antennas take on many different forms, but in all cases they are essentially tapered waveguides. *Sectoral horns* are only flared in one dimension. *Pyramidal* or *conical horns* are flared in both dimensions. See Figure 9-32 for diagrams of typical horn antennas.

FIGURE 9-32 Typical horn antennas: pyramidal and conical

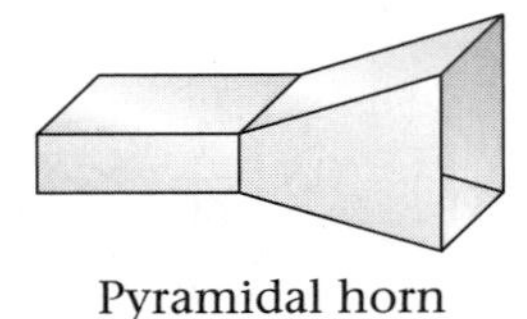

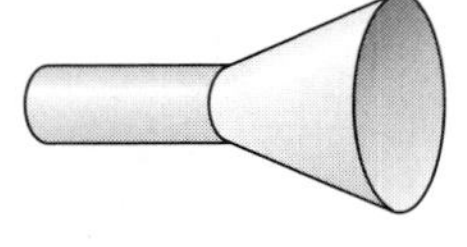

A typical pyramidal horn-antenna field pattern is shown in Figure 9-33. Figure 9-34 on page 360 shows a millimeter-wave sectoral horn antenna. The electric and magnetic field patterns are different due to the different flares in those planes.

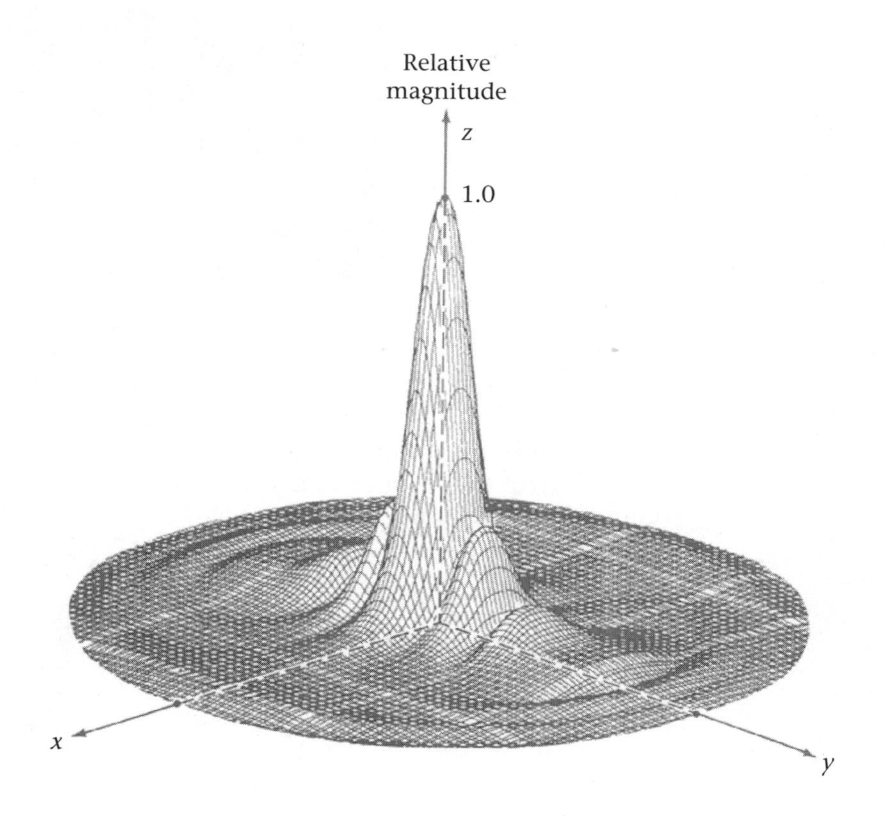

FIGURE 9-33 A typical horn-antenna field pattern

9.10 Microstrip Patch Antennas and Arrays

Microstrip patch antennas belong to a class of antenna known as *printed antennas* because they are constructed using printed circuit techniques. These antennas are popular due to their low cost, ease of construction, and ability to be configured in special shapes. A **microstrip patch antenna** is basically two parallel conductors separated by a sheet of dielectric material. One conductor acts as a ground plane, while the other conductor acts like a resonant circuit that radiates EM energy away from the structure.

FIGURE 9-34 Typical millimeter-wave horn antenna

Figure 9-35 shows a typical structure for a microstrip patch antenna. Note the antenna feed is itself a microstrip. The direction of maximum radiation is normal to the patch.

Microstrip patch antennas can be fashioned into arrays, as shown in Figure 9-36. Refer back to Figure 9-8b for an actual photograph of such an array.

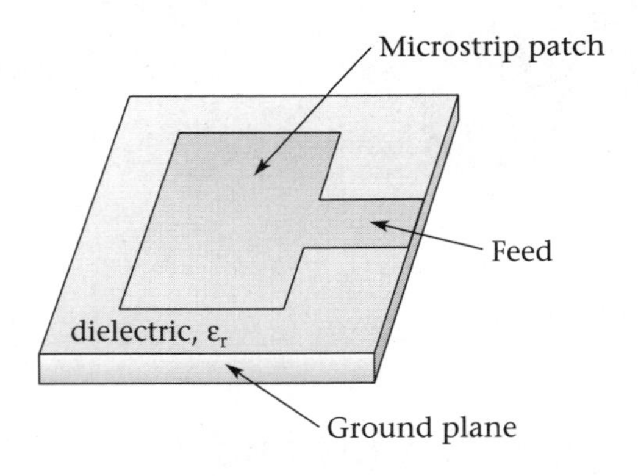

FIGURE 9-35 Typical microstrip patch antenna

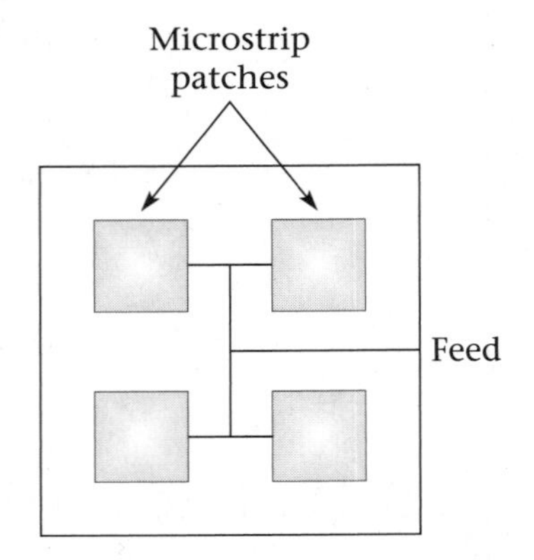

FIGURE 9-36 Four-element microstrip patch antenna array

9.11 Cell-Site Antenna Technology

Cell site antennas are appearing everywhere as a result of the popularity of cellular-telephone service. Typically, cell sites are broken into three sectors of 120 degrees each. This is accomplished with cell-site antennas that consist of arrays of printed half-wave dipoles that provide a 120-degree beamwidth. Figure 9-37 shows a typical cell-site antenna setup. Figure 9-38 shows a closer view of the actual antenna arrays. The antenna elements are protected from the elements by plastic *radomes*. As cellular technology improves, so-called smart antennas with adjustable radiation patterns will be used at the cell site to effectively increase the amount of spatial multiplexing employed.

FIGURE 9-37 Typical cell-tower antennas (each level is a different service provider)

FIGURE 9-38 Typical cellular antennas

9.12 Software Resources for Antenna Theory

There have been many advances in the field of computational electromagnetics (CEM) that allow one to plot antenna radiation patterns. Some of these programs are available free of charge over the World Wide Web. A software program called *APV* (**A**ntenna **P**attern **V**isualization) can be particularly useful for dipoles and arrays and the *SABOR* software program can be helpful for horn-antenna and reflector-antenna analysis. Other readily available and fairly low-cost mathematical software packages offer graphics capabilities that allow one to plot both two-dimensional and three-dimensional diagrams of antenna field patterns from mathematical equations. Additionally, there are numerous commercial EM simulation programs that have antenna field-pattern plotting capabilities.

Summary

Antennas provide the interface between TLs and EM waves. For transmission, power applied to an antenna is converted into an EM wave that propagates away from the antenna structure. An antenna is typically about ¼ to ½ of a wavelength long. The antenna radiation pattern depends upon the antenna's shape, electrical length, and spacing from the ground. Antennas have many characteristics of operation besides their radiation patterns. Radiation resistance, beamwidth, bandwidth, efficiency, capture area, and polarization are all antenna parameters of operation.

Antennas can be organized into arrays of basic elements. An antenna array allows for the enhancement of various antenna parameters such as directivity and bandwidth. As the frequency of operation increases, it is possible to use reflecting surfaces or simple aperture structures to obtain good performance characteristics. Parabolic antennas yield extremely large values of gain and narrow beamwidth values. Newer antennas such as microstrip patch antennas are also very useful at high frequencies because they are easily matched to printed circuit components.

Questions and Problems

Section 9.1

1. What is the basic definition of an *antenna*?

Section 9.2

2. Define an *isotropic source*.

3. What is a dipole?
4. What is meant by the term *radiation pattern*?

Sections 9.3 to 9.6

5. Determine the length of a half-wave dipole for 1030 kHz. Repeat the calculation for 38 GHz.
6. Define the *beamwidth* of an antenna.
7. Define the *gain* of an antenna.
8. Define the *input impedance* of an antenna.
9. Define *antenna polarization*.
10. Determine the antenna height and wavelength for the antenna of an AM radio station in your area (call the station). Is the antenna height what was expected?

Section 9.7

11. Determine the antenna directivity for an antenna with equal E and H beamwidths of 7.5 degrees.
12. What is an antenna *null*?
13. What special characteristics does a folded dipole have?
14. Why are vertical antennas needed for long wavelength signals?
15. How can a loopstick antenna be used for "radio location" functions?

Section 9.8

16. Describe the Yagi-Uda array antenna.
17. Describe the log-periodic array antenna.
18. A certain antenna has a gain of 8.5 dB. Determine a total system gain for a stack of eight of these basic antennas.
19. Check the local AM radio stations in your area. Determine if any of them use monopole arrays to direct their radiation patterns in any special way. Report your findings.
20. What is meant by the term *parasitic element* when referring to array antennas?

Sections 9.9 to 9.12

21. Determine the gain of a 3-m parabolic dish operating at 6 GHz. Give your result as a number and in dB.
22. Determine the beamwidth of the antenna described in question 21.
23. Determine the approximate size of a half-wave dipole antenna element for a cellular telephone antenna at about 890 MHz.
24. Explain the difference between an active feed and a Gregorian feed for a parabolic dish antenna.
25. Do an Internet search for the antenna software programs mentioned on page 362 of the text. A good starting point with links to other sites is **www.ieeeaps.org**. Download these programs and try them out.

Chapter Equations

$$P_D = \frac{P_T}{4\pi r^2}\ \text{W/m}^2 \qquad 9.1$$

$$\text{Directivity } (D) = \frac{32{,}400}{\theta_1 \times \theta_2} \qquad 9.2$$

$$D = \frac{4\pi A_{\text{eff}}}{\lambda^2} \tag{9.3}$$

$$\eta = \frac{P_{\text{rad}}}{P_{\text{in}}} \tag{9.4}$$

$$\frac{1}{\tau} = \frac{l_1}{l_2} = \frac{l_2}{l_3} = \frac{l_n}{l_{n+1}} \quad \textit{and} \quad \frac{1}{\tau} = \frac{d_1}{d_2} = \frac{d_2}{d_3} = \frac{d_n}{d_{n+1}} \tag{9.5}$$

$$\text{Total Gain in dB} = \text{Gain of Basic Antenna} + 3\text{ dB} \times \log_2 N \tag{9.6}$$

$$\text{Gain} = \frac{\pi^2 D^2}{\lambda^2} \tag{9.7}$$

$$\theta = \frac{70\lambda}{D} \tag{9.8}$$

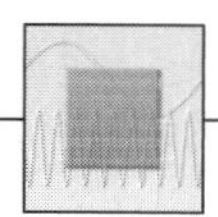

A

absorption The process by which the energy of a propagating signal decreases as it travels through the transmission medium.

access technologies A term used to describe various forms of multiplexing methods used with the wireless or air interface channel.

adaptive DPCM Adaptive differential pulse-code modulation. A form of pulse-code modulation that transmits a signal that represents the difference between successive samples.

ADSL Asymmetrical digital subscriber line. A form of high-speed digital modulation used over copper pairs.

A-law A standardized form of companding used in Europe with pulse-code modulation.

aliasing An effect peculiar to the sampling process. If an insufficient number of samples are taken to encode a signal, the spectral components contained in the samples will overlap one another and cause distortion during signal recovery.

amplitude modulation (AM) A form of analog modulation. The amplitude of a high-frequency sine wave is varied instantaneously in accordance with the message signal.

analog A signal that within certain limits has a continuous variation of possible values.

analog-to-digital converter (ADC) A device that converts an analog signal to a digital code.

angle modulation This term refers to forms of analog modulation where either the phase or frequency of a high-frequency sine wave is varied instantaneously in accordance with the message signal.

antenna A device used for the conversion of a high-frequency electrical current into a propagating electromagnetic wave or vice versa.

antenna element Usually, one of the wire-like metal conductors used to fabricate a wire antenna array. However, any basic antenna building block type of structure (i.e., a microstrip patch or slot) can be considered to be an antenna element.

aperture antenna An antenna structure that achieves its directional characteristics through the use of patterns of slots or holes in a ground plane.

array A physical arrangement of antenna elements used to achieve some optimum electrical characteristic (typically gain).

array antennas Antennas constructed from several to many basic antenna elements.

attenuation The process whereby the intensity of an electromagnetic wave or an electrical signal decreases.

audio frequency (AF) Frequencies in the normal range of human hearing. Often quoted as 50 to 15,000 Hz or 20 to 20,000 Hz.

audio mixer A device that linearly adds audio signals together. Typically used by vocalists or band members to create a desired composite sound.

azimuthal The horizontal angular distance relative to some standard reference direction (typically north).

B

balanced modulator A device that accepts two input signals of different frequencies and produces an output consisting of the sum and difference of these input frequencies.

balun A device that matches a balanced transmission line or antenna to an unbalanced transmission line, or vice versa.

bandwidth The range of frequencies in a signal. Usually calculated by subtracting the lowest frequency in the signal from the highest frequency. Other measures of bandwidth specify the range of signals above a certain power level.

bandwidth efficiency The factor by which the rate of data transmission per hertz exceeds a system that transmits one bit per bit time.

baseband A signal with frequency components starting close to or at 0 Hz and extending upward. Voice would be considered to be a baseband signal.

beamwidth The angle between the minus 3-dB points in the radiation pattern of an antenna.

Bessel functions Functions that are solutions to certain types of partial differential equations (typically encountered when dealing with physical phenomena having cylindrical symmetry).

binary phase-shift keying (BPSK) A form of digital modulation where two different phase values are transmitted to represent 0 and 1.

bipolar When referring to the polarity of a signal, this term indicates that both positive and negative voltage values are present.

bit time The basic length of time used to transmit a bit of information.

BlueTooth™ A form of wireless protocol that is used for short range (0–10 m) wireless communication.

Bode plotter A virtual instrument, similar to a network analyzer, that is available in the MultiSIM circuit-simulation program.

boundary conditions The values of electric- and magnetic-field intensity at or on the interfaces or boundaries between two different transmission media.

broadside An antenna radiation pattern that is perpendicular to the plane formed by parallel antenna elements in an array.

C

cable modem A device used by the cable-TV industry to allow a subscriber to transmit and receive digital data over the cable-TV system, allowing the cable-TV subscriber to obtain Internet access over the system.

capture area The effective area of an antenna that intercepts and converts incident electromagnetic energy into a high-frequency electric current.

capture effect The tendency of an FM receiver to lock on to a more powerful FM station and reject others.

carrier frequency The output frequency of an unmodulated radio transmitter, usually specified by the FCC in tables of channel assignments for a particular service.

channel The physical means by which a signal propagates. Also, any switching fabric used to direct the signal to the proper destination that it will pass through.

characteristic impedance For transmission media, this is the ratio of the electric-to-magnetic field intensity of a propagating electromagnetic wave.

circuit switching A type of transmission channel where the switching fabric physically completes a particular point-to-point connection from many possible connections.

circularly polarized A type of electromagnetic wave that, as it propagates, has rotating electric- and magnetic-field components.

coarse wavelength-division multiplexing (CWDM) A proposed standard for the implementation of a 10-GHz ethernet system.

code-division multiple access (CDMA) A wireless multiplexing system that uses different codes to differentiate signals with the same carrier frequency.

collinear array An antenna array with equally spaced parallel antenna elements.

communications The act of sending a signal from one point to another.

companding The process of compressing and expanding a signal, usually related to the pulse-code modulation process.

constellation diagrams A means of diagramming the signals that make up a digital modulation scheme, usually plotted on the *xy* plane.

coplanar A type of waveguide that is formed by the elimination of a strip of metal from a ground plane.

cross-modulation products The resulting frequencies produced by the higher-order, nonlinear mixing of signals.

cylindrical symmetry This occurs when an object looks the same regardless of the azmuthal angle that one observes it from.

D

dB The logarithmic ratio of two values.

de-emphasis A process used in FM systems whereby the demodulated message signal is passed through an LPF.

delta modulation (DM) A form of pulse-code modulation where only one bit (a 0 or 1) is transmitted every bit time, depending upon whether the signal amplitude has decreased or increased.

demodulation The process that is the reverse of modulation, normally taking place at the receiver and converting the modulated signal back to the original message signal.

demultiplexing The process that is the reverse of multiplexing and by which signals that have shared the same transmission media are separated from one another.

dense wavelength-division multiplexing (DWDM) The use of many different wavelength signals over the same fiber-optic cable; used to increase the data transmission rate.

deregulation The process of reducing the legal restrictions upon an entity, business, or industry.

detector The subsystem in a receiver that performs signal demodulation or decoding.

deviation sensitivity The amount of frequency shift (hertz per volt) of applied signal for an FM transmitter.

dielectric constant The relative permittivity of a material.

differential PCM A digital-modulation scheme where only the difference between successive samples is transmitted.

diffraction The process by which a propagating electromagnetic wave appears to bend around the corner or sharp edge of a structure or some type of obstruction in its path.

diffuse reflection A reflection of an electromagnetic wave (typically light) that scatters the rays of light in many different directions.

digital Refers to a signal that has a finite number of values.

digital companding A form of companding that uses an algorithm to encode and decode the pulse-code modulated samples.

digital cross-connect switches (DCS) A form of switching fabric used to process overhead signals, groom and consolidate payloads, and cross-connect digital and optical signals.

digital modulation (DM) A modulation scheme that employs a limited number of transmitted signals (usually a power of two).

digital multitone A form of digital modulation that uses numerous parallel channels to send information, in effect raising the data rate.

digital subscriber line (DSL) A type of transmission technique used to transmit digital data over a pair of copper wires.

digital-to-analog converter (DAC) A device that converts a digital code word into an analog voltage level.

direct digital frequency synthesis The synthesis of a periodic signal using read-only memory (ROM) digital code words, a DAC, and filtering.

direct FM The production of an FM signal by the direct application of a message signal to a system with V/F conversion characteristics.

directivity The characteristic of an antenna that makes it more receptive to signals coming from a particular direction.

Doppler radar A form of radar that allows one to measure relative speed between the radar system and a reflective target.

double conversion A superheterodyne receiver that uses two intermediate-frequency stages to obtain better image rejection and selectivity characteristics.

double-sideband full-carrier AM (DSBFC-AM) A modulation technique whereby the transmitter outputs a carrier signal and a pair of sidebands.

double-sideband suppressed-carrier AM (DSBSC-AM) A modulation technique whereby the transmitter outputs a pair of sidebands and a suppressed carrier.

downstream In a cable-TV system, the direction from the headend to the subscriber.

driven element For an antenna array, the element connected to the transmission line.

duty cycle For a repetitive pulse signal, the ratio of on time to the total pulse period.

dynamic range The range of signal values that a system can handle without distortion or overload.

E

effective area For an antenna, the equivalent physical size of an antenna that is 100% efficient at converting incident electromagnetic energy into an electrical signal.

electrically small dipole (ESD) A very short piece of electrical conductor used as an antenna.

electromagnetic spectrum The range of all possible electromagnetic waves.

electronic filters Electronic systems that are frequency selective.

elevation angle The angle between the horizon and a particular vertical direction.

end-fire For an antenna array, a radiation pattern in the direction parallel to the plane formed by the array elements.

envelope For an AM wave, the positive and negative limits of the waveform amplitude.

equipotential A surface with the same voltage at each point.

extremely high frequency (EHF) The range of frequencies from 30 GHz to 300 GHz.

extremely low frequency (ELF) Frequencies in the range of 30 Hz to 300 Hz.

eye diagram An oscilloscope display that superimposes received pulses on top of one another.

F

fan-beam An antenna with different beamwidths in the electric- and magnetic-field planes.

feed point The transmission-line connection point for an antenna structure or the illumination center for a reflecting plane-type antenna.

folded dipole A two-wire half-wave dipole antenna.

Fourier analysis A mathematical method used to determine the frequency content of a complex signal.

free-space impedance The ratio of electric field intensity (V/m) to magnetic field intensity (A/m) for a propagating electromagnetic wave in free space. Equal to 377 Ω.

frequency A measure of the rate of change of a periodic signal; typically, the number of cycles per second (hertz).

frequency deviation The amount of change in the rest frequency of an FM transmitter.

frequency-division multiplexing (FDM) A common multiplexing technique that allows the sharing of a common communications channel by assigning different carrier frequencies.

frequency domain Displaying a signal in terms of its power and frequency; typically, the type of display used by a spectrum analyzer.

frequency modulation (FM) An analog-modulation technique. For FM, the frequency of a high-frequency sine wave is varied instantaneously in accordance with the modulating signal.

frequency reuse group A term used in cellular radio to indicate the repeating pattern of cell-site channel assignments.

frequency shift keying (FSK) A digital-modulation technique that employs two different frequencies to represent the symbols 0 and 1.

G

geographic multiplexing A form of multiplexing that uses the spacing between transmitters to allow for the sharing of the same electromagnetic spectrum.

gigascale integration (GSI) Integrated circuits with billions of transistors per chip.

grand-scale integration (GSI) Integrated circuits with millions of transistors per chip.

ground plane A reflecting surface of good conductivity.

ground waves Propagating, vertically polarized, low-frequency electromagnetic waves; usually the transmitting range for these signals is dictated by power level.

guided waves Electromagnetic waves that are guided by some form of transmission-line structure.

H

half-wave dipole A standard elementary antenna that serves as a building block for numerous antenna arrays.

headend In cable-TV systems, the source of over-the-air TV signals and satellite feeds from various network providers.

helical antenna A type of antenna that is formed in the shape of a helix or spring.

hertzian antenna An ungrounded resonant antenna.

high frequency (HF) Frequencies in the range of 3 MHz to 30 MHz.

I

I channel For a digital-modulation system, the signal path of one of two orthonormal signals. For this channel, the so called in-phase signal does not undergo any processing.

image frequency An interfering signal frequency that is converted to the intermediate frequency along with the desired signal in a superheterodyne receiver.

image rejection (IR) The ability of a superheterodyne receiver to reject image-frequency signals.

image signal An interfering signal that is applied to the detector of a superheterodyne receiver along with the desired signal as a result of the mixing action of the receiver.

index of modulation A measure of the amount or depth of modulation for analog modulation schemes like AM or FM.

indirect FM A type of frequency modulation derived from phase modulation.

input impedance The ratio of input voltage to input current for an electronic system.

interference Any signal that masks or interferes with the proper reception of a desired signal.

intermediate frequency (IF) The fixed frequency used in a superheterodyne receiver to provide signal gain and selectivity to the desired signal after mixing with the local oscillator signal.

Internet A collection of computer networks that are networked together.

interoperability The ability of complex telecommunications systems, manufactured by different vendors, to operate together without compatibility problems.

intersymbol interference (ISI) Results from the ringing tails of other transmitted pulses interfering with the desired pulse.

ISM bands The instrumentation, medical, and scientific bands located at 2.45 GHz, 5.8 GHz, and several other frequencies.

isotropic radiator An antenna that radiates equally in all directions.

isotropic source A source that radiates equally in all directions.

L

large-carrier amplitude modulation (LCAM) Standard AM modulation.

large-scale integration (LSI) Integrated circuits with thousands to tens of thousands of transistors per chip.

line codes Types of electrical signals used to encode 0s and 1s for transmission over copper-wire pairs.

local access territory A term used to describe an area within which calls must be carried by a local exchange carrier (LEC) according to the Modified Final Judgment of 1984.

local oscillator A subsystem of a superheterodyne receiver. This subsystem produces a high-frequency signal that is applied to the mixer of the superheterodyne along with the tuned signal.

log-periodic array A type of antenna array that has extremely wideband characteristics.

loopstick antenna An antenna used for low frequencies in portable equipment.

low frequency (LF) Frequencies in the range of 30 kHz to 300 kHz.

M

medium frequency (MF) Frequencies in the range of .3 MHz to 3 MHz.

medium-scale integration (MSI) Integrated circuits with hundreds to thousands of transistors per chip.

microstrip patch antenna A type of antenna constructed from microstrip patches.

microwaves Generally considered to be frequencies above 3 GHz.

millimeter waves Generally considered to be frequencies with wavelengths in the millimeter range and frequencies in the range of 30 GHz to 300 GHz.

minimum shift keying (MSK) A form of digital modulation that does not undergo abrupt phase shifts.

mixer A device that has a nonlinear input-output transfer characteristic.

modulation The process by which a signal is modified instantaneously in accordance with a message signal to make it compatible with the transmitting channel.

modulation index See *index of modulation*.

modulation sensitivity A term used with FM transmitters to characterize the amount of frequency deviation per volt.

modulator A device that modulates a signal. See *modulation*.

monopole A type of antenna that usually consists of a single vertical element.

Moore's Law A rule proposed by Moore stating that the number of devices on an integrated circuit can double every 18 to 24 months.

Motion Pictures Expert Group (MPEG) An ad hoc industry group that works on signal-compression techniques and standards.

multimedia Combined signals of video, text, audio, and data.

multiplexing The process by which a transmission media (such as a coaxial cable) is shared by numerous individual signals.

MXY multiplexers A device that combines several standard T-carrier digital bit streams into another standard, but faster, digital bit stream.

N

narrow-band frequency modulation (NBFM) A form of frequency modulation for which the index of modulation is less than 0.5. This type of signal usually only has one pair of significant sidebands.

network A collection of computers wired together and able to share the same resources.

network analyzer A type of test and measurement equipment that can display the input/output transfer characteristics of a system or device under test versus frequency.

noise Any type of signal that interferes with the reception of a desired signal.

noise figure The figure of merit of an active device like a field-effect transistor. It indicates the amount of noise introduced by the device and is usually expressed as a dB value.

noise ratio The same as noise figure but expressed as a numerical ratio.

noise temperature This term refers to the same figure of merit as *noise figure*; however, the units are in degrees Kelvin.

non-return-to-zero A type of line code where the signal does not return to zero between symbols.

O

oblique incidence Incidence of an electromagnetic wave on a surface at an angle other than perpendicular.

offset QPSK A type of digital modulation that uses two phase-shifted signal constellations to encode data.

orthogonal FDM Frequency-division signals that use orthonormal signals.

orthonormal signals Signals that are orthogonal to one another. Mathematically, these types of signals have no cross-correlation (similarity) and have unit energy.

P

packet switching A transmission scheme in which part of the data being transmitted (usually called the header) represents the address of the destination.

parasitic element An element of an array antenna that is electrically and magnetically coupled to the driven element.

party line A telephone line that is shared by several subscribers.

passband The range of frequencies allowed through a system; typically, a radio-frequency signal like that produced by a cell phone.

patch antennas See *microstrip patch antennas*.

pencil-beam A narrow beam formed by an antenna array or reflecting surface antenna.

periodic A repeating signal.

permeability of free space The ease with which free space can support a magnetic field.

personal-area networks Networks that are physically small and usually restricted to personal computer equipment (PC, laptop, printer, and so forth).

phase-division multiplexing A form of multiplexing where two data channels (the I and Q) modulate the same carrier.

phase shift keying (PSK) A form of digital modulation that employs various phase shifts relative to a reference to encode binary bits.

phasor A representation of a signal that includes both amplitude and phase information.

pilot subcarrier A low-level signal transmitted by a system to serve as a reference for the receiver in the process of demodulation.

plane wave An electromagnetic wave where the electric and magnetic fields are perpendicular to one another and also to the direction of propagation.

point source A conceptual source of electromagnetic radiation that emits energy equally in all directions.

polarization The physical orientation of the electric field of an electromagnetic wave.

polar plot A two-dimensional graph representing the cylindrical coordinates, r (distance from the center) and θ (theta, the angular coordinate). Typically used to show antenna radiation patterns.

power density The amount of energy contained in an electromagnetic wave, given in watts/square meter.

pre-emphasis A means of combating noise in an FM transmitting system. The message signal is high-pass filtered before modulating the transmitter, which has the effect of reducing high-frequency system noise.

product detector A type of detector used in a receiver to recover a single sideband signal.

pseudonoise A digital code overlaid on a signal to accomplish code-division multiple access (CDMA).

pulse repetition rate (PRR) The number of pulses per second in a periodic pulse train.

Q

Q The quality factor of a resonant circuit. This value is a measure of the selectivity of a bandpass filter.

Q channel For a digital-modulation system, the signal path of one of two orthonormal signals. For this channel, the quadrature signal modulates the carrier after the carrier has undergone a 90-degree phase shift.

quadrature A term that indicates a 90-degree phase shift.

quadrature amplitude modulation (QAM) A form of digital modulation that employs both phase shifts and amplitude changes to encode binary combinations.

quadrature phase-shift keying (QPSK) A type of digital modulation that uses four phase angles to represent two binary bits of data.

Quality of Service (QoS) A term that refers to how a telecommunications system deals with different types of traffic. Different types of traffic require different treatments (QoS) for correct transmission.

quantizing levels The number of levels into which the analog-to-digital converter (ADC) process encodes a sample.

quarter-wave matching transformer A type of transmission-line matching scheme using a quarter-wave section of transmission line.

quasi-optical antennas Antennas that use lenses to focus extremely high-frequency electromagnetic waves.

quieting sensitivity A measure of the effectiveness of an FM receiver in reducing noise in the presence of a signal.

R

radiation pattern Typically, a polar plot of an antenna's gain characteristics.

radio frequency (RF) Generally considered to be frequencies above 3 MHz.

radio-frequency amplifier An amplifier used to raise the signal level of a radio-frequency signal.

rate-adaptive DSL (RADSL) An *ADSL* system that adapts itself to present line conditions.

Rayleigh fading A type of signal fading common to cell-phone systems due to multipath propagation.

receiver A device used to recover a modulated message signal.

refarming The reassignment of portions of the frequency spectrum for use by different radio services.

reflection The process by which a propagating electromagnetic wave bounces off of a different transmission media.

reflection coefficient The amount of reflection a propagating electromagnetic wave undergoes when it encounters a new transmission media. Typically given by the ratio of reflected-to-incident signal amplitude.

refraction The process by which a propagating electromagnetic wave undergoes a change in direction when it encounters a new transmission media at an oblique angle.

regenerator A device that reconstitutes a signal back to its original shape.

regulated monopoly A sole provider of some type of goods or service.

regulation The process by which the government controls the operations of various industries.

relative permittivity The dielectric constant of a material.

repeater In a telecommunications system, a device that receives an analog signal, demodulates it, remodulates it, and retransmits it.

resolution The fineness of detail available with a digital-to-analog converter (DAC).

rest frequency The output frequency of an FM transmitter without any modulation.

roll-off rate (ROR) The amount of attenuation introduced by either an LPF or an HPF. Usually specified in dB/decade or dB/octave.

S

sample-and-hold (S/H) A device that when triggered samples an input signal and holds the sample voltage constant so that it can be digitized.

sampling theorem A theorem that states that if one samples a signal periodically at a fast enough rate, one will be able to reconstruct the sampled signal perfectly from the samples.

scalar signals Signals that are modulated by changing only one signal parameter.

selectivity The ability of a radio receiver to select between two signals that are located close together in the frequency spectrum.

shape factor A measure of the sharpness of a bandpass filter.

sidebands Frequencies that are created symmetrically around the carrier frequency of a transmitter during the process of analog modulation.

signal A description of some type of physical phenomena; typically, an electrical signal.

signal-space diagram A means by which one can display the discrete signals of a digital-modulation scheme on the *xy* plane.

signal-to-noise ratio (SNR) A measure of the quality of a signal.

signal-to-quantizing noise ratio (SQNR) A measure of the noise introduced during the analog-to-digital converter (ADC) quantizing process.

sinc pulse A particular type of signal that has desirable properties for the transmission of pulses (low intersymbol interference).

single-sideband AM (SSB) A type of amplitude-modulation signal that has eliminated the carrier-frequency component and one of the sidebands.

skin depth The depth into a material at which a propagating electromagnetic wave has had its field intensities reduced to approximately 37% of their initial values.

skin effect The phenomena whereby a high-frequency electrical current flows mostly near the surface of a conductor.

sky waves Medium- to high-frequency electromagnetic waves that propagate by reflecting off of the ionospheric layers.

small-scale integration (SSI) Integrated circuits with tens to hundreds of transistors per chip.

software radio A radio receiver that digitizes the received signal and can be reprogrammed to change its demodulation characteristics.

space-division multiplexing (SDM) A form of multiplexing that relies on the physical separation of transmitters operating on the same frequency.

space waves Electromagnetic waves at high frequencies and above that propagate in a straight line and are limited in range by the radio horizon.

spectra The frequency content of a signal.

spectral fold-over An adverse effect encountered with the under sampling of signals.

spectrum analyzer A piece of test equipment that is able to display the individual frequency components in a signal.

specular reflection A reflection of an electromagnetic wave from a mirror-like surface.

spherical wave A wave emitted by a point source. When close to the source, the wavefront is spherical in shape.

split band For cable-TV systems, the entire useable bandwidth is split apart for two different services.

spread spectrum A technique whereby the bandwidth of a signal is spread out to achieve noise immunity.

standards Documented agreements containing rules, guidelines, or definitions of characteristics.

standing wave The result of a transmission-line mismatch. Incident and reflected electromagnetic waves add vectorially to form standing waves.

standing-wave ratio (SWR) The ratio of incident-to-reflected electromagnetic wave intensity on a transmission line with a mismatch.

stripline A form of transmission line used at very high frequency with planar components.

super group Typically, 60 frequency-division multiplexed (FDM) telephone channels.

superheterodyne A type of radio receiver that converts the received signal to a standard intermediate frequency (IF) before demodulation.

super-high frequency (SHF) Frequencies in the range of 3 GHz to 30 GHz.

super master group Typically, 600 to 1800 (by multiples of 300) frequency-division multiplexed (FDM) telephone channels.

switching fabric The means by which a signal is directed to its correct destination.

symbol A signal that represents one of a discrete number of messages.

symbol time The time required to transmit one symbol.

synchronous optical network (SONET) An optical transport mechanism for digital data transmission.

T

telecommunications The transmission of voice or data (including multimedia content) over a distance using the public-switched telephone network (PSTN) or privately owned networks.

threshold effect An effect experienced by frequency-modulation receivers when the input signal exceeds a certain level. The signal-to-noise ratio at the receiver output increases rapidly for a small-input signal-to-noise-ratio improvement.

time-division duplex (TDD) The transmission of data over a channel by the assignment of time slots for different transmission directions.

time-division multiple access (TDMA) A type of multiplexing scheme that shares the transmission channel by assigning time slots to different users.

time-division multiplexing (TDM) A common multiplexing technique that allows the sharing of a common communications channel by assigning different time slots to each user.

time domain The display of a signal in terms of its amplitude variation with time; typically, the display used by an oscilloscope.

time-slot interchange switch (TSIS) A type of switching fabric that allows the switching of user time slots for time-division multiplexing systems.

time-variant signals Vector signals that use time and/or code variations to access the transmission channel.

trans-ionospheric A term that refers to the propagation of signals across the ionospheric layers.

transmission coefficient The ratio of transmitted-to-incident signal amplitude for a propagating electromagnetic wave when it encounters a new transmission media.

transmission-control protocol/Internet protocol (TCP/IP) A transport-control protocol used in conjunction with Internet protocol.

transmission lines (TLs) Electrical conductors or guiding structures arranged in such a fashion as to efficiently transport electrical or electromagnetic signals from one point to another.

transmitter The subsystem of a telecommunications system that modulates or encodes a signal to make it acceptable to the transmission channel.

transverse This term refers to the direction of the electric field of an electromagnetic wave in relation to the direction of propagation. They are at right angles.

transverse electromagnetic wave (TEM) Another term for a plane wave.

Trellis code-modulation See *Trellis coding*.

Trellis coding A form of encoding that provides an effective coding gain in the signal-to-noise ratio of a channel.

tuned radio-frequency receiver (TRF receiver) An early receiver design that utilized several cascaded radio-frequency amplifiers.

U

μ-law A standard form of companding used with pulse-code modulation in North America and Japan.

ultra-high frequency (UHF) Frequencies between .3 GHz and 3 GHz.

ultra large-scale integration (ULSI) Integrated circuits with millions to tens of millions of transistors per chip.

unipolar Signals that have only one polarity.

upstream In a cable-TV system, signals from the subscriber sent back toward the headend.

V

vector signals Signals that are modulated by changing two signal parameters.

vertical monopole See *monopole antenna.*

very-high frequency (VHF) Frequencies in the range of 30 to 300 MHz.

very large-scale integration (VLSI) Integrated circuits with tens of thousands to hundreds of thousands of transistors per chip.

vestigial sideband AM (VSB) A form of amplitude modulation (AM) that reduces the frequency components and amplitude of one sideband, thus reducing the overall signal bandwidth.

video Signals that extend from close to 0 Hz to several MHz (wideband baseband signals).

video amplifier Generally considered to be an amplifier that operates from close to dc to several MHz.

voice frequencies (VF) Generally considered to be frequencies in the range of 30 Hz to 3 kHz.

W

Walsh codes Orthogonal pseudorandom noise sequences used in code-division multiple access (CDMA) systems.

wavelength This is defined for an electromagnetic wave as the distance between identical points on a propagating wave.

wavelength-division multiplexing (WDM) The use of different wavelength signals over a fiber-optic cable to increase data-transmission rates.

wideband frequency modulation Frequency modulation with a high index of modulation.

wireless A form of communications that uses the free-space channel.

wireline Refers to the propagation of electrical signals through copper-wire pairs or coaxial cable.

Y

Yagi-Uda array A particular type of antenna array typically used at very-high frequencies (VHF).

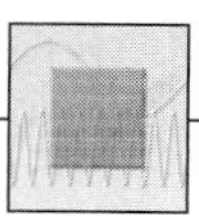

REFERENCES

Chapter 1

Bondyopadhyay, Probir K. (1998). Moore's Law Governs the Silicon Revolution. *Proceedings of the IEEE, 86*(1), 78–81.

Bowers, Brian. (2000). The First Atlantic Telegraphs. *Proceedings of the IEEE, 88*(7), 1131–1133.

Brittain, James E. (1997). Ralph Bown and the Golden Age of Propagation Research. *Proceedings of the IEEE, 85*(9), 1511–1513.

———. (1996). Reginald A. Fessenden and the Origins of Radio. *Proceedings of the IEEE, 84*(12), 1852–1853.

Cannon, Robert. (2001). Where Internet Service Providers and Telephone Companies Compete: A Guide to the Computer Inquiries, Enhanced Service Providers, and Information Service Providers. Presented at the 2001 Telecommunications Policy Research Conference and found online at http://www.tprc.org/Agenda00.htm.

Carne, Bryan E. (1999). *Telecommunications Primer: Data, Voice, and Video Communications* (2nd ed.). Upper Saddle River, New Jersey: Prentice Hall PTR.

Cole, Marion. (2002). *Introduction to Telecommunications* (2nd ed.). Upper Saddle River, New Jersey: Prentice Hall.

Contanis, Nicolae. (1997). The Radio Receiver Saga: An Introduction to the Classic Paper by Edwin H. Armstrong. *Proceedings of the IEEE, 85*(4), 681–684.

Corazza, Gian Carlo. (1998). Marconi's History. *Proceedings of the IEEE, 86*(1), 1307–1311.

Davis, Jeffery, et al. (2001) Interconnect Limits on Gigascale Integration (GSI) in the Twenty-First Century. *Proceedings of the IEEE, 89*(3), 305–324.

Farley, Tom. Privateline's Telephone History, 1892 to 1921. http://www.privateline.com.

Morton, David. (1999). A Snapshot of Telephony at the Turn of the Century. *Proceedings of the IEEE, 87*(4), 691–694.

National Inventor's Hall of Fame. A History of Morse. http://www.invent.org.

Nebeker, Frederik. (1992). An oral history of Harold Beverage. IEEE History Center, Rutgers University, New Brunswick, New Jersey.

Stephan, Karl D. (1996). The French Cable Station in Orleans, Massachusetts. *IEEE Antennas and Propagation Magazine, 37*(5).

Vermeulen, Dirk J. (1998). The Remarkable Dr. Hendrik van der Bijl. *Proceedings of the IEEE, 86*(12), 2445–2454.

Chapter 3

Brittain, James E. (1996). John S. Stone and the Professionalization of Communications Engineering. *Proceedings of the IEEE, 84*(10), 1573–1574.

———. (1996). Reginald A. Fessenden and the Origins of Radio. *Proceedings of the IEEE, 84*(12), 1852–1853.

Brock, Darren K., Mukhanov, Oleg A., Rosa, Jack. (2001). Superconductor Digital RF Development for Software Radio. *IEEE Communications Magazine, 39*(2).

Contanis, Nicolae. (1997). The Radio Receiver Saga: An Introduction to the Classic Paper by Edwin H. Armstrong. *Proceedings of the IEEE, 85*(4), 681–684.

De Forest, Lee. (1913). Recent Developments in the Work of the Federal Telegraph Company. *Proceedings of the IRE, 1*(1), 37–51.

Maas, Stephen A. (1993). *Microwave Mixers* (2nd ed.). Norwood, Massachusetts: Artect House, Inc.

Chapter 4

Arfken. (1985). *Mathematical Methods for Physicists* (3rd ed.). Orlando, Florida: Academic Press, Inc.

De Forest, Lee. (1913). Recent Developments in the Work of the Federal Telegraph Company. *Proceedings of the IRE, 1*(1), 37–51.

Morton, David. (1999). Radio Broadcasting in the Electrical Century. *Proceedings of the IEEE, 87*(5), 929–932.

Chapter 5

Haykin, Simon. (2001). *Communications Systems* (4th ed.). New York: John Wiley & Sons, Inc.

Tomasi, Wayne. (2001). *Electronic Communications Systems: Fundamentals Through Advanced* (4th ed.). Upper Saddle River, New Jersey: Prentice Hall.

Chapter 6

Shannon, Claude E. (1949). Communication in the Presence of Noise. *Preceedings of the IRE, 37*(1), 10–21.

Chapter 7

Black, Uyless. (1996). *Wireless and Mobile Networks* (2nd ed.). Upper Saddle River, New Jersey: Prentice Hall.

Rauschmayer, Dennis. (2000). *ADSL/VDSL Principles*. Macmillan Technology Series.

State University of New Jersey. Thomas A. Edison Papers Web Site. http://www.rci. edu/~taep.

Tomasi, Wayne. (2001). *Electronic Communications Systems: Fundamentals through Advanced* (4th ed.). Upper Saddle River, New Jersey: Prentice Hall.

Chapter 8

Alexanderson, Ernst F. (1920). Trans-Oceanic Radio Communication. *Proceedings of the IRE, 8*(4), 263–285.

Carne, Bryan E. (1999). *Telecommunications Primer: Data, Voice, and Video Communications* (2nd ed.). Upper Saddle River, New Jersey: Prentice Hall PTR.

Hioki, Warren. (2001). *Telecommunications* (4th ed.). Upper Saddle River, New Jersey: Prentice Hall.

Marconi, Guglielmo. (1922). Radio Communication. *Proceedings of the IRE,* 10, 215–238.

———. (1929). Radio Communication. *Proceedings of the IRE*, 16, 40–69.

Richards, Carol. (2001). New 1-GHz Channel-Spacing Technology Narrows in the Last Mile. *Lightwave*.

Trueman, Christopher W. (2000). Interactive Transmission Line Computer Program for Undergraduate Teaching. *IEEE Transactions on Education, 43*(1).

———. (1999). Teaching Transmission Line Transients Using Computer Animation. Presented at the 29th ASEE/IEEE Frontiers in Education Conference, San Juan, Puerto Rico.

Chapter 9

Stutzman, Warren L. and Thiele, Gary A. (1998). *Antenna Theory and Design* (2nd ed.). New York: John Wiley & Sons, Inc.

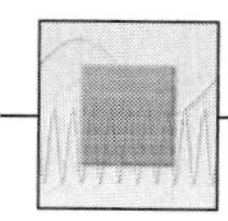

INDEX

C

D

N

Q

R

S

T